# MODELING AND SIMULATION IN CHEMICAL ENGINEERING

## ROGER G. E. FRANKS

SENIOR CONSULTANT: ENGINEERING COMPUTATION AND ANALYSIS
ENGINEERING DEPARTMENT
E. I. DU PONT DE NEMOURS & CO., INC.
1972

## WILEY–INTERSCIENCE,

A DIVISION OF JOHN WILEY & SONS, INC.
NEW YORK | LONDON | SYDNEY | TORONTO

*Library of Congress Cataloging in Publication Data*

Franks, Roger G E
  Modeling and simulation in chemical engineering.

  Includes bibliographical references.
  1. Chemical engineering—Mathematical models.
  2. Digital computer simulation.    I. Title.

TP155.F724     660'.2'0184     72-39717
ISBN 0-471-27535-2

Printed in the United States of America.

10 9 8 7 6 5 4 3 2

TO MY WIFE BARBARA
AND MY DAUGHTERS
BARBARA ANN, DEBORAH LOUISE,
AND JENNIFER ELIZABETH

# PREFACE

The forerunner to this book, published in 1967, was titled *Mathematical Modeling in Chemical Engineering*. It was written at a time (1964–1966) when analog computation in industry was phasing out, to be replaced by digital simulation. Digital computer technology at that time was in an early stage of evolution. Thus, other than a brief description of the MIMIC dynamic simulator, the emphasis of that book was almost entirely on presenting the analytical techniques that lead to models suitable for simulation by either analog, hybrid, or digital computers. During the past seven years, considerable progress has been made in the development of digital programs for the chemical engineering field, and some common procedures are now emerging. Consequently, this book will present two FORTRAN IV computer programs (INT and DYFLO) that are designed for solving sets of differential equations, thereby simulating the dynamic behavior of chemical processes. The primary reason for offering these programs in FORTRAN is that it is the most common language for computers today. The program (developed for the UNIVAC 1108) as listed in this text can be adapted to any computer with FORTRAN with only minor modifications.

This book addresses itself to those chemical engineering systems that, when analysed, lead to models involving ordinary and/or partial differential equations. Consequently, it does not concern itself with steady-state energy and material balance, nor does it cover programs directed to detailed engineering design of process equipment.

In order to make the simulation programs meaningful, they are described in a framework of engineering applications, the various parts of the program being introduced in the chapters throughout the book as each area of chemical engineering is covered. If these programs are to be used effectively a knowledge of elementary numerical methods is recommended. For this reason chapters II and III cover the basic concepts of numerical iteration and integration. This leads to the development of the INT program, which consists of a set of subroutines for solving simultaneous differential equations and associated algebraic equations. The INT program forms the basis of a higher-level set of subroutines (DYFLO) that simulate the dynamic behavior of most of the common unit operations in chemical processes. These are introduced successively in the following chapters and deliberately are designed for simplicity in order to gain execution speed and to facilitate an understanding of the internal computation sequence. It is hoped that this understanding will encourage the reader to develop additional subroutines tailored to his specific needs.

The computer programs provide a ready means for executing the calculations of the mathematical models created in the analytical portions of the text, demonstrating that simulation is a practical approach to complex process problems, typical of present day practice in industry. As in the first book, model formulation methods are stressed. Many engineers have the latent ability to describe physical reality in mathematical symbology. This requires imagination, coupled with experience that is accumulated from frequent practice. Development of these talents leads to greater analytical insights into process systems and a more thorough grasp of fundamentals. This book is intended to stimulate the analytical approach by removing outdated obstacles and presenting a straightforward, unified procedure for constructing mathematical models of complex process systems.

Compared with the complexities of the computer simulations commonly done in industry, and occasionally described in the literature, most of the examples discussed in this book are elementary. For this reason this text should be considered as an introduction to computer simulation. However, most of the general principles inherent in the complex models have been deliberately included in the text examples. The subject matter progresses from lumped systems through staged operations to distributed systems that give rise to partial differential equations. The last chapter shows how nonlinear control problems can be tackled by simulation without detailed knowledge of advanced control theory.

It is assumed that the reader is familiar with the fundamentals of energy and mass transfer, chemical kinetics, and vapor/liquid equilibrium and that he has a working knowledge of FORTRAN. The logical candidates for this book, therefore, are senior undergraduates or first-year graduate students and practicing engineers who have not been exposed to computer modeling.

I am grateful to my colleagues, Mr. R. L. Buchanan, Mr. D. Culver, Mr. T. Keane and Mr. A. Auster for their assistance in the preparation of this text. Mr. D. Culver has been particularly helpful in the development of the DYFLO program. The assistance of Mrs. J. Schweikert and Miss B. Franks in typing the manuscript is gratefully acknowledged.

*Wilmington, Delaware*                                           R. G. E. FRANKS
*December 1971*

# CONTENTS

# MODELING AND SIMULATION
# IN CHEMICAL ENGINEERING

# CHAPTER I

# INTRODUCTION

Over the last fifteen years in the chemical and petroleum industries there has been a gradual trend toward a more quantitative approach to problems in the design and operation of processes. This trend has been made possible by the increasing use of powerful electronic computers in the solution of complex systems of mathematical equations. This analytical approach to engineering problems allows both considerably wider scope in investigating alternative designs and more efficient operation of commercial batch and continuous processes. In addition, the computerized analytical approach provides a deeper understanding of the internal mechanisms of the processes studied.

Today powerful computers have become almost universally available in both schools and industry. The future promises the rapid expansion of time-sharing and more sophisticated terminal connections, such as graphic displays and automatic plotting equipment. During the period when the analog computer was used to solve process dynamics problems (1955–1965), there existed, by necessity, a specialized staff at each computer installation that performed the programming and computer operation chore for clients with problems to be solved. The analysis of the problem, that is, its definition in mathematical terms, was performed by either the client or the computer specialist, or sometimes both in collaboration. During the middle and late 1960s, the digital computer, because of its increase in speed and size, and the accumulation of library routines, became the favored computer for chemical process simulations. As a result, the analog computer suffered a decline in usage and has now been phased out from most industrial computer installations. The advent of digital simulation with user access to library routines that perform a wide variety of calculations has enabled many analysts to bypass the computer specialist, by constructing their own

1

programs. In other words, it has encouraged the trend to open-shop operation with less reliance by the user on the specialist. One objective of this book is to provide the reader with a set of programs that will enable him to program his own problem with a minimum of effort, specifically in the area of process simulation.

As processes become more complex, incorporating ever-increasing degrees of automation, there will be a greater need for the analytical approach to problems associated with their design and operation. Modern analysis of process problems usually involves some form of mathematical modeling, and, in one sense, this should appeal to chemical engineers because modeling of processes, either on a bench or pilot scale, has long been a favored preliminary to a commercial plant. There are various mathematical models for the same system, each one suited to solve a particular problem associated with the system. The two broad classifications constitute steady-state and dynamic models, and in either type the degree of detail required depends on the problem to be solved as well as the amount of basic data available. A very precise description of a chemical process system will often lead to a large set of unwieldy equations. Although they can be solved, it is advisable for the analyst to use engineering judgment to reduce the equations to a less complex set that for all practical purposes will yield an engineering solution within the accuracy of the basic data provided.

One important aspect of mathematical modeling is the arrangement of the equations. It has been found by experience that if the equations are arranged in a logical or cause-and-effect sequence the computer model is stable. This sequence is termed the "natural" order, for it invariably closely parallels the cause-and-effect sequence found in nature. It will soon be realized that the key to understanding the internal mechanism lies in being able to define this natural cause-and-effect sequence.

The educational background required for modern analysis is an increasing problem, for it is clear that computers have changed the emphasis placed on this subject. Before computer usage became popular, instruction in engineering analysis was (and still is in some places) restricted to simple systems, and most of the effort was devoted to solving the few elementary equations that were derived. These cases were mostly of academic interest and, because of their simplicity, were of little practical value. To this end, a considerable amount of time was devoted to acquiring skills in mathematics, especially to methods of solving differential equations. In fact, most chemical engineers are given courses in differential equations, but experience shows that very little of this knowledge is retained by the engineer after graduation for the simple reason that mathematical methods are not adequate to solving most systems of equations encountered in

**Table 1-1. Classification of Mathematical Problems\* and Their Ease of Solution by Analytical Methods**

| | Linear Equations | | | Nonlinear Equations | | |
|---|---|---|---|---|---|---|
| Equation | One Equation | Several Equations | Many Equations | One Equation | Several Equations | Many Equations |
| Algebraic | Trivial | Easy | Essentially impossible | Very difficult | Very difficult | Impossible |
| Ordinary differential | Easy | Difficult | Essentially impossible | Very difficult | Impossible | Impossible |
| Partial differential | Difficult | Essentially impossible | Impossible | Impossible | Impossible | Impossible |

\* Courtesy of Electronic Associates, Inc.

industry, and any advanced mathematics is forgotten through sheer disuse. Table 1-1 shows the various classes of mathematical equation and the limited class amenable to analytical solution. Most industrial process problems fall into the category of nonlinear differential equations, which can be solved only by a computer. The classes of equation that are amenable to analytic solution are of a trivial nature and restricted to very few cases of industrial interest. The table shows a heavy border that separates the possible from the impossible, and by these standards practical problems defy *analytical* solution or yield extremely cumbersome, essentially useless answers.

The three broad areas that are fertile fields of analytical studies are the following:

1. Research and process development
2. Process design
3. Improvement of process operations

The areas of analysis cover these categories:

1. Fluid flow
2. Mass transfer
3. Heat transfer
4. Kinetics
5. Dynamics and control

The approach to mathematical model building outlined in this book assumes that the student is familiar with the fundamentals of these areas. A model of a typical process will often involve all five categories and the

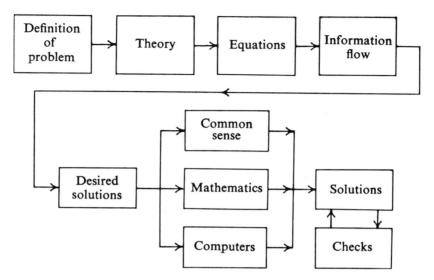

FIG. 1-1. Procedure for Analytical Approach

general procedure for conducting an analytical study can be grouped into seven stages as shown in Figure 1-1

The first step, perhaps the most important one, is that of problem definition, yet it is not possible to establish rules for problem definition that are sufficiently general to be useful. Technical problems are so diverse that it is up to the analyst to state clearly the nature of the individual problem. This will establish a definite objective for the analysis and is invaluable in outlining a path from the problem to the solution.

The second step is a definition of the theory that governs the phenomena of the problem. This theory is usually available from a variety of sources, both published and unpublished, but for those isolated cases in which there is no theory available it is worthwhile to postulate one, or several, and to test its validity later by comparing the solution of the mathematical model with experimental results. One of the advantages of the computerized approach is the facility of rapidly obtaining solutions to various cases; this makes comparisons between alternate theories possible.

Next, the theory, as applied to the problem, is written in mathematical symbology, a necessary step that forces the analyst to a clear unambiguous definition of the problem. The physical systems to be studied in this book are always described by a set of simultaneous algebraic and differential equations, which must be written in the most direct form possible; no manipulations are required at this stage. It is worthwhile at this point.

however, to simplify the equations, whenever possible, by omitting insignificant terms. Care is needed, though, to ensure that any terms omitted are indeed insignificant during the entire course of the problem run. Often it is possible to eliminate entire equations by merely neglecting minor fluctuations in certain intermediate variables; for example, suppose the specific heat of a multicomponent mixture required for a heat balance varies only 1% of its value because of expected variations in composition; rather than include an equation in the model, to compute a value continuously, an average constant number could be substituted.

When the equations are assembled, a procedural method for solving them as a simultaneous set is required. This is sometimes referred to by mathematicians as "equation ordering" but is called "natural arrangement" in this book and consists of placing each equation in an information-flow block diagram which shows how each equation is to be used, that is, the variable it solves for and the interrelationship between the equations.

This technique is merely an extension of the classical linear transfer function notation to systems of nonlinear equations. Such an arrangement, paralleling the logical cause-and-effect relationship in the physical system, presents a clear picture of the postulated mechanism and sometimes reveals interrelationships between variables that were not apparent during previous stages. A consideration of the solutions required from the model is a necessary step preliminary to the computation phase. A list of the various cases required and the information that is expected in each case will reveal possible redundant situations and will be helpful in the programming of the computation phase.

The computation phase that follows offers several alternate routes to the solution. The method selected depends on the complexity of the equations to be solved. There are three general levels, the most elementary being common sense; that is, the solutions desired can be obtained from the model by inspection if the equations or the solutions required are sufficiently simple. It should be realized that this technique cannot be extrapolated to more complex cases without requiring increasing amounts of pure guesswork. The next level, again restricted to systems of modest complexity, solves the equations by analytical techniques. As pointed out in the preceding discussion, a considerable amount of skill is required to solve even some of the simplest sets of nonlinear equations, and such a level is usually beyond the reach of the average process engineer. Fortunately, the third alternative of automatic computation offers the most fruitful path and is the only expedient method for problems of even fair complexity.

The last phase is the study and verification of the solution obtained from the mathematical model. Any unexpected solution should be rationalized to ensure that no errors have occurred in the computation; also, some of

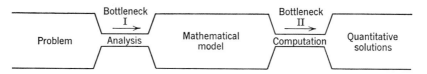

FIG. 1-2.  Difficulties of Analytical Approach before Computers

the computer runs should be specifically designed to check the validity of the mathematical model.

The procedure for obtaining quantitative solutions from an analytical approach has changed significantly in recent years. Figure 1-2 is a symbolic representation of the situation that existed before computers became readily available.

Basically, there were two bottlenecks, the analysis of the problem and then the solution of the equations resulting from the analysis. This latter bottleneck was generally impassable, so that efforts made at mathematical modeling were of no practical use since the equations could not be solved anyway. As a result, the effective way to solve technical problems was by laboratory or pilot-plant experimental procedures. The disciplines of the rational/analytical approach were regarded as purely academic and consequently were not practiced as a way of life in industry.

Over the last 15 to 20 years computers became more widely used, and the recent impact of software languages and especially simulation languages has made computers almost completely accessible to the average engineer if he can get past the first bottleneck of analysis. It can be safely stipulated, then, that today the second bottleneck does not exist, as shown in Figure 1-3.

Another purpose of this book, then, is to encourage engineers, having once eliminated their reservations toward computers, to develop their analytical abilities by adopting as a start the simple approach presented in the following pages. It will be helpful for the student to review a few basic concepts in mathematics, covered in the next sections.

## 1-1  EQUATIONS

Equations can be classified into two broad groups: algebraic and integral/differential equations. Generally, an algebraic equation does not contain

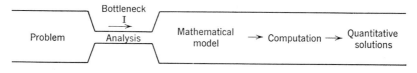

FIG. 1-3.  Difficulty of Analytical Approach Today

a variable expressed as a derivative. For example, the equation $x = ay + bz$ is algebraic, whereas $dx/dt = ay + bz$ is a differential equation and $dx/dt$ is the derivative.

### 1-1-1  Linearity

The concept of linearity in equations is important. An example of a linear equation would be the definition of the pressure at the bottom of a vessel containing liquid (Figure 1-4):

$$P = h\phi + P_0$$

where $P_0 = $ pressure above surface (lb/ft²)
$P = $ pressure at depth $h$ (ft)
$\phi = $ density (lb/ft³)

The relationship between $P$ and $h$ is shown as a straight line on the graph; that is, at any level $h$ a given change in level ($\Delta h$) will produce a corresponding proportional change in pressure $\Delta P$. An example of a nonlinear equation would be the relationship between flow and pressure drop through a valve.

$$Q = C_v\sqrt{P_1 - P_2}$$

where $Q = $ flow rate (gal/min)
$C_v = $ valve constant
$(P_1 - P_2) = $ pressure difference across valve

In this expression the incremental change in flow $Q$ is not proportional to a given change in the pressure drop $(P_1 - P_2)$. Note, however, that although the relationship between flow and pressure drop is nonlinear that between flow and $C_v$ is linear.

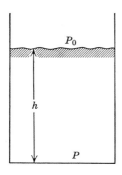 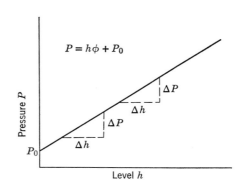

FIG. 1-4.  Example of Linear System

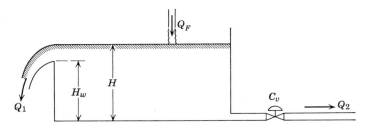

FIG. 1-5. Example of Nonlinear System

### 1-1-2 Implicit and Explicit Equations

Relationships between variables in an equation can be either explicit or implicit. An example of an explicit relationship was shown in the equation for flow, $Q = C_v\sqrt{(P_1 - P_2)}$; that is, given $P_1$, $P_2$, and $C_v$, $Q$ can be established directly. An example of an implicit equation relates to a tank with an outflow at the base and a weir overflow at the side (Figure 1-5). If the total flow to the tank is $Q_F$, then, when steady-state conditions are achieved, it can be shown that $Q_F = 3.336(H - H_w)^{1.5} + C_v(H_w)^{0.5}$. If $Q_F$, $C_v$, and $H_w$ (weir height) are known, $H$ cannot be determined from the above equation *directly*, and a series of manipulations is required before $H$ can be established.

### 1-1-3 Simultaneous Equations

The concept of simultaneity is now described. Consider the system shown in Figure 1-6. A pump supplies a constant flow rate $Q_F$ to two locations through two valves, discharging into pressures $P_1$ and $P_2$, respectively. The equations for the system are

$$Q_F = Q_1 + Q_2$$
$$Q_1 = C_{v1}\sqrt{P_F - P_1}$$
$$Q_2 = C_{v2}\sqrt{P_F - P_2}$$

where $C_{v1}$ and $C_{v2}$ are the value constants and $P_F$ is the discharge pressure of the pump. There are three unknown values for given values of $P_1$, $P_2$, $C_{v1}$, $C_{v2}$, and $Q_F$—the two flows $Q_1$ and $Q_2$ and the pressure $P_F$. None of these unknown values can be determined by solving any one of the above equations by itself. They can be determined only by solving all three simultaneously.

In a broad sense variables in a set of simultaneous equations are implicitly defined, for none can be established *directly* by solving any one equation.

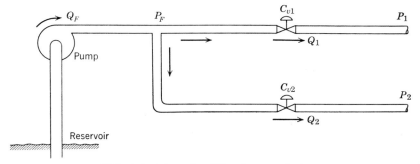

FIG. 1-6. Example of Simultaneous System

## 1-1-4 Sufficiency and Redundancy

In order to obtain a solution to a set of equations, it is necessary to specify as many independent equations as there are dependent variables. Independent equations mean that no redundant equation derived from the other equations can be used as an independent equation; for example, in the set

$$x + y + 2z = 5 \tag{a}$$

$$3x + y + 2z = 3 \tag{b}$$

$$2x + y + 2z = 4 \tag{c}$$

(c) is redundant because it is merely the sum of equations (a) and (b) divided by two and as such it is not a separate statement, and they cannot be solved to give values of $x$, $y$, and $z$. These comments also apply to simultaneous differential equations.

For systems in which there are more variables than equations, there exists an infinite set of solutions, but if a maximizing or minimizing condition is stated one optimum solution can be selected. This general problem area is termed "linear or nonlinear programming." The other, more common, situation of having more equations than unknown variables requires finding a solution that best fits all the equations with a minimum of error. This is the general area of data-fitting problems.

## 1-1-5 Differential Equations

To acquire confidence in formulating differential equations it is essential to understand clearly the meaning of a derivative. The symbol $dv/dt$ merely states "the rate of change of $v$ with respect to $t$." If $v$ is related to $t$, as shown

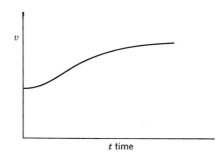

FIG. 1-7

in Figure 1-7, then $dv/dt$ is the slope of the curve at any point $t$. If, for example, a vessel is being filled at a rate $F(t)$ (this symbol means that the feed rate $F$ is not necessarily constant but varies or is a function of time $t$), then, as stated in the equation,

$$\frac{dv}{dt} = F$$

that is, the rate of change of volume $v$ with respect to time is equal to the feed rate $F$. This could also be stated as an integral equation by integrating both sides of the equation

$$v = \int_0^t F \, dt$$

which says, "the volume $v$ at any time $t$ is the accumulation of the flow $F$ over the time period $0 \rightarrow t$ plus the volume at time 0."

If the vessel being filled has a constant cross sectional area $A$, the volume $v = AH$, where $H$ is the height of the surface above a datum level. Now, in general,

$$\frac{dv}{dt} = \frac{d}{dt}(AH) = A\frac{dH}{dt} + H\frac{dA}{dt}$$

but because $dA/dt = 0$ ($A$ is constant)

$$\frac{dv}{dt} = A\frac{dH}{dt}$$

that is, the cross-sectional area $A$ times the rate of change of height $dH/dt$ is equal to the feed rate. This manipulation is called "differentiating by parts" and is sometimes convenient for simplifying equations.

### *Order of Differential Equations*

The order of a differential is the number of times a dependent variable has been differentiated. In the equation considered in the preceding section, the differential was first-order because the volume ($V$) was only differentiated once ($dv/dt$). Sometimes a differential equation is written as a higher-order differential; for example, the classic acceleration equation

$$M \cdot \frac{d^2x}{dt^2} = F$$

mass × acceleration = force

This is an example of a second-order differential, but one interesting aspect is that generally all higher-order differentials can be broken down into a series of simultaneous first-order differentials involving intermediate variables—in this case, the velocity $v$.

$$M \frac{dv}{dt} = F \quad \text{and} \quad \frac{dx}{dt} = v$$

When a computer is to be used to solve the equations in a model, there is nothing to be gained in deriving a higher-order differential equation by a process of substitution, because in any case the equations must be programmed as *first*-order equations. A more important aspect is that the fundamental relationship is invariably a first-order relationship. For instance, the example defined as a second-order relationship (mass × acceleration = force) is really a special case in which the mass is constant. The more fundamental relationship is that

$$\text{force} = \text{rate of change of momentum} \qquad F = \frac{d}{dt}(Mv)$$

and

$$\text{velocity} = \text{rate of change of distance} \qquad v = \frac{dx}{dt}$$

A computer mechanization of these relationships is identical to the cause-and-effect relationship that binds these functions in physical reality, starting with a known force and mass, neither of which need necessarily be constant with time; a computer program would be as follows:

1. Integrate force → momentum.
2. Divide momentum by the mass → velocity.
3. Integrate velocity → distance.

This program is shown in Figure 1-8 in symbolic form.

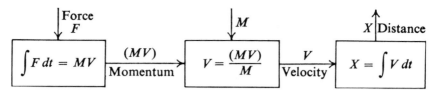

FIG. 1-8. Breakdown of Second-Order Equation

The analyst is able to visualize each of these steps clearly—the *force* changing the *momentum* and the *velocity* changing the *distance*. This mental grasp of the situation is the key factor in ensuring success for an analytical study. The overall relationship for this system, that is,

$$\frac{d}{dt}\left(M\frac{dx}{dt}\right) = F$$

is not so easily visualized and cannot be programmed in this form. For these reasons it serves no useful purpose today. This philosophy, demonstrated by the above simple example, is the basis of the approach to analytical modeling adopted in this book. Experience has shown that a mental visualization of the physical meaning of analytical expressions and statements (i.e., equations) on the part of the analyst and programmer is absolutely invaluable. Because of this, it is worthwhile to break down any incomprehensible complex expression into its components, which are more easily understood and simpler to program.

### Boundary Conditions

A complete definition for differential equations must include numerical values for the *boundary* conditions. An example would be the equation for the volume of liquid in the tank considered above: $dv/dt = F$ (Figure 1-9). This equation defines the volume $V$ at any time $t$ but the initial volume

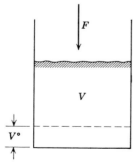

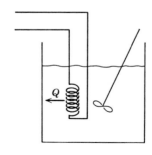

FIG. 1-9. Boundary Condition

FIG. 1-10. System Described by Ordinary Differential Equation

$V^0$ at time $t = 0$ must also be stated. This initial volume is called a boundary condition, and in order to solve the differential equation a value for this condition must be supplied. In any system of equations representing a mathematical model that contains a number of differential equations specific values of the dependent variables are required at particular values of the independent variable for all those variables for which derivative terms are included in the equations; for example, the following simultaneous set of equations defines the variables $X$, $Y$, $Z$:

$$\frac{dY}{dt} = X^2 - Y^2 + 3Z \tag{1-1}$$

$$\frac{dZ}{dt} = Y - 2Z + X \tag{1-2}$$

$$X = 5Z^2 - Y + 6 \tag{1-3}$$

Values of $Y$ and $Z$ are required at a particular value of the independent variable $t$, and because (3) is algebraic $X$ is thereby automatically defined. Usually, boundary values for the dependent variables are all specified for the initial value of the independent variable and are called "initial conditions." In a few situations, however, they are specified at different values of the independent variable and are termed "split boundary value" problems.

Whenever equations include higher-order differentials, boundary values corresponding to the order of the equation are required; for example, the equation

$$\frac{d^2X}{dt^2} = 3 - X$$

requires a value $X^0$ and $(dX/dt)^0$ at $t = 0$ as initial conditions.

### 1-1-6 Partial Differential Equations

In all the differential equations we have considered the derivatives have been defined with respect to only one independent variable time. A large and important class of differential equations involves derivatives with respect to several independent variables. The two contrasting examples following demonstrate the difference between ordinary and partial differential equations.

EXAMPLE 1-1

Consider a well-agitated liquid in a tank being heated by an immersion heater emitting a flow of heat, $Q$ heat units/sec (Figure 1-10). The equation defining the temperature of the fluid is

$$\frac{d}{dt}(WCT) = Q$$

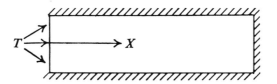

FIG. 1-11. System for Partial Differential Equation

where $W$ is the mass of fluid, $C$ is the unit mass heat capacity, and $T$ is the temperature. The equation assumes that the temperature is the same at all points in the fluid and follows from the assumption of good agitation. For this case, then, there is only one temperature $T$ to be considered, and it leads to the simple ordinary differential equation with only one independent variable, time ($t$).

EXAMPLE 1-2

The case of a solid bar, heated at one end and insulated at all the other sides (Figure 1-11), will produce a partial differential equation if considered during a transient state. The equation relating temperature, time, and distance is

$$\frac{\partial T}{\partial t} = +K \frac{\partial^2 T}{\partial X^2}$$

where $K$ is the thermal diffusivity. This equation defines the temperature as a function of both time $t$ and distance $X$; that is, at any particular time $t_i$ the temperature varies with the distance $X$ or, conversely, at any particular location $X_i$ the temperature will vary with time. This equation, although a brief, correct, and elegant definition of the relationship between the variables, is difficult to visualize in terms of physical reality. Some later chapters in this text offer a more pragmatic approach to this problem area by using the finite difference approach. This permits the definition of a physical system directly in terms of approximate but understandable ordinary differential equations.

## 1-2 CONTINUOUS SYSTEM SIMULATORS

Following the introduction of FORTRAN, there have been a succession of programs (Ref. 3) designed to duplicate the function of analog computers by providing a means for solving sets of ordinary differential equations by digital simulation. These have finally evolved into the current continuous system simulation programs, the most popular being MIMIC (for CDC and UNIVAC computers) and CSMP (for IBM computers). Several modified versions of these programs are available at various computer installations; however, only where the usage is frequent will the computer staff provide software support. These programs often compete with local library routines for performing the integration of sets of differential equations. These routines, combined with the functions and procedures available in the FORTRAN library, essentially duplicate the features of the simulation programs. The advantage of library routines is that they do not require the specialized systems service that is necessary for the more structured programs. Furthermore, they can be readily adapted to any computer operating with FORTRAN, making them virtually universal. For this reason, the policy adopted for this book is to provide the reader with a complete listing of a set of FORTRAN subroutines that fulfill most of the functions of a simulation program. The routines are short and direct, avoiding sophistication and thereby gaining in execution speed. The system, to be referred to as "INT," comprises about a dozen subroutines, and it is hoped that the reader will obtain a grasp of how they function from the explanations in the text and will thereby be stimulated to develop other programs fulfilling his specialized needs.

## 1-3 DYNAMIC PROCESS SIMULATORS

The availability of differential equation simulation programs has led to the development of several higher-level programs that are designed to simulate the dynamic behavior of chemical process units, either singly or combined in a system. They are the dynamic equivalent of the steady-state material and energy-balance programs such as PACER (Ref. 12), FLOWTRAN (Ref. 13), CHESS (Ref. 14), and others. Figure 1-12 shows schematically the supporting structure for these programs and also demonstrates the fact that a process simulation problem can be programmed in any of the three levels.

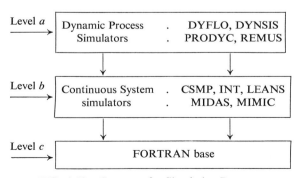

FIG. 1-12.   Structure for Simulation Programs

Use of a higher level merely facilitates the coding effort required; for example, programming at the continuous system level (*b*) eliminates coding most of the numerical procedures required for integrating differential equations, generating arbitrary functions, and converging algebraic equations. Programming at level *c*, using an available dynamic process simulator, in addition to eliminating the coding of numerical procedures, offered by level *b*, also greatly simplifies the effort for defining the system to be simulated. This is accomplished by the use of subprograms that simulate entire unit operations such as fractionation columns, heat exchangers, reactors, and so on. This book provides the listing and describes the use of one of these programs, called "DYFLO," that offers a library of several dozen unit-operation subroutines. It has been in use for several years, and although it was originally conceived for use in schools at the undergraduate level, it has been used successfully for the solution of large, complex industrial problems.

The "INT" program is described in Chapters 2, 3, and 11, and the "DYFLO" program is described and its use demonstrated in Chapters 4, 5, and 8. The bibliography at the end of the chapter provides references on other programs that are currently available.

## REFERENCES

### A.  Continuous System Simulation Programs

1. CSMP: *System/360 Continuous System Modeling Program Uses Manual*, Form H20-036703, Program Number 360A-cx-16x, I, B.M., 1967.
2. MIMIC: *A Digital Simulator Program* H. G. Petersen and F. J. Sansom, SESCA Internal Memo.
3. LEANS: *Lehigh Analog Simulator*, S. M. Morris, Lehigh University, Bethlehem, Penna.
4. Reference to predecessor programs no longer in use may be found in *Mathematical Modeling in Chemical Engineering* 1st ed., R. G. E. Franks, Wiley, New York, 1967; and *The Evolution of Digital Simulation Programs* R. G. E. Franks and W. E. Schiesser, A.I.Ch.E. 59th Annual Meeting, Detroit, December 1966.

### B.  Dynamic Systems Simulators

1. DYNSYS: *Digital Computer Programs for Studying the Transient Behaviour of Systems using a Modular Approach*, S. Babrow, et al., Department of Chemical Engineering, McMaster University, Hamilton, Ontario, Canada, October, 1969.
2. PRODYC: *A Simulation Program for Chemical Process Dynamics and Control*, D. M. Ingels and R. L. Motard, University of Houston, Houston, Texas, RE 4-70, August 1970.
3. REMUS: "Routine for Executive Multi Unit Simulation," *REMUS Users Manual*, P. G. Ham, University of Pennsylvania, Philadelphia, October 1969.
4. FLEX: D. R. Shern, Proctor and Gamble, Cincinnatti.
5. KARDASZ: *A High Level Structure Oriented Simulation Language for Chemical Plants*, J. H. Kardasz, University of Pisa.

6. OSUSIM: E. J. Freeh, Ohio State University, Columbus.
7. EARLYBIRD: R. E. Weaver, Tulane University, New Orleans.

## C. Steady State Energy and Material Balance Programs

1. PACER: P. T. Shannon, Dartmouth College, Hanover, N.H.
2. FLOWTRAN: Monsanto Co., Computerized Engineering Applications Department, St. Louis, Missouri.
3. CHESS: *System Guide*, Technical Publishing Co., Houston, Texas.
4. GIFS: Service Bureau Corporation, New York.

# CHAPTER II

# NUMERICAL SOLUTION OF ALGEBRAIC EQUATIONS

This chapter develops the basic concepts underlying the iterative procedures used in digital computers for solving complex systems of algebraic equations. Several simple, but powerful, methods are described and packaged as subroutines to simplify the programming effort for larger systems. The objective here, and indeed throughout this text, is to build a library of subroutines that perform a variety of tasks common in chemical engineering calculations. There is today a proliferation of systems similar to this, some proprietory, some public—developed and used in industrial corporations and universities—in addition to some packages available from computer manufacturers. The availability of such systems encourages the analyst or problem formulator to construct his own program, and experience has shown that generally this happens. It is hoped that by greatly simplifying problem programming by frequent use of macrosystems, the analyst/programmer can devote his attention to the more important aspects of problem solution, namely analysis and interpretation of results.

Use of macroprograms eliminates the necessity for repeated detailed coding of common procedures, but this does not eliminate the need for the programmer to understand the sequence of events, or at least the basic principle of a particular program he is using, as well as to appreciate its limitations. Thus it is advisable for the reader to understand the concepts described in the following sections.

## 2-1 EXPLICIT AND IMPLICIT EQUATIONS

This subject is covered briefly in Chapter 1, but it is worthwhile to reconsider the meaning of explicit and implicit equations especially as they

**18**

apply to sets of simultaneous equations. Given the equation

$$X = A + B$$

with $A$ and $B$ known, $X$ can be established directly; that is, $X$ is defined explicitly by $A$ and $B$. Now consider the equation

$$X = A^X + B$$

Here, even though $A$ and $B$ are known, $X$ can no longer be established directly without some manipulations. In this sense, a pair of simultaneous algebraic equations such as

$$3Y + 2X = 2 \tag{2-1a}$$
$$2Y + 3X = 4 \tag{2-1b}$$

is *implicit* in that $X$ and $Y$ cannot be established directly without some manipulations. Most physical situations, when described as a set of equations, are nonlinear and simultaneous; that is, the variables are defined implicitly, and the only practical method of obtaining a solution is by numerical methods that, although unsophisticated, are quite successful. Some simple methods will now be demonstrated using equations 2-1a and 2-1b as an example.

These particular equations can be solved by elementary manipulations; in fact, the solution is apparent almost by inspection. However, they will be used here to demonstrate several methods for converging to the solution by iteration. These methods are used for more complex situations not amenable to straightforward solution by mathematical manipulation. The most elementary procedure is termed "direct substitution" and is as follows:

1. Estimate a trial value for $X$ (say 3).
2. Solve equation 2-1a for $Y$ using $X = 3$.
3. Substitute the value of $Y$ obtained in Step 2 in equation 2-1b and solve for $X$.
4. Compare this new value for $X$ with the original $X$, and if not within a prescribed tolerance limit, return to Step 2 using the new value of $X$.
5. When the tolerance limit in Step 4 is satisfied, continue (i.e., either print the result or go on to other calculations).

To assist in visualizing the way the numbers change from cycle to cycle, the following table shows the values of $X$ and $Y$ for each cycle.

| CYCLE | 1 | 2 | 3 | 4 | 5 | 6 | 7 | 8 |
|---|---|---|---|---|---|---|---|---|
| $X$ | 3 | 2.22 | 1.87 | 1.72 | 1.67 | 1.62 | 1.61 | 1.60 |
| $Y$ | $-1.33$ | $-0.81$ | $-0.58$ | $-0.51$ | $-0.44$ | $-0.41$ | $-0.40$ | $-0.40$ |

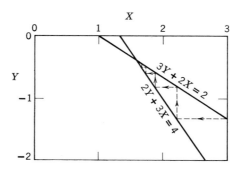

FIG. 2-1. Stepwise Iteration

As will be observed from the table, the values of $X$ and $Y$ "converge" to their final values of 1.6 and $-0.4$ in eight cycles. Starting with other initial guesses for $Y$, the computation will always converge on these values. If the student is not convinced, he should satisfy himself with a trial calculation. A graphical demonstration of these cycles is shown in Figure 2-1. The two equations are shown as straight lines, intersecting at the point $Y = -.4$, $X = +1.6$. The stepwise procedure of direct substitution finally converges on this point of intersection. A simple FORTRAN program that would implement this procedure is as follows:

```
  X = 3.
6 Y = (2. − 2. * X)/3.
  XC = (4. − 2. * Y)/3.
  IF (ABS ((XC − X)/X). LT. .0001) Go to 5
  X = XC
  Go to 6
5 Continue
```

A more effective way of viewing the problem of algebraic convergence is suggested by the above program, which is to realize that for each trial value of $X$, there will result a calculated value ($XC$). The objective of the procedure is to reduce the difference between these two values below a prescribed tolerance limit. Usually this difference is expressed as a fraction of the variable itself ($X$). For most purposes a tolerance of 0.01% is adequate, though it can be changed if necessary.

For most cases of typical complexity, the equations involved are nonlinear, and there may be a large number of equations between $X$ and $XC$. This situation can be summarized by the expression $XC = f(X)$, where $f(X)$

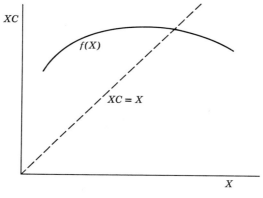

FIG. 2-2

could be a series of explicit equations such as

$$Y = f_1(X)$$
$$Z = f_2(Y)$$
$$X = f_3(Z)$$

The calculation sequence would then be:

$\xrightarrow{\ X\ }$ $\boxed{Y = f_1(X)}$ $\xrightarrow{\ Y\ }$ $\boxed{Z = f_2(Y)}$ $\xrightarrow{\ Z\ }$ $\boxed{XC = f_3(Z)}$ $\longrightarrow XC$

$\xrightarrow{\ X\ }$ $\boxed{\text{Equations}}$ $\longrightarrow XC$

If a plot is made of $X$ versus $XC$, a curve, such as that shown in Figure 2-2, would result. The diagonal line represents all values that satisfy $XC = X$. Therefore, the point where the curve $f(X)$ cuts the diagonal is the solution point where $X = f(X)$.

This diagram can be used to demonstrate several typical functions and to characterize the efficiency of convergence for the direct substitution method.

### Case 2-1  Satisfactory Convergence

Figure 2-3 shows a typical case where satisfactory convergence can be achieved in a relatively few iterative cycles (start at $X_1 \rightarrow X_2$, etc.).

### Case 2-2  Slow Convergence

Figure 2-4 shows a situation where many cycles are required to find the solution.

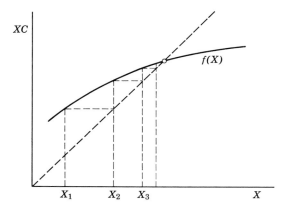

FIG. 2-3.  Efficient convergence

*Case* 2-3  *Oscillatory Instability*

Figure 2-5 shows a typical case of instability where the value of *XC* will continue to oscillate between two values $(X_1, X_2)$ on either side of the true solution.

*Case* 2-4  *Progressive Divergent Instability*

Some situations result in equations which behave as shown in Figure 2-6; that is, direct substitution carries the value of *XC* away from the solution point.

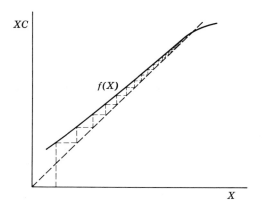

FIG. 2-4.  Poor convergence

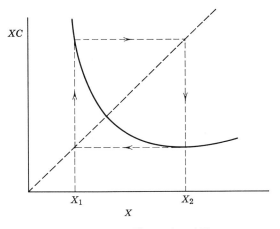

FIG. 2-5. Oscillatory instability

*Case* 2-5 *Oscillatory Divergent Instability*

This situation is characterized by fluctuation of both $X$ and $XC$ around a mean value, but the amplitude of the fluctuations becomes progressively larger and will continue to a limit, as in Case 2-3, or to a computer overflow. Figure 2-7 shows a typical function that produces this effect.

It should be realized by now that although direct substitution is a simple method, it is only satisfactory for well-behaved situations such as Case 2-1, but is inadequate or fails completely for more complex situations such as Cases 2-2, 2-3, 2-4, and 2-5. A modification of the direct substitution method

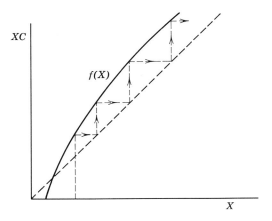

FIG. 2-6. Divergent instability

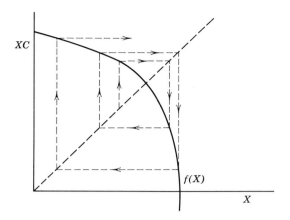

FIG. 2-7.   Divergent instability

can be used that allows the programmer greater control over the convergence procedure in cases of difficulty. This method is described in the next section.

### 2-2   PARTIAL SUBSTITUTION

This method is sometimes used for situations that display oscillatory instability as in Case 2-3. It simply consists of establishing the new trial value somewhere between the old trial value and the resulting calculated value. The formula for this would be

$$X = X_0 + (XC - X_0) * R$$

where $X_0$ = old trial value

$X$ = new trial value

The ratio $R$ can be adjusted by the programmer to achieve stable convergence. If $R = 0.5$, the mean value between $X_0$ and $XC$ is to be the next trial value. If $R = 1$, direct substitution is achieved, but reducing $R$ provides greater stability. Figure 2-8 shows the effect of $R = 0.5$ for Case 2-3, where rapid convergence is achieved.

In a similar manner, it will be effective for oscillatory divergence (Case 2-5). It will be realized, though, that this method of partial substitution aggravates situations of slow convergence shown in Case 2-2 and is equally useless for progressive divergent instability (Case 2-4). However, since it is sometimes useful for special situations, a subroutine encapsulating the method is provided in Figure 2-9.

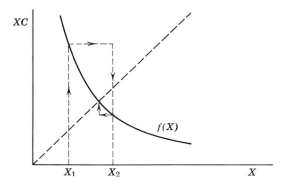

FIG. 2-8.   Partial substitution

where X = trial value,
   XC = calculated value,
   R = partial substitution,
   NC = converge logic control (NC = 1 is converged).

An example of how this subroutine would be used is as follows.
   Suppose the equation to be solved for $X$ is

$$X = (5Y^2 + 3\sqrt{X} - 8X^{.8})^{.36}$$

where the value for $Y$ is provided elsewhere in the program.
   The program for this equation would be

```
      X= 5.                          (initial estimate)
      CONTINUE
   5  XC = (5. *Y**2 + 3. * SQRT(X) − 8.*X**.8) **.36
      CALL CPS(X,XC,.5,NC)
      IF (NC.NE.1 ) GO TO 5
   6  CONTINUE
```

```
1 *       SUBROUTINE CPS(X,XC,R,NC)
2 *       NC=2
3 *       IF(ABS((X-XC)/(X+XC)).LT..0001) NC=1
4 *       X=X+(XC-X)*R
5 *       RETURN
6 *       END
```

FIG. 2-9.   Listing for Subroutine CPS
   X = Trial value
   XC = calculated value
   R = partial substitution ratio
   NC = converge index (NC = 1, converged)

The only advantage of the partial substitution method is that it offers the programmer the opportunity to intervene manually in the convergence sequence by adjusting the substitution ratio $R$. This is most useful for those cases that exhibit oscillatory instability. For general usage, however, a more universal method of convergence is required, one that will achieve convergence for all usual situations. There are several such methods, but only two will be discussed here, namely the Wegstein and Newton-Raphson methods.

### 2-3 WEGSTEIN METHOD FOR ALGEBRAIC CONVERGENCE (Ref. 2)

This method is based on a projection technique and is similar to the mathematical technique called "false postion." The idea is to project from two known points on the $f(X)$ line and thereby determine the next trial value for $X$. This procedure is continued, always using the latest two values of $X$ to project for the next trial value. Figure 2-10 demonstrates this technique.

Figure 2-10 shows a function $f(X)$ for which the solution $X = f(X)$ is to be found. Since two points on the curve are required for the projection, the first two points are established by direct substitution. Starting with $X_1$, the calculated value of $XC1$ is obtained. Substituting $XC1 (= X_2)$ directly in $f(X)$, the second point is established as $XC2$. We now know the coordinates of two points on the curve, namely $(X1, XC1)$ and $(X2, XC2)$. Projecting these two points to the intersection with the diagonal $XC = X$ requires solving the

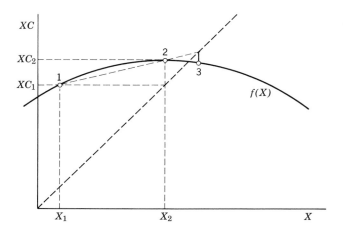

FIG. 2-10.  Wegstein method for convergence

equation

$$X_3 = \frac{X_1 * XC_2 - XC_1 * X_2}{X_1 - X_2 + XC_2 - XC_1}$$

This expression is derived by solving for $X$ in two equations, one for the diagonal $XC = X$ and the other for a line passing through the points 1 and 2.

The projected value $X_3$ is now substituted in $f(X)$ to calculate $XC3$, and the points 2 and 3 are used to project to the diagonal $XC = X$ to establish $X4$. This procedure is repeated until convergence is obtained, which is usually achieved quite rapidly, compared with the more elementary methods. It will also be apparent that the method is effective for all the cases shown previously (Cases 2-1 to 2-5) where direct or partial substitution failed. The reader should sketch a few curves with the characteristic shapes shown in Figure 2-7 and by graphically duplicating the Wegstein method should assure himself of the effectiveness of the method.

The Wegstein method is sufficiently involved to justify a subroutine for repeated use. The listing for this subroutine, called CONVerge, is shown in Figure 2-11. The first item in the argument list, X, is the trial value of the variable, while the second argument XC is the calculated value. The next argument NR is the numeral identifying each call on the subroutine, which is used as a subscript to specify storage locations in two arrays XA(NR), XCA(NR). The last item in the argument list (NC) is an integer that has a

```
 1 *        SUBROUTINE CONV(X,Y,NR,NC)
 2 *        DIMENSION XA(10) , YA(10)
 3 *        IF(ABS((X-Y)/(X+Y)).LT..0001) GO TO 6
 4 *        IF(NC.LE.1) GO TO 5
 5 *        XT = (XA(NR)*Y-YA(NR)*X)/(XA(NR)-X+Y-YA(NR))
 6 *        XA(NR) = X
 7 *        YA(NR) = Y
 8 *        X = XT
 9 *        RETURN
10 *      5 XA(NR) = X
11 *        YA(NR) = Y
12 *        X = Y
13 *        NC= 2
14 *        RETURN
15 *      6 X=Y
16 *        NC=1
17 *        RETURN
18 *        END
```

FIG. 2-11.   Listing for subroutine CONV
Argument list;
  X = trial value
  XC = calculated value
  NR = routine call number
  NC = converge index (NC = 1, converge)

value of either 1 or 2, indicating convergence or nonconvergence. It is used in the main program to direct the computation to either continue (NC = 1) or to recycle for an additional evaluation of the function. After convergence, on the next pass through the routine a direct substitution is made, since this will again be the first pass of a new convergence cycle (lines 10–14). The projection formula is on line 5 and is used for the succeeding estimates for X after the first direct substitution cycle. The test for convergence is on line 3 and is set for 0.01 % which should be adequate for most cases. The preceding values of X and XC are stored internally in the arrays XA and XCA in the location identified by the call number NR.

The reader should realize that if a situation calls for a converged solution XC = 0, the test for convergence will break down. It is only valid for finite values of XC other than zero. The denominator of the convergence criterion was made X + XC in order to avoid the possibility that if either X or XC are temporarily zero during a series of iterations to convergence, the criterion would numerically overflow.

An example of how this subroutine would be used in a main program is as follows (see Example 2-2):

```
    X = 5
    CONTINUE
  5 XC = (5.* Y ** 2 + 3. * SQRT(X) − 8 * X ** .8) ** .36
    CALL CONV (X,XC,1,NC)
    GO TO (6,5),NC
  6 CONTINUE
```

The number of times this subroutine can be called in any program is determined by the dimensions of the internal arrays XA and XCA. These are nominally set to 10 in Figure 2-11 but can be increased if necessary.

## 2-4 EXAMPLE: BEATTIE-BRIDGEMAN EQUATION

The Beattie-Bridgeman equation of state is expressed as

$$V = \left(RT + \frac{\beta}{V} + \frac{\gamma}{V^2} + \frac{\delta}{V^3}\right)\frac{1}{P}$$

where $\beta = RTB_0 - A_0 - RC/T^2$,
$\gamma = -RTB_0b + aA_0 - RB_0C/T^2$,
$\delta = RB_0bc/T^2$,
$R$ = gas constant (atm liter/°K g-mole) (0.08206),
$T$ = temperature (°K),
$P$ = pressure (atm),
$V$ = volume (liters).

```
1*          DATA A,B,C,AO,BO/.11171,.07697,3.E6,16.6037,.2354/
2*          DATA R,T,P/.0827,408.,36./
3*      100 FORMAT(4E12.5)
4*    C **PRELIMINARY CALCULATION**
5*          BT=R*T*BO-AO-R*C/T**2
6*          GM=-R*T*BO*B+A*AO-R*BO*C/T**2
7*          DEL=R*BO*B*C/T**2
8*          V=R*T/P
9*    C **ITERATION SECTION**
10*       5 VC=(R*T+BT/V+GM/V**2+DEL/V**3)/P
11*          PRINT 100,V,VC
12*          CALL CONV(V,VC,1,NC)
13*          GO TO (6,5),NC
14*       6 CONTINUE
15*          END
```

FIG. 2-12.   Use of CONV Subroutine

The constants $A_0$, $B_0$, $a$, $b$, $c$ are available from tabular data (Ref. 1) for a number of common gases. The values for isobutane are

$$A_0 = 16.6037$$
$$B_0 = .2354$$
$$a = .11171$$
$$b = .07697$$
$$c = 300.*10^4$$

For particular values of temperature $T$ and pressure $P$ the value of the specific volume $V$ cannot be obtained explicitly from this equation, it would require iteration to converge on the correct solution. Selecting $T = 350°K$ and $P = 1$ atm, the program shown in Figure 2-12 will perform the necessary iterations using the Wegstein subroutine CONV. A starting value for $V$ can be obtained as shown on line 8 from the ideal gas law (also the first term on the Beattie-Bridgeman equation); that is, $V = RT/P$. The trial values of $V$ and the calculated values ($VC$) at each cycle are shown in Figure 2-13 both in numerical form and also plotted, giving an estimate of the rapidity of convergence.

This example is used in the next section to demonstrate the Newton-Raphson method for convergence.

## 2-5   NEWTON-RAPHSON CONVERGENCE

This method is particularly useful for those cases where the derivative of the function can be obtained analytically. A more convenient way of expressing the functional relationship is $f(X) = 0$. The value of $X$ that satisfies this relationship is the root of the equation. The method uses the derivative of the function with respect to $X$ $df(X)/dX$ to determine the next trial value

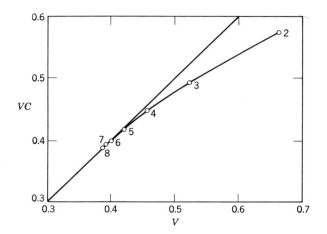

FIG. 2-13.  Results of Program in Figure 2-12

| V | VC |
|---|---|
| .93727 + 00 | .66555 + 00 |
| .66555 + 00 | .57210 + 00 |
| .52312 + 00 | .49408 + 00 |
| .45888 + 00 | .44828 + 00 |
| .42196 + 00 | .41824 + 00 |
| .40200 + 00 | .40080 + 00 |
| .39242 + 00 | .39212 + 00 |
| .38923 + 00 | .38919 + 00 |

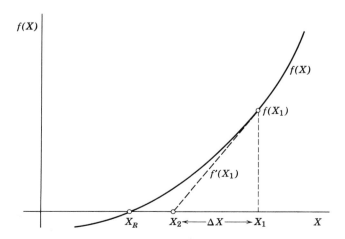

FIG. 2-14.  Newton Raphson Convergence

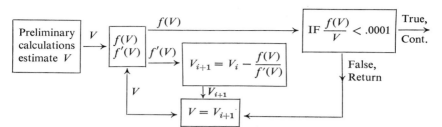

FIG. 2-15.  Signal flow for Newton-Raphson procedure

by the relationship

$$x_{i+1} = x_i - \frac{f(x_i)}{f'(x_i)}$$

Figure 2-14 shows a $f(X)$ plotted versus $X$. At the first trial using $X_1$ the function is evaluated as $f(X_1)$, and the derivative will be $f'(X_1)$. It will be evident that since the derivative $f'(X_1) = f(X_1)/\Delta X$, the change in the trial value $\Delta X = f(X_1)/f'(X_1)$; that is, $X_2 = X_1 - f(X_1)/f'(X_1)$. This new trial value will be closer to the root $X_R$. This procedure is repeated until $X$ is sufficiently close to $X_R$ to reduce $f(X)$ below a prescribed tolerance.

## 2-6  EXAMPLE: NEWTON–RAPHSON METHOD

The example in the previous section showed how the specific volume was calculated by iteration using the Beattie-Bridgeman equation of state. The Newton-Raphson method will now be used to obtain the specific volume. The B-B equation, expressed as $f(V)$ is

$$f(V) = \left( RT + \frac{\beta}{V} + \frac{\gamma}{V^2} + \frac{\delta}{V^3} \right) \frac{1}{P} - V = 0$$

Differentiating with respect to $V$

$$f'(V) = \frac{d}{dV} f(V) = -\left( \frac{\beta}{V^2} + \frac{2\gamma}{V^3} + \frac{3\delta}{V^4} \right) \frac{1}{P} - 1$$

The procedure followed in the program is shown in signal flow form in Figure 2-15. The iteration section of the program is shown in Figure 2-16. The preliminary calculation section is the same as in Example 2-6 (Figure 2-12). The numerical results are shown in Figure 2-17 and are also plotted. When comparing the convergence sequence for the Wegstein and Newton-Raphson methods (Figures 2-13 and 2-17), conclusions should be reached only as to their efficacy for this particular example. Other situations will show one method significantly superior over the other. For example, a useful

```
 9*   C **ITERATION SECTION**
10*       5 FV=(R*T+BT/V+GM/V**2+DEL/V**3)/P-V
11*         FPV=-(BT/V**2+2.*GM/V**3+3.*DEL/V**4)/P-1.
12*         ERR=ABS(FV/V)
13*         PRINT 100,V,ERR,FV,FPV
14*         V=V-FV/FPV
15*         IF(ERR.GT..0001) GO TO 5
16*         CONTINUE
17*         END
```

FIG. 2-16.   Newton-Raphson Procedure

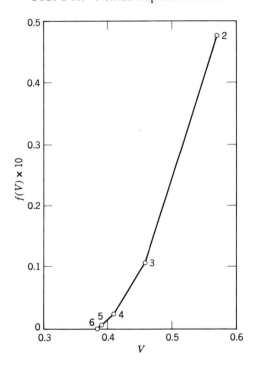

FIG. 2-17.   Results of Newton-Raphson Iteration

| V | ERR | FV | FPV |
|---|---|---|---|
| .93727 + 00 | .28991 + 00 | −.27172 + 00 | −.74215 + 00 |
| .57114 + 00 | .83264 − 01 | −.47555 − 01 | −.42288 + 00 |
| .45869 + 00 | .23017 − 01 | −.10558 − 01 | +.22443 + 00 |
| .41164 + 00 | .56219 − 02 | −.23142 − 02 | −.12520 + 00 |
| .39316 + 00 | .92682 − 03 | −.36439 − 03 | −.85916 − 01 |
| .38892 + 00 | .48267 − 04 | −.18772 − 04 | −.77075 − 01 |

application of Newton-Raphson is in converging multicomponent vapor-liquid equilibrium, inevitably involved in separation processes. In fact, it is used in Chapter 5 as part of a general routine for an equilibrium calculation (EQUIL).

## 2-7 IMPLICIT SYSTEMS OF HIGHER ORDER

In this chapter we are concerned with methods for handling implicit equations that may occur in the algebraic sections of a mathematical model. Typically, a single loop situation can occur such as $X = f(X)$, where $f(X)$ could be a series of equation steps. For this case it is necessary to converge on $X$ by one of the methods described in the previous sections. It is also possible to have to deal with more than one implicit loop. For example:

$$X = f_1(Y, Z) \qquad (2\text{-}7a)$$

$$Y = f_2(X, Z) \qquad (2\text{-}7b)$$

$$Z = f_3(X, Y) \qquad (2\text{-}7c)$$

The recommended method for such a system is as follows:

Step 1   Estimate $Z$
Step 2   Estimate $Y$
Step 3   Calculate $X$ from equation 2-7a
Step 4   Calculate YC from equation 2-7b
Step 5   CONVerge YC and $Y$ (estimated) by returning to 2
Step 6   When Step 5 is satisfied calculate ZC from equation 2-7c
Step 7   CONVerge ZC and $Z$ by returning to Step 1

The important point of the above procedure is to converge the inner loop $(Y = YC)$ at each iteration of the outer loop; otherwise the system may fail to converge. These comments apply to the Wegstein or partial-substitution methods. Newton-Raphson, of course, is particularly suited for these situations, since each variable would be converged individually by adjusting according to its own derivative. An example of this method as applied to a multielement case is described in Chapter 5.

Another recommendation is to take advantage of any possibilities of rearranging the equations to eliminate the inner loops. For example, the equations 2-7a, 2-7b, 2-7c could be reduced to

$$X = f_1'(Y)$$

$$Z = f_2'(Y, X)$$

$$Y = f_3'(X, Y, Z)$$

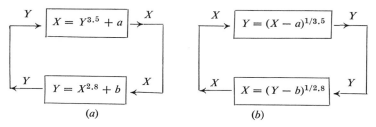

FIG. 2-18. Alternate Implicit Sequence

This set now requires converging only on $Y$. An example of this reduction is given in Example 6-5.

Another recommendation is to arrange the sequence of calculations within the implicit loop to be converged so that the calculation proceeds to lower powers of the variables involved. This will in most cases ensure a greater measure of stability for the convergence. For example, Figure 2-18 shows two arrangements for a pair of simultaneous algebraic equations.

Arrangement $a$ is more likely to be unstable than $b$ since each trial value is raised to a higher power in order to calculate the succeeding variable.

## 2-8 ARBITRARY FUNCTION GENERATION (FUN1)

An invaluable addition to a library of subroutines is a procedure for generating an arbitrary function. The opposite of an arbitrary function is an analytic function, such as $Y = e^x$ or $Y = \sin x$. Arbitrary functions cannot be expressed so elegantly; in fact, the relation can only be expressed as a series of tabulated values or a curve (Figure 2-19).

It is possible, by regression methods, to adjust coefficients in a high-order equation to fit the function to a prescribed accuracy. However, by describing the function as a series of coordinate points and by using an interpolation technique it is possible to represent the function directly. This avoids having to develop an analytic expression by regression methods. The only restriction is that the function must be single valued for all values of the input variable. The accuracy achieved will be established by the number of coordinate points employed.

Figure 2-19 shows an arbitrary function relating $Y$ to $X$. This function can be described by a series of straight-line segments connecting a set of points located on the curve. The greater the number of points, the closer the series of straight lines approximates the curve. Also, since the points do not have to be equally spaced, they can be crowded together in regions of sharp curvature change and spaced further apart in relatively linear regions. Most relationships in chemical engineering are usually simple monotonic curves; thus

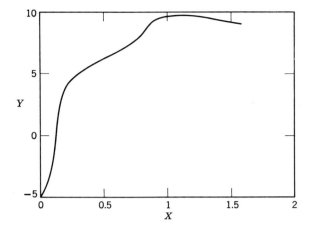

FIG. 2-19. Example of an Arbitrary Function

| X | Y |
|---|---|
| 0 | −5 |
| 0.1 | −2.6 |
| 0.2 | +3.8 |
| 0.4 | +5.6 |
| 0.7 | +7.2 |
| 0.9 | +9.3 |
| 1.3 | +9.4 |
| 1.6 | +9.0 |

10 to 20 points should be sufficient for reasonably accurate representations, but more points can be used if required. The coordinates of each point are stored in an X and Y array in the main program. A subroutine FUN1 can now be used to calculate a value of $Y$ for a particular value of $X$. It will search through the array and locate the adjacent coordinate points around $X$, that is, $X_i$, $Y_i$ and $X_j$, $Y_j$ (see Figure 2-20). It then obtains $Y$ by linear

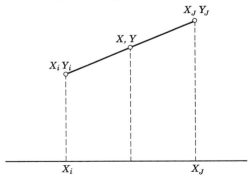

FIG. 2-20. Interpolation between Adjacent Points

```
 1 *          C ARBITRARY FUNCTION SUBROUTINE Y VERSUS X
 2 *                  FUNCTION FUN1(A,N,X,Y)
 3 *                  DIMENSION X(2),Y(2)
 4 *                  IF (A-X(1)) 5,5,6
 5 *                6 IF (A-X(N)) 1,2,2
 6 *                2 FUN1 = Y(N)
 7 *                  RETURN
 8 *                5 FUN1 = Y(1)
 9 *                  RETURN
10 *                1 DO 3 I = 2,N
11 *                  IF (A .LT. X(I)) GO TO 4
12 *                3 CONTINUE
13 *                4 FUN1 = Y(I-1) + (A-X(I-1))*(Y(I)-Y(I-1))/(X(I)-X(I-1))
14 *                  RETURN
15 *                  END
```

FIG. 2-21.   Listing for Subroutine FUN1
Argument list:
A = input variable
N = total number of coordinate points
X = X array
Y = Y array

interpolation between the points

$$Y = Y_i + \frac{X - X_i}{X_j - X_i} (Y_j - Y_i)$$

The function subroutine also provides for the case where the input is less than the first point or greater than the last point. In either case the output will be the first value of $Y$ (i.e., $Y_1$) or the last value $Y_n$ respectively. Figure 2-21 shows a listing of the function subroutine FUN1, where A is the input variable, N the total number of coordinate points, X and Y the arrays designating the coordinates of each point. The first two lines in the subroutine (lines 4 and 5) test the imput against the X coordinate of the first

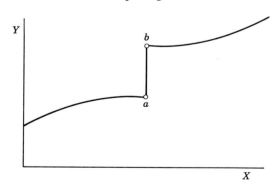

FIG. 2-22.   Discontinuous Function

and last point. The sequential search through the array to find the adjacent points to A is conducted in the DO loop starting on line 10. The interpolation formula is calculated on line 13.

It should be evident from the above search procedure that the coordinate points must be listed in the X and Y arrays in ascending values of $X$. The $Y$ coordinates will be established by the function and may have any value. If it is ever necessary to represent a sharp discontinuity such as shown in Figure 2-22, an approximation can be made by providing the $X$ coordinate of the point $b$ with a slightly greater value than that of point $a$, to give the segment an almost vertical slope. An example of how this subroutine is to be used will now be demonstrated.

## 2-9   USE OF FUN1 SUBROUTINE

An arbitrary function representing the measured wall temperature of an electrically heated pipe is to be generated in a computer simulated as a continuous function of length. The function is shown in Figure 2-23.

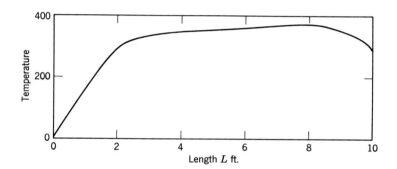

FIG. 2-23.   Arbitrary Function

| L | T |
|---|---|
| 0 | 10. |
| 1 | 150. |
| 2 | 290. |
| 2.5 | 320. |
| 3 | 330. |
| 4 | 345. |
| 5.3 | 350. |
| 7.4 | 370. |
| 8 | 375. |
| 8.5 | 370. |
| 9.7 | 325. |
| 10 | 290. |

A suitable series of 12 points are shown marked on the curve, and the coordinates of each point are shown in the adjoining table. This data can be entered into an array in several ways, depending on the programmers' preference. One suitable method is by using the following data statement:

$$\text{DATA (AL (N), N} = 1, 12)/\ 0., 1., 2., 2.5, 3., 4., 5.3,$$
$$7.4, 8., 8.5, 9.7, 10./$$

$$\text{DATA (AT (N), N} = 1, 12)/\ 10., 150., 290., 320., 330.,$$
$$345., 350., 370., 375., 370., 325., 290.,/$$

Having entered the data into the X (AL) and Y (AT) array, the subroutine can be called any time the temperature is required at any position down the tube by the following statement

$$\text{T} = \text{FUN1 (L, 12, AL, AT)}$$

where L is the distance down the tube and T is the temperature.

This subroutine can be used in various ways. For example, suppose an iterative solution procedure is required to solve a particular problem. Each iteration produces a function F(L) that has to be used in the following iteration. This can be achieved quite simply by storing the function F(L) and the corresponding value of L in two storage arrays, and on the following iteration specifying these two arrays as imputs to FUN1 and thus obtaining F(L). An example of this procedure is discussed later.

### 2-10   TWO-DIMENSIONAL FUNCTION  (FUN2)

The interpolation technique programmed in the single-dimension function generator FUN1 can be extended to the two-dimensional function. In this situation a variable $Y$ is a function of two imput variables $X$ and $Z$. A typical method of defining such a function is a family of curves as shown in Figure 2-24.

The interpolation procedure for this two-dimensional function is programmed in the subroutine FUN2 shown in Figure 2-25. Most of the statements in this subroutine are tests to determine whether the input variables

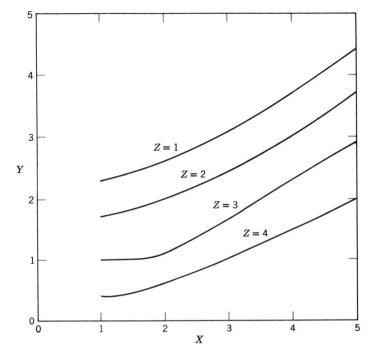

FIG. 2-24.   Two-Dimensional Function

$X$ and $Z$ are beyond the bounds of the data supplied. If such a situation does occur, the closest point on the border of the data field is determined by proper extrapolation and passed out as the value of the function. Under normal circumstances the $X$ and $Z$ values (symbolized in the argument list as A and B) will be within the data limits. The search procedure (starting on line 30) first establishes the adjacent points to the value of A on the X axis, then searches along the Z axis for the adjacent points to the B input. The interpolation formulas are on lines 47, 52, and 53.

The preceding brief description of the procedure is adequate for an appreciation of what is involved. The data preparation for the arrays is restricted by one simple rule: there must be the same number (M) of coordinate points in the Z direction for each X coordinate point. An example of how the data would be entered in the X, Y, and Z arrays will make this clear. The family of curves shown in Figure 2-23 can be expressed as a series of triplet values X, Z, and Y) representing the coordinates of each point. The X, Z, and Y

```
  1 *    C ARBITRARY FUNCTION SUBROUTINE  Y VERSUS X AND Z
  2 *           FUNCTION FUN2(A,B,N,M,X,Z,Y)
  3 *           DIMENSION X(2), Z(2), Y(2)
  4 *           IF ((A .LE. X(1)) .AND. (B .LE. Z(1))) GO TO 13
  5 *           IF ((A .GE. X(N)) .AND. (B .GE. Z(N))) GO TO 14
  6 *           IF ((A .LE. X(1)) .AND. (B .GE. Z(M))) GO TO 15
  7 *           IF ((A .GE. X(N)) .AND. (B .LE. Z(N-M+1))) GO TO 16
  8 *           IF (A .LE. X(1)) GO TO 19
  9 *           IF (A .GE. X(N)) GO TO 23
 10 *           MP = M+1
 11 *           GO TO 17
 12 *        23 I = (N-M+1)
 13 *           I2 = N
 14 *           GO TO 22
 15 *        19 I = 1
 16 *           I2 = I+M-1
 17 *        22 DO 20 J = I,I2
 18 *        20 IF (B .LE. Z(J)) GO TO 21
 19 *        21 J = J - 1
 20 *           FUN2 = Y(J)+ (B-Z(J))/(Z(J+1)-Z(J))*(Y(J+1)-Y(J))
 21 *           RETURN
 22 *        13 FUN2 = Y(1)
 23 *           RETURN
 24 *        14 FUN2 = Y(N)
 25 *           RETURN
 26 *        15 FUN2 = Y(M)
 27 *           RETURN
 28 *        16 FUN2 = Y(N-M+1)
 29 *           RETURN
 30 *        17 DO 3 I = MP ,N,M
 31 *           IF (A .LT. X(I)) GO TO 4
 32 *         3 CONTINUE
 33 *         4 IF (B .LT. Z(I-M)) GO TO 9
 34 *           IF (B .GT. Z(I-1)) GO TO 11
 35 *           GO TO 12
 36 *         9 YT1 = Y(I-M)
 37 *           YT2 = Y(I)
 38 *           GO TO 18
 39 *        11 YT1 = Y(I-1)
 40 *           YT2 = Y(I+M-1)
 41 *           GO TO 18
 42 *        12 J1=I-M
 43 *           J2=I-1
 44 *           DO 5 J = J1,J2
 45 *           IF (B .LT. Z(J)) GO TO 6
 46 *         5 CONTINUE
 47 *         6 YT1 = Y(J-1) + (B-Z(J-1))/(Z(J) - Z(J-1))*(Y(J)-Y(J-1))
 48 *           I2=I+M-1
 49 *           DO 7 J = I,I2
 50 *           IF (B .LT. Z(J)) GO TO 8
 51 *         7 CONTINUE
 52 *         8 YT2 = Y(J-1) + (B-Z(J-1))/(Z(J)-Z(J-1))*(Y(J)-Y(J-1))
 53 *        18 FUN2 = YT1 + (A-X(I-1))/(X(I)-X(I-1))*(YT2-YT1)
 54 *           RETURN
 55 *           END
```

FIG. 2-25.   Listing for FUN2
Argument list:
A = X input variable
B = Z input variable
N = total number coordinate points
M = number coordinate points in groups
X = X array
Z = Z array
Y = Y array

arrays would then be as follows:

| X | Z | Y |
|----|----|-----|
| 1. | 1. | 2.3 |
| 1. | 2. | 1.7 |
| 1. | 3. | 1.0 |
| 1. | 4. | 0.4 |
| 2. | 1. | 2.6 |
| 2. | 2. | 2.0 |
| 2. | 3. | 1.1 |
| 2. | 4. | 0.6 |
| 4. | 1. | 3.7 |
| 4. | 2. | 3.0 |
| 4. | 3. | 2.3 |
| 4. | 4. | 1.5 |
| 5. | 1. | 4.4 |
| 5. | 2. | 3.7 |
| 5. | 3. | 2.9 |
| 5. | 4. | 2.0 |

The third item in the argument list of FUN2 is N, the total number of coordinate points. In this example $N = 16$. The fourth item is M, the number of points in each group (i.e., having a common X coordinate), which in this case is 4. The last three items symbolize the data arrays X, Z, and Y. The data would be entered into the arrays as shown in the previous section for FUN1, and the subroutine would be used in a manner similar to that shown for FUN1.

It will be found that FUN2 enjoys frequent use in chemical engineering, since two-dimensional functions are quite common. However, the use of the routine is computationally time-consuming and should not be used for cases where the data can be approximated adequately by analytical expressions.

**Problems**

1. The Benedict-Webb-Rubin (Ref. 1) equation of state is expressed as

$$Pv = RT + \frac{\beta}{v} + \frac{\sigma}{v^2} + \frac{\zeta}{v^4} + \frac{\omega}{v^5}$$

$$\beta = RTB_o - A_o - C_o/T^2$$
$$\delta = bRT - a_2 + (c/T^2)e^{-v/v^2}$$
$$\eta = cye^{-v/v^2}/T^2$$
$$\omega = a\alpha$$

For isobutane the eight constants have the following values:

$$A_o = 10.2326$$
$$B_o = 0.137544$$
$$C_o = 0.84994$$
$$a = 1.9376$$
$$b = 0.042435$$
$$c = 0.286 * 10^6$$
$$\alpha = 1.0741 * 10^{-3}$$
$$\gamma = 3.4 * 10^{-2}$$

where $R = 0.08207$ (gas constant atm liters/g-mole °K),

$$P = \text{pressure (atm)},$$
$$v = \text{volume (liters)},$$
$$t = \text{temperature (°K)}.$$

Calculate the volume $v$ for a pressure of 36 atm for temperature ranging from 300 to 410°K (every 10°) and determine the % deviation from the values determined by the Beattie-Bridgeman equation of state (Example 2-4).

2. A centrifugal pump is located 5 ft above a tank car 8 ft in diameter containing a liquid with a density of 85 lb/ft³ (Figure 2-26a). The pump has to deliver the contents of the tank car to a storage tank through a horizontal delivery pipe.

The pressure rise across the pump is specified as a function of the flow rate, as shown in Figure 2-26b. The total pressure drop due to friction in the

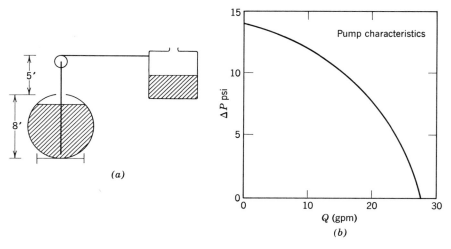

(a)

(b)

FIG. 2-26

suction and delivery pipes is $P_f = 0.012Q^2$, where $Q$ = gpm and $P$ = psi. The pressure at the tank car and the storage tank is atmospheric. Calculate how the delivery rate to the storage tank varies as the level in the tank car falls through 8 ft from full to empty.

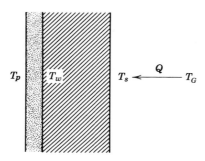

3. A furnace gas at a temperature $T_G$ radiates heat $Q$ to the outer surface of a pipe whose temperature is $T_s$. The heat is conducted through the pipe wall to the inner surface $T_w$ and then through a film to a process stream having a temperature $T_p$. Analyzing this system, the following relations are established:

Radiant heat flux $Q = a(T_G{}^4 - T_s{}^4)$
Conduction through pipe wall $Q = b(T_s - T_w)$
Conduction through film $Q = c(T_w - T_p)$
$$a = 1.2 * 10^{-9}$$
$$b = 70 + 0.07(T_s + T_w)$$
$$c = 6$$

Calculate $Q$, $T_s$, and $T_w$ for $T_p = 900°K$ and $T_G$ ranging from 1200 to 2000°K every 50°K.

4. A continuous flow chemical reactor operating as a steady-state system can be described by the following set of equations:

$$F * (XA) = FA - R$$
$$F * (XB) = FB - R$$
$$F = FA + FB - R$$
$$R = 60 * (XA) * (XB)^{.6}$$

Solve the above set of equations for $XA$, $XB$, $F$, and $R$ assuming $FA = 5$ and $FB = 7$. The physical limits on the reaction rate $R$ are 0 and $FA$.

5. The arbitrary function generator subroutine FUN1 described in this chapter suffers from a minor inefficiency; namely, on every call the search

procedure starts from the beginning of the array. What modification to the subroutine would be required in order to start the search for the coordinate points straddling the imput not from the beginning but from the straddling points found on the preceding use of the subroutine for that particular call?

## REFERENCES

The following references provide a more advanced discussion of the topics covered in this chapter.
1. *Material and Energy Balance Computation*, E. J. Henley and E. M. Rosen, Wiley, New York, 1969.
2. "Accelerating Convergence of Iterative Processes" J. H. Wegstein, *Comm. Assoc. Computing Machinery*, 1, 9.
3. *Computational Techniques for Chemical Engineers*, H. H. Rosenbrock and C. C. Storey, Pergamon Press, New York, 1967.
4. *Applied Numerical Methods*, B. Carnahan, H. A. Luther, and J. O. Wilkes, Wiley, New York, 1970.

CHAPTER III

# NUMERICAL SOLUTION OF DIFFERENTIAL EQUATIONS

Many methods are available for performing numerical integrations of differential equations. Experience with a wide range of typical chemical engineering problems has shown that, with almost no exceptions, the fourth-order method (Runge-Kutta) is quite adequate for any situation and that in many cases the second- and first-order methods will provide reasonable accuracy coupled with greater efficiency. This chapter starts at an elementary level and proceeds to explain and demonstrate the three methods of numerical integration. A system of FORTRAN subroutines is developed that incorporate all three methods, allowing the programmer to make a suitable selection for any particular application.

## 3-1 ORDINARY DIFFERENTIAL EQUATIONS

An ordinary differential equation is one in which there is only one independent variable. In a set of simultaneous ordinary differential equations the independent variable, usually time or distance, is common to all the equations. The purpose of these equations is to relate to the independent variable system variables that are interdependent, and the solution to the equations is the values of the system variables related to the independent variable. This relationship is commonly expressed numerically or in graphical form. An example should clarify the terms just quoted. Figure 3-1 shows a tank into which a liquid is flowing at a rate $Q_i$. This flow could be a constant rate. More typically, it would be varying with time, in which case it is written as $Q_i(t)$. This means "$Q_i$ is a function (i.e., varies with) of time $t$." The liquid flows out from the tank through a restriction at a rate $Q_o(t)$. The rate of outflow $Q_o(t)$ is determined by the hydraulic height in the tank $H$ and is

**45**

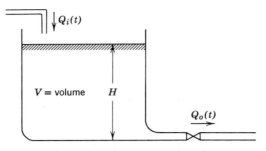

FIG. 3-1.   Continuous Flow Tank

expressed by the relationship

$$Q_o = C_v\sqrt{H}$$

where $C_v$ is a characteristic of the restriction. A differential equation can be formed for this system by symbolizing the statement

rate of accumulation = inflow − outflow

$$\frac{dV}{dt} = Q_i - Q_o \tag{3-1}$$

where $V$ = volume = $A * H$      $A$ = tank area.

Equation 3-1 is a nonlinear ordinary differential equation with time $t$ as the *independent* variable and $V$ as the dependent variable. The derivative of $V$ is the expression on the righthand side of the equation. The *parameters* of the system are $A$ and $C_v$, while the input $Q_i$ is sometimes referred to as a *forcing function*. If this equation is integrated on both sides with respect to time $t$ the following *integral* equation results:

$$\int \frac{dV}{d_\theta}\, dt = V = \int (Q_i - Q_o)\, dt \tag{3-2}$$

Equation (3-2) is the integral form of the differential equation and is a direct parallel of the natural process involved. It says that the volume $V$ is the *integral* of the difference between the inflow and outflow. The natural process equivalent says that the volume in the tank is the *accumulation* of the difference between the inflow and outflow. Clearly, then, integration is the mathematical equivalent of accumulation. It should be realized that natural processes can only accumulate (i.e., integrate) but not differentiate. A differential equation is an elegant way of stating a relationship, but it is invariably solved by an integrating procedure that is equivalent to the natural process. At this point it will be realized that all physical systems could be described in terms of integral equations rather than differential equations.

While this is certainly true, it would be a radical departure from traditional analysis. It is necessary, then, for the analyst to be able to work comfortably with both forms of expressions and to realize that they are equivalent to each other.

A precaution is necessary when changing from the differential to the integral form—the specifications of integration limits an initial condition (sometimes referred to in mathematics as the constant of integration). For example, "at any time the rate of change of volume $V$ is the difference between the inflow and outflow" (equation 3-1) is a complete statement requiring no further qualification. The integral equivalent that "the volume is the integral of the difference between the inflow and outflow" requires a definition of the integration limits, that is, from when to when, and a specification for the initial value of $V$, that is, how much volume was present at the start of the integration. So a complete integral statement would be:

Total volume at $t = $ initial volume at $t_1 +$ accumulation from $t_1$ to $t$

$$V(t) = (V)_{t_1} + \int_{t_1}^{t} (Q_i - Q_o)\, dt$$

The following two rules, then, must be observed by the analyst/programmer whenever one or more equations are to be solved by numerical integration. (a) An initial condition or starting value for each integration variable must be provided. (b) The integration limits, that is, initial and final values of the independent variable, must be specified.

The remainder of this chapter will introduce the reader to the three basic methods of numerical integration, followed by a discussion of the precautions to be observed when solving large systems of nonlinear differential (or integral equations.

## 3-2  FIRST-ORDER METHOD  (Simple Euler)

This method, commonly referred to as the simple Euler, is the most elementary method available.

Referring to equation 3-1, this can be restated simply as:

$$V = \int V'\, dt$$

where $V'$ is the derivative or rate of change of $V$, symbolizing the expression involving $Q_i$ and $Q_o$. Suppose the solution for $V(t)$ at a particular set of conditions is as shown in Figure 3-2a. The derivative of $V$ is the slope of the $V$ curve and is also shown in Figure 3-2a. Consider now a narrow section $Dt$ along the time axis, shown greatly magnified in Figure 3-2b. Within this

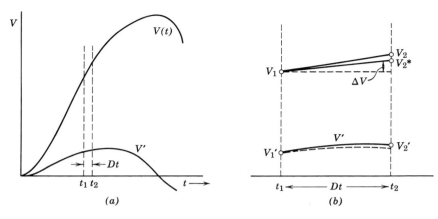

FIG. 3-2. Simple Euler

interval $Dt$, the derivative $V'$ changes from $V'_1$ to $V'_2$, with a corresponding change of $V$ from $V_1$ to $V_2$.

The first-order method evaluates the derivative $V'_1$ at the beginning of the time interval $t_1$, assumes that it is constant across the interval $Dt$ to time $t_2$, and calculates the corresponding change in $V$ as

$$\Delta V = V'_1 \, Dt$$

and

$$V^*_2 = V_1 + V'_1 \, Dt$$

that is, the value of $V$ at $t_2$ is $V_1$ plus the slope of $V$ at $t_1$ multiplied by the increment $Dt$, a relation that follows from elementary trigonometry.

$V^*_2$ calculated in this manner is actually an approximation to the correct value $V_2$ since it is based on the assumption that the derivative $V'$ is constant at the initial value $V'_1$ at the beginning of the interval. It will be realized, of course, that there is a small change in $V'$ from $V'_1$ to $V'_2$ that will give rise to the *truncation error* $V_2 - V^*_2$.

The procedure for the entire integration is to specify an independent variable increment $Dt$, to evaluate the derivative at the start of the integration, $i$, and to step forward a distance $Dt$ to $i + 1$, calculating the dependent variable as

$$V_{i+1} = V_i + V'_i * Dt$$

The derivative $V'_{i+1}$ is reevaluated at $t = t_{i+1}$, and the stepping procedure is repeated. This continues until the entire traverse of the integration (i.e., from lower limit to upper limit) has been completed.

The truncation errors mentioned previously may accumulate throughout the course of the integration, resulting in a significant difference between calculated and true values. Fortunately, because of the nature of chemical engineering problems, there is usually a tendency for the errors to cause slight changes in the calculated derivatives that in turn will diminish the errors. This is due to the inherent self-stabilization typical of natural phenomena. By way of explanation, referring to the tank problem described previously, an error (as shown in Figure 3-2b) $V_2 - V_2^*$ will result in a lower $H$, causing a slightly lower outflow $Q_0$. This in turn will increase the derivative $(Q_i - Q_o)$ which on the next step will increase $V$ by a greater amount, thus decreasing the error.

## 3-3  RELATIONSHIP BETWEEN ERROR AND INCREMENT SIZE

Referring to Figure 3-2b, it will be evident that if the step size is reduced, a closer approximation to the correct values is obtained.

This effect is shown in Figure 3-3 where a single step from $t_1$ to $t_2$ is compared with two half steps $t_1 \rightarrow t_a$ and $t_a \rightarrow t_2$.

For the latter case, starting with the derivative $V_1'$ at $t_1$, only a half step is made to $t_a$, where the derivative $V_a'$ is reevaluated, then the second half step $t_a$ to $t_2$ is made with this new derivative $V_a'$. It will be seen from Figure 3-3 that the final result $V_{2a}^*$ is closer to the correct $V_2$ than the original $V_2^*$ obtained with one full step. In fact, it would appear that the error $V_2 - V_2^*$ has been halved (approximately) by halving the interval size, leading to the generalization that for a first-order method of integration, the numerical errors are directly proportional to the step size. This can be proved mathematically by defining the function $V$ as a Taylor series in terms of its derivatives $V'$ and

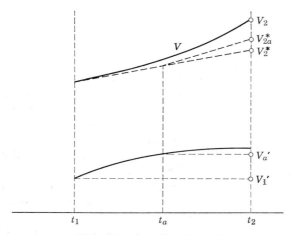

FIG. 3-3.  Step Size Reduction

noting that only the first two terms are used for the first-order method; that is, Taylor's expansion for $dV/dt = f(V, t)$ is

$$V(t_1 + Dt) = V(t_1) + Dtf(V(t_1), t_1) + \frac{(Dt)^2}{2!} f'(V(t_1), t_1) + \cdots$$

Eulers method is the first approximation to this series, namely:

$$V(t_1 + Dt) \approx V(t_1) + Dtf(V(t_1), t_1)$$

The sum of the remaining terms of the series will be the error involved in the approximation; that is, the series is truncated after the first two terms (hence truncation error). Since a Taylor series converges fairly rapidly, the first term of the neglected series of terms can be considered to represent the bulk of the error, that is,

$$\epsilon \approx \frac{(Dt)^2}{2!} f'(V(t_1), t_1) \tag{3-3}$$

The function $f'$ can also be expanded into a Taylor series and can be approximated by

$$f' = \frac{f(V(t_1) + Dt), t_1 + Dt) - f(V(t_1), t_1)}{Dt}$$

$$= \frac{f_1 - f_2}{Dt}$$

Substituting this result in equation 3-3 for $\epsilon$ we obtain

$$\epsilon = \frac{Dt}{2!} (f_1 - f_2)$$

showing that the error involved in the first-order method is proportional to the increment size $Dt$.

The desired tolerance, that is, maximum error permissible, determines the step size to be used. If a particular step size gives rise to errors greater than the specified tolerance, the step size is decreased, which reduces the errors to acceptable levels within the tolerance. Errors can be specified as either a % of the current value of the variable (i.e., $\epsilon\% = 100 \, |\epsilon_v|/V$) or as an absolute value $|\epsilon_v|$. The reason for this alternate definition is that the fractional or % error definition breaks down when $V \approx 0$. Mathematically, one can be quite precise with these definitions. However they can be misleading, resulting in costly computer running times. For example, Figure 3-4 shows two solutions to a set of differential equations, an exact solution and an approximate solution, obtained, perhaps, with a first-order integration method. Visual inspection of these two solutions would show that for

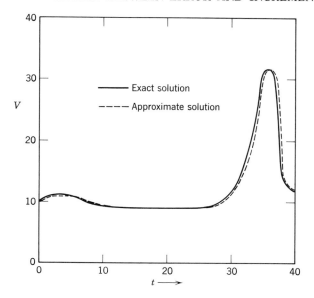

FIG. 3-4.    Exact and Approximate Solutions

practical engineering purposes, the approximate solution can be considered quite adequate, especially for scouting work. However, mathematically the fractional error of the numerical solution is as high as 40% (around $t = 38$), which is a theoretically unacceptable level when compared with a typical tolerance of, say, 1%. Specified error tolerances, then, can be deceiving and costly, which leads to the following practical advice to the programmer when deciding on a suitable integration step size:

1. Solve the equations numerically by using a nominal step size based on a knowledge of the problem equations.

2. If the results from the previous step appear stable (see Section 3-13), repeat the calculation, halving the step size used in 1.

3. Compare the solutions from 1 and 2 and, using engineering judgement as in the example demonstrated by Figure 3-4, decide whether the step size used in 1 was adequate. If so, repeat the calculation using twice the nominal step size.

4. Continue increasing the step size until the solution deteriorates beyond acceptable limits, or decrease the step size until sufficient accuracy is achieved.

The above procedure need only be followed when very large problems are being solved that require long and costly computer runs. Usually only a few trial runs are required to find a suitable optimum step size. For small problems requiring a few seconds of computer time, only two trial runs are required at some small step size to establish the authenticity of the results.

### 3-4 FORTRAN PROGRAM

The model describing the level change in the tank ($H$) as a function of the inflow $Q_i$ and outflow $Q_o$ described in the previous section can be summarized by the following equations

$$\frac{dH}{dt} = \frac{Q_i - Q_o}{A} \qquad \text{mass balance}$$

$$Q_i = f(t) \qquad \text{inflow}$$

$$Q_o = C_V\sqrt{H} \qquad \text{outflow}$$

It is convenient to retain the identity of each of the three variables $H$, $Q_i$, $Q_o$; thus substitution into one final equation will be avoided. The following data is required for a solution to these equations:

1. $C_V = $ ft³/(ft½ min) valve constant
2. $A = $ ft² tank area
3. $H = o$ ft at $t = o$ initial condition
4. Calculate $H$ for $t = o$ to $t = 10$

The remaining specification is the inflow $f(t)$. This will be an arbitrary function of time as shown in Figure 3-5a and will be approximated by a series of linear segments connecting 12 points on the curve. The coordinates of the points are specified in the table shown in Figure 3-5b.

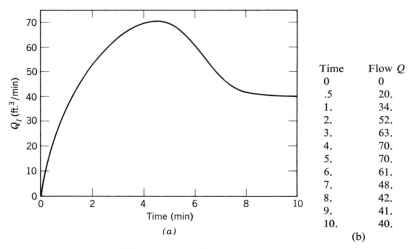

| Time | Flow $Q$ |
|------|----------|
| 0    | 0        |
| .5   | 20.      |
| 1.   | 34.      |
| 2.   | 52.      |
| 3.   | 63.      |
| 4.   | 70.      |
| 5.   | 70.      |
| 6.   | 61.      |
| 7.   | 48.      |
| 8.   | 42.      |
| 9.   | 41.      |
| 10.  | 40.      |

(a)

(b)

FIG. 3-5. Feed Flow Rate

The program for this set of equations can be classified into the three following sections.

### Initiation Section

This section contains the preliminary calculations and necessary housekeeping statements as follows:

```
 1*       100 FORMAT(1F10.4)
 2*       101 FORMAT(6E12.5)
 3*           DIMENSION AT(15),AQ(15)
 4*    C **INITIATION SECTION**
 5*           DATA(AT(N),N=1,12)/0.,,5,1.,2.,3.,4.,5.,6.,7.,8.,9.,10./
 6*           DATA(AQ(N),N=1,12)/0.,20.,34.,52.,63.,70.,70.,61.,48.,42.,41.,40./
 7*         8 READ 100,DT
 8*           T=0.
 9*           H=0.
10*           TPRNT=0.
```

Statements 5 and 6 enter the data for the coordinates of the points on the inflow function. Statements 8 and 9 set the initial condition for time and tank level H. TPRNT is a print index whose use is explained below, which is initiated to zero in this section.

### Derivative Section

This section contains the algebraic expressions whose final result is the calculation of the derivative, or derivatives if there is more than a single equation to be solved. It is important that the logical FORTRAN sequence be observed here; that is, no variable can be used until it has been established by a previous statement. For example, the equation for the derivative $dH$ must succeed the definition of $Q_i$ and $Q_o$. The sequence then is:

```
11*    C **DERIVATIVE SECTION**
12*         7 QO=17.*SQRT(H)
13*           QI=FUN1(T,12,AT,AQ)
14*           DH=(QI-QO)/25.
```

NOTE. The imput flow QI is obtained from FUN1, which is the arbitrary function generation subroutine described in Chapter 2.

At this point, all the integrated variables (T, H) have been specified, and the derivative (DH) and intermediate or other dependent variables ($Q_i$ and $Q_o$) have been calculated. This, then, is the logical point to print out information on the relative state of all the variables of interest. The print section, therefore, comes at the end of the derivative section. Generally, this information is desired at specified intervals, in this case, every 1 or 5 min; thus it is necessary to include a test to compare the time T with a print index

(TPRNT) that specifies the next print time. This index is updated on line 17. This section of the program will be as shown below.

```
15*    C    TEST FOR PRINT AND FINISH
16*            IF(T.GE.TPRNT) PRINT 101,T,H,DH,QI,QO,DT
17*            IF(T.GE.TPRNT) TPRNT=TPRNT+1.
18*            IF(T.GE.10.) GO TO 8
```

## Integration Section

This is the last of the three sections and contains the procedure for stepping forward along the axis of the independent variable in an orderly fashion. For the first-order method this procedure is simply

```
19*    C **INTEGRATION SECTION**
20*            T=T+DT
21*            H=H+DH*DT
22*            GO TO 7
23*            END
```

For multiple differential equations, the integration steps may be listed in any order. After completion of this stepwise integration, the calculation is directed to the first line of the derivative section to cycle through again, reevaluate the derivatives, and so on. The cycle continues repetitively until the independent variable T reaches a prescribed limit. This test is made immediately following the print statement and is simply

$$IF \ (T. \ GE. \ 10.) \ GO \ TO \ 8$$

When this statement is satisfied, the computer is directed to a new value of DT as specified in the data section corresponding to the READ card.

```
1*        100 FORMAT(1F10.4)
2*        101 FORMAT(6E12.5)
3*            DIMENSION AT(15),AQ(15)
4*    C **INITIATION SECTION**
5*            DATA(AT(N),N=1,12)/0.,.5,1.,2.,3.,4.,5.,6.,7.,8.,9.,10./
6*            DATA(AQ(N),N=1,12)/0.,20.,34.,52.,63.,70.,70.,61.,48.,42.,41.,40./
7*          8 READ 100,DT
8*            T=0.
9*            H=0.
10*           TPRNT=0.
11*   C **DERIVATIVE SECTION**
12*          7 QO=17.*SQRT(H)
13*            QI=FUN1(T,12,AT,AQ)
14*            DH=(QI-QO)/25.
15*   C    TEST FOR PRINT AND FINISH
16*            IF(T.GE.TPRNT) PRINT 101,T,H,DH,QI,QO,DT
17*            IF(T.GE.TPRNT) TPRNT=TPRNT+1.
18*            IF(T.GE.10.) GO TO 8
19*   C **INTEGRATION SECTION**
20*            T=T+DT
21*            H=H+DH*DT
22*            GO TO 7
23*            END
```

FIG. 3-6.   FORTRAN Program for Tank Simulation

| Time | $\Delta T$ | | | | | | |
|------|------|------|------|------|------|------|-------|
|      | 0.5 | 0.2 | 0.1 | 0.05 | 0.02 | 0.01 | 0.005 |
| 0 | 0.0 | 0.0 | 0.0 | 0.0 | 0.0 | 0.0 | 0.0 |
| 1 | 0.4000 | 0.4405 | 0.4663 | 0.4795 | 0.4877 | 0.4905 | 0.4919 |
| 2 | 1.4088 | 1.4789 | 1.5113 | 1.5279 | 1.5382 | 1.5416 | 1.5434 |
| 3 | 2.7090 | 2.7800 | 2.8113 | 2.8275 | 2.8374 | 2.8408 | 2.8425 |
| 4 | 4.1116 | 4.1723 | 4.1994 | 4.2134 | 4.2221 | 4.2250 | 4.2265 |
| 5 | 5.4755 | 5.4963 | 5.5098 | 5.5171 | 5.5217 | 5.5233 | 5.5241 |
| 6 | 6.5516 | 6.4960 | 6.4841 | 6.4788 | 6.4759 | 6.4750 | 6.4745 |
| 7 | 7.0981 | 6.9640 | 6.9260 | 6.9077 | 6.8971 | 6.8935 | 6.8918 |
| 8 | 7.1430 | 6.9847 | 6.9447 | 6.9230 | 6.9102 | 6.9060 | 6.9039 |
| 9 | 7.0000 | 6.8622 | 6.8216 | 6.8020 | 6.7904 | 6.7866 | 6.7847 |
| 10 | 6.8360 | 6.7125 | 6.6754 | 6.6582 | 6.6478 | 6.6443 | 6.6426 |

FIG. 3-7.  Tabulated Results of Program in Figure 3-6

Actually, any other problem condition could be used to stop the computation and start a new run. For example, if the level H reaches a prescribed level, a suitable finish statement would be

$$\text{IF (H. GE. 6.5) GO TO 8}$$

The entire program is shown in Figure 3-6. Runs were made, each with a different integration step size, varying from 0.5 min to 0.005 min. The results for tank level H are tabulated in Figure 3-7.

### 3-5  SECOND-ORDER  INTEGRATION

The level of sophistication above the first-order method is suggested by the limitations inherent in the first-order method. This requires making two derivative evaluations in either of the following two ways:

1. The derivatives are evaluated at the beginning of the step $(t_1)$ and with these derivatives a step is made *halfway* across the interval to $(t_1 + Dt/2)$. At this point the derivatives are reevaluated and are considered to be the average derivatives for that interval. So, starting back at $t_1$, the entire step is made for a full interval $Dt$ with the average derivative found at $(t_1 + Dt/2)$. This method is commonly termed the modified Euler method.

2. The second variation of the second-order method is actually the first of the Runge-Kutta methods. The step procedure is as follows:
(a) Evaluate the derivatives $dV/dt$ at $t_1$.
(b) With the derivatives from step a, integrate across the full increment to

$t_2 \ (= t_1 + Dt)$ by the simple Euler method

$$V)_2 = V)_1 + \left\{\frac{dV}{dt}\right\}_1 Dt$$

3. Reevaluate the derivatives $\{dV/dt\}_2$ at $t_2$.

4. Calculate an average derivative from the two derivatives established at $t_1$ and $t_2$

$$\frac{\overline{dV}}{dt} = \frac{1}{2}\left(\left\{\frac{dV}{dt}\right\}_1 + \left\{\frac{dV}{dt}\right\}_2\right)$$

5. Starting again at $t_1$, integrate across the full increment to $t_2$ using the average derivative from step 4.

$$V)_2 = V)_1 + \frac{\overline{dV}}{dt} \cdot Dt$$

In this method, two derivative evaluations are required for each integration increment; hence the term *second order*.

The procedure described above is rather tedious; thus it will be included in a subroutine system that will automatically take care of all necessary housekeeping details. The objective of the system is to allow the programmer merely to state the differential equations and to call on the subroutine to do the integration. The entire array of subroutines for solving ordinary differential equations that are developed throughout this text are called the INT program.

### 3-6  SUBROUTINE INT

There are two key subroutines in this system, called INT and INTI. They must always be used together since they share information through a COMMON statement.

The first subroutine to be described is the INT subroutine that is called from a main program by the following statement:

### CALL INT(X, DX)

where X is the name of the integrated variable and DX is the derivative calculated in the preceding derivative section. It is capable of performing any one of the three orders of integration, that is, first, second, or fourth. The procedure for the first and second orders have already been described; thus only the section of the subroutine dealing with these two methods is described now. The section in the subroutine dealing with the fourth-order method is described following the section describing the fourth-order procedure.

```
1*          SUBROUTINE INT(X,DX)
2*          COMMON/CINT/T,DT,JS,JN,DXA(500),XA(500),IO,JS4
3*          JN=JN+1
4*          GO TO (9,8,3,3),IO
5*        9 X=X+DX*DT
6*          RETURN
7*        8 GO TO (1,2),JS
8*        1 DXA(JN)=DX
9*          X=X+DX*DT
10*         RETURN
11*       2 X=X+(DX-DXA(JN))*DT/2,
12*         RETURN
```

FIG. 3-8. Listing for Subroutine INT (First- and Second-Order Sections Only)
Argument list:
X = integrated variable
DX = derivative

A listing of the first- and second-order sections of INT is shown in Figure 3-8. The argument list has two variables, the variable to be integrated (X) and its derivative (DX). The COMMON/CINT/ statement contains the independent variable T, the integration interval DT, and an integer JS that has values of either 2 or 1 specifying whether the calculation is performing its first or second derivative evaluation, that is, step 1 or step 3 in the foregoing list. The next item in the COMMON list is the integer JN that acts as an internal index to identify a storage location in the array DXA. The values of T, DT, JS, and the initial value of JN = 0 are established in the control subroutine INTI that integrates the independent variable T and is described later.

The first line of computation (3) increases JN by unity, thus identifying each particular CALL. In a typical problem, several, even many, differential equations may have to be integrated, requiring an equal number of calls on the integration subroutine. Since they will be listed in sequence in the main program, JN will automatically index each successive CALL. The GO TO statement on line 4 directs the computation to one of 3 locations controlled by the integer IO symbolizing *I*ntegration *O*rder and available from the COMMON statement. The first location, address 9, is for first order (IO = 1) and is self evident. The second location, address 8, is for the second-order method where a second GO TO statement directs the computation to address 1 or 2 depending on the value of JS. At address 1 (line 8) the derivative from the first derivative evaluation is stored in the array DXA(JN) and the variable X is increased by the quantity DX * DT, the first simple Euler step. On the second pass (JS = 2), the value of X is recalculated from the expression

$$X = X + (DX - DXA(JN)) * \frac{DT}{2}$$

This can be explained by recalling that from the basic definition of second-order integration, the corrected value of X at $t_2$ is

$$X_2 = X_1 + (DX + DXA(JN)) * \frac{DT}{2}$$

However, in order to reduce storage requirements, $X_1$ is not available at $t_2$; instead the X that is available is the value of $X_2$ obtained by simple Euler:

$$X = X_1 + DXA(JN) * DT$$

Eliminating $X_1$ from these last two equations results in the expression on line 11.

### 3-7   SUBROUTINE INTI

The corresponding sections in the control subroutine INTI for the first and second order are shown in the partial listing of INTI in Figure 3-9. The three items in the argument list are, in order, TD, the name of the independent variable; DTD, the integration increment; and IOD, the integration order required. This subroutine must always be the first call in the integration section; that is, it must precede all the calls on the subroutine INT. Since the three items in the argument list correspond to the T,DT. and IO in the COMMON/CINT statement they have to be "dummied" to avoid confusion of identity. The variables in the argument list have to be transferred to the common statement at the appropriate time. For example, line 3 transfers the integration order to the common statement, which incidentally permits changing integration methods during a computation run.

```
 1*          SUBROUTINE INTI(TD,DTD,IOD)
 2*          COMMON/CINT/T,DT,JS,JN,DXA(500),XA(500),IO,JS4
 3*          IO = IOD
 4*          JN=0
 5*          GO TO (6,5,1,1),IO
 6*        6 JS=2
 7*          GO TO 7
 8*        5 JS=JS+1
 9*          IF(JS.EQ.3)JS=1
10*          IF(JS.EQ.2)RETURN
11*        7 DT=DTD
12*        3 TD=TD+DT
13*          T=TD
14*          RETURN
```

FIG. 3-9.   Listing for Subroutine INTI (First- and Second-Order Sections Only)
Argument list:
  TD = independent variable
  DTD = integration step size
  IOD = integration order

Line 4 resets the call counter JN to 0, and line 5 directs the computation to one of three places, depending on IO. Address 6 is the first line of the section for first order integration and sets the pass index JS to 2, signifying a "legitimate" pass, since for first order there are no "dummy" or intermediate passes. At address 7 the integration interval DTD is transferred to common as DT, a step that permits operating this subroutine with a *variable* interval size. The next line (12) integrates the independent variable TD, then transfers this to common as T (line 13).

The second-order procedure starts at address 5 (line 8) where the pass index is increased by 1, then tested. If it is 3, it is reset to 1 (line 9). If JS = 1 (i.e., the computation has completed its first pass for the present increment through the derivative section) then the independent variable is integrated (line 12). If the second pass has been completed (JS = 2) then no further action is required here, and a return is made to the main program (line 10).

Before showing an example of how these routines are used to program a solution to a set of differential equations, the fourth-order method of integration is described, since the implementation of this method will complete the subroutines INT and INTI.

## 3-8  FOURTH-ORDER RUNGE-KUTTA METHOD

This method is the most common higher-order method in use and requires four derivative evaluations per increment. The double computational effort, as compared with the second-order method, is compensated by the fact that since the integration errors are proportional to the fourth power of the integration interval, that is, $\epsilon \propto (DT)^4$, larger intervals can be used for a specified accuracy. This comment is applicable only when the maximum interval size is not limited by other factors, such as stability or print interval (see Section 3-13).

The procedure for the fourth-order method is to evaluate the derivatives at the beginning, halfway, and end of the integration interval. The final step is then made across the interval using a weighted average of all these derivatives. In detail the procedure follows (see Figure 3-10):

1. The derivative $V_1'$ is evaluated at $t_1$, and, using a Simple Euler step, the value of the function is calculated at $(t_1 + DT/2) \rightarrow V_2$.
2. The derivative is evaluated at $t_1 + DT/2 \rightarrow V_2'$.
3. Starting at $V_1$, $t_1$ the function at $t_1 + DT/2$ is recalculated using the derivative $V_2'$ to give point $V_3$.
4. The derivative $V_3'$ is reevaluated at $t_1 + DT/2$.
5. Starting at $V_1$, $t_1$ the function is calculated at $t_2(= t_1 + DT)$ using the derivative $V_3'$ to give point $V_4$.
6. The derivative $V_4'$ is evaluated at $t_2$.

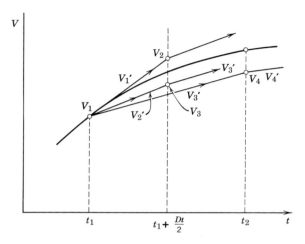

FIG. 3-10. Fourth order Runge-Kutta Procedure

7. Using the derivatives $V_1' \to V_4'$ obtained in steps 1–6, the final value of the function $V(t_2)$ is calculated using a weighted average of the derivatives, that is,

$$V(t_2) = V(t_1) + \frac{(V_1' + 2 * C_2' + 2 * C_3' + V_4') * DT}{6}$$

The above procedure is programmed in the second part of the subroutine INT, shown in Figure 3-11. The COMMON statement contains an additional storage array XA and a pass counter JS4. The dimensions of the two arrays XA and DXA, nominally set for 500, determine the total number of differential equations that can be solved by the program.

When IO is set to 4 in the argument list of INTI, the computation is

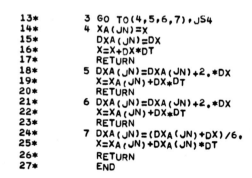

```
13*        3 GO TO(4,5,6,7),JS4
14*        4 XA(JN)=X
15*          DXA(JN)=DX
16*          X=X+DX*DT
17*          RETURN
18*        5 DXA(JN)=DXA(JN)+2.*DX
19*          X=XA(JN)+DX*DT
20*          RETURN
21*        6 DXA(JN)=DXA(JN)+2.*DX
22*          X=XA(JN)+DX*DT
23*          RETURN
24*        7 DXA(JN)=(DXA(JN)+DX)/6.
25*          X=XA(JN)+DXA(JN)*DT
26*          RETURN
27*          END
```

FIG. 3-11. Fourth-Order Section of Subroutine INT

```
15*      1  JS4=JS4+1
16*         IF(JS4.EQ.5)JS4=1
17*         IF(JS4.EQ.1) GO TO 2
18*         IF(JS4.EQ.3) GO TO 4
19*         RETURN
20*      2  DT=DTD/2.
21*         GO TO 3
22*      4  TD=TD+DT
23*         DT=2.*DT
24*         T=TD
25*         RETURN
26*         END
```

FIG. 3-12.   Fourth-Order Section of Subroutine INTI

directed to the fourth-order section, where the 7 steps outlined above are programmed as follows:

Step 1.  (JS4 = 1) Lines 14 → 16, X and DX are stored in the arrays. Also, DT from COMMON is half the integration interval.
Step 2.  Accomplished in the main program.
Step 3.  (JS4 = 2) Lines 18, 19. Note, 'DT' at this stage is still half the total interval DT).
Step 4.  Accomplished in the main program.
Step 5.  (JS4 = 3) Lines 21, 22. The interval DT is now back to full size.
Step 6.  Accomplished in the main program.
Step 7.  (JS4 = 4) Lines 24, 25. The various derivatives in their proper ratios are accumulated in the array DXA on lines 15, 18, 21, and 24. The final averaged value is calculated on line 24.

The corresponding section in the control subroutine INTI is shown in Figure 3-12. The first line (15) increments JS4, and the second line resets it to 1 if it is at 5. If JS4 is either 2 or 4, no further action is taken. If it is 1, the integration interval (DTD) from the argument list is halved (line 20) and transferred to COMMON as DT. Then the computation continues to address 3 where T is integrated to T + DT. When JS4 is 3, T is further integrated to the end of the interval (line 22) and the DT is increased back to its full size (line 23).

## 3-9  GENERAL ARRANGEMENT OF MAIN PROGRAM

The description of the two subroutines INT and INTI is now complete. The subroutines are shown as complete listings in Appendices B-3 and B-4, together with a definition of the items in the argument list. The program for the integration procedure is now reduced to a list of CALL INT statements. The CALL INT statement can, of course, be in a DO loop, and the argument list can contain ordinary or subscripted variables. Use of these is shown in later examples.

The main program arrangement for solving a set of differential equations is similar to the layout shown in Figure 3-6 for the first-order method. It is shown in general form in Figure 3-13. Following a preliminary housekeeping section containing FORMATS, TYPE, DIMENSION, COMMON statements, and so on, the initiating section is laid out. Here all the calculations and specifications for the problem parameters and coefficients are made. These

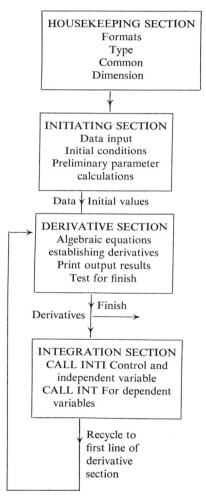

FIG. 3-13. General Arrangement of Main Program for Use of INT Subroutines

are the values that are invariant during the course of the iterative calculation. In this section, the initial conditions for the integrated variables must also be specified, and new data for repetitive runs is introduced via a READ statement.

The next section is the derivative section that contains all the algebraic equations specifying the derivatives of the integrated variables. The sequence of equations must be properly ordered; that is, no variable can be used unless specified in a prior statement. At the end of this section, a test for printing and end of run is made. A subroutine PRNTF will be described in Section 3-10 to accomplish this purpose.

The last section is the integration section, where a listing of the CALL INT statements is made for each integrated variable. One cardinal rule must be observed here; namely the first call must be for the independent variable subroutine INTI, since this routine sets the call and pass counters in COMMON (CINT) controlling the other routines to their proper values. The listings for the CALL INT subroutines following the control subroutine INTI can be in any order except for the following rule: No derivative in an INT argument can be specified by the output of a preceding INT routine.

As an explanation of this rule, suppose for example the following two equations are to be integrated.

$$DX = \int D^2X$$

$$X = \int DX$$

The correct sequence of CALL statements in the integration section must be in the reverse order, that is

$$CALL\ INT(X,DX)$$

$$CALL\ INT(DX,D2X)$$

This reverse sequence will correctly coordinate the derivative with the independent variable.

After all the CALL INT statements have been listed, the last statement directs the computation to the first line of the derivative section. This connection directs the calculation to cycle through the derivative and integration sections continuously until the end of the run.

## 3-10  SUBROUTINE  PRNTF

The purpose of this subroutine is twofold. It controls the frequency of printing the variables of interest, and it also tests for the end of the run. The logical place for printing out results is at the end of the derivative section, as shown in Figures 3-6 and 3-13. It is necessary here to print out values of selected variables at specified intervals called the print interval and designated as PRI. These values can only be printed after the previous legitimate step has been completed and the new derivatives calculated. This is the pass where the pass index for the first- and second-order method JS is 2. For the fourth-order method the pass index JS4 is 4. The subroutine that takes care of these details is shown in Figure 3-14. The first item in the argument list is PRI, the print interval, which is referred to the independent variable T available from COMMON/CINT/. The next item FNR is the FiNish-Run value of T specifying when the run is complete. This is followed by the index NF that indicates whether the run is Not Finished (NF = 1) or complete (NF = 2). This index is used in the main program to direct the computation to continue (NF = 1) or stop (NF = 2).

The following 10 items in the argument list A → Q represent the variables to be printed. If less than 10 variables are required, the remaining locations must contain dummy numbers, such as 0. or 1. The COMMON/CPR statement on line 3 contains the index NPR that controls a secondary PRNTRS subroutine (Figure 3-15) that will print an additional set of 10 variables. When the master print routine PRNTF prints its variables, NPR is made equal to 1, and this will automatically print the variables listed in the succeeding PRNTRS routines.

The test for printing is on line 7 and depends on the print index TPRNT (time to print) and JS the pass index. If the conditions are satisfied, the print

```
 1  •         SUBROUTINE PRNTF(PRI,FNR,NF,A,B,C,D,E,F,G,O,P,Q)
 2  •         COMMON/CINT/T,DT,JS,JN,DXA(500),XA(500),IO,JS4
 3  •         COMMON/CPR/NPR
 4  •   100   FORMAT(10E12.5)
 5  •         NPR = 0
 6  •         IF(TPRNT.LT.PRI) GO TO 4
 7  •         IF((T.GE.FNR-DT/2.).AND.((JS.EQ.2).OR.(JS4.EQ.4))) GO TO 6
 8  •         IF((T.GE.TPRNT-DT/2.).AND.((JS.EQ.2).OR.(JS4.EQ.4))) GO TO 5
 9  •         RETURN
10  •     4   NF = 1
11  •     5   TPRNT = TPRNT+PRI
12  •     8   PRINT 100,A,B,C,D,E,F,G,O,P,Q
13  •         NPR = 1
14  •         RETURN
15  •     6   T = 0.
16  •         TPRNT = 0.
17  •         NF = 2
18  •         DO 7 J=1,500
19  •     7   XA(J) = 0.
20  •         GO TO 8
21  •         END
```

FIG. 3-14.   Listing for Subroutine PRNTF
Argument list:
    PRI = print interval
    FNR = end of run
      NF = finish index (NF = 2 is finish)
  A → Q   are the variable to be printed

```
1  *        SUBROUTINE PRNTRS(IS,A,B,C,D,E,F,G,O,P,Q)
2  *        COMMON/CPR/NPR
3  *    100 FORMAT(10E12.5)
4  *        IF(NPR.EQ.1)GO TO 5
5  *        RETURN
6  *      5 PRINT 100,A,B,C,D,E,F,G,O,P,Q
7  *        IF(IS.EQ.1)PRINT 100
8  *        RETURN
9  *        END
```

FIG. 3-15.   Listing for Subroutine PRNTRS
Argument list:
   IS = for space after print line (IS = 1 for space)
   A → Q variable to be printed

index TPRNT is increased by the print interval PRI (line 10), which specifies the succeeding print time. The variables are printed, and the print index NPR is switched to 1 in case of further printing requirements from PRNTRS.

The test for FINISH is made on line 13, and if positive, it resets to zero the independent variable T (line 15), and also resets TPRNT to zero, ready for the next run. The array XA is cleared (line 19) and the finish index is switched to 2.

## 3-11  SUBROUTINE PRNTRS

This is the subroutine used in situations where more than 10 variables have to be printed. The excess has to be handled by this *Print Repeat Space* routine shown in Figure 3-15. As explained previously, the test for printing is made in the master routine PRNTF, and the print index is relayed to this routine through the COMMON/CPR/NPR statement. The remainder of the coding should be self-evident. This subroutine offers the option of specifying a space following the line of printing, accomplished by assigning the integer 1 as the first item in the argument list, thus separating the last line of output at each print interval with the first line of output of the succeeding interval.

The three routines INTI, INT, and PRNTF described in this chapter form a framework for the solution of nonlinear, ordinary differential equations. In order to appreciate the simplicity of programming equations using this system, a simple example will be demonstrated in the next section.

## 3-12  PROGRAMMING EXAMPLE USING INT SYSTEM

The same example will be used here to demonstrate the use of the INT program that was used at the beginning of the chapter and described in Section 3-3. The INT program is shown in Figure 3-16 where the equations describing the change in tank level H are solved for a succession of values of

```
26*     C **HOUSEKEEPING SECTION**
27*     100 FORMAT(1H1,1F10.5)
28*         DIMENSION DTA(15),AT(15),AQ(15)
29*     C **INITIATION SECTION**
30*         DATA(DTA(N),N=1,8)/1.,.5,.2,.1,.05,.02,.01,.005/
31*         DATA(AT(N),N=1,12)/0.,,5,1.,2.,3.,4.,5.,6.,7.,8.,9.,10./
32*         DATA(AQ(N),N=1,12)/0.,20.,34.,52.,63.,70.,70.,61.,48.,42.,41.,40./
33*         DO 8 J=1,8
34*         T=0.
35*         H=0.
36*         DT=DTA(J)
37*         PRINT 100,DT
38*     C **DERIVATIVE SECTION**
39*       7 QO=17.*SQRT(H)
40*         QI=FUN1(T,12,AT,AQ)
41*         DH=(QI-QO)/25.
42*     C     TEST FOR PRINT AND FINISH
43*         CALL PRNTF(1.,10.,NF,T,H,DH,QI,QO,DT,0.,0.,0.,0.)
44*         GO TO (5,8),NF
45*     C **INTEGRATION SECTION**
46*       5 CALL INTI(T,DT,1)
47*         CALL INT(H,DH)
48*         GO TO 7
49*       8 CONTINUE
50*         END
```

FIG. 3-16. Example of a Main Program using INT Subroutines

the integration interval DT contained in the array DTA. This is accomplished by including the three sections, initiation, derivative, and integration, inside a DO loop (line 33 to 49). On each pass of the DO loop a new value of DT is used. The integration starts at $T = 0.$ and continues to $T = 10.$ min as specified by the second argument in PRNTF. Results are printed every 1 min as specified by the first argument in PRNTF. Line 39 (address 7) is the first line of the derivative section. On completion of the CALL INT(H,DH) (line 47) the computation is recycled to this point. The reader should study the program in Figure 3-16 and recognize the recommendations made for the general arrangement, which to summarize are:

1. Division of the program is into four functional sections.
2. Location of the PRNTF routine is at the end of the derivative section.
3. The first call in the integration section is for the independent variable routine INTI.
4. The initial conditions $(T=0, H=0)$ are specified in the initiation section.
5. The recycle loop is closed to the first line of the derivative section.

When the above precautions are observed, this system will readily solve all types of ordinary nonlinear differential equations. The total number is limited only by the dimensions of the storage arrays in COMMON/CINT or, ultimately, by the size of the computer available. Mathematical models

| | | | | | |
|---|---|---|---|---|---|
| .00000 | .00000 | .0000n | .00000 | .00000 | .5₀000-n2 |
| .10050+01 | .49631+00 | .88454+00 | .34090+n2 | .11976+02 | .5₀000-02 |
| .20050+01 | .15495+01 | .12357+01 | .52055+02 | .21162+02 | .5₀000-02 |
| .30050+01 | .28493+01 | .13736+01 | .63035+02 | .28696+02 | .50000-02 |
| .40050+01 | .42335+01 | .14009+01 | .70000+n2 | .34978+02 | .50000-02 |
| .50050+01 | .55301+01 | .11991+01 | .69955+02 | .39977+02 | .50000-02 |
| .60050+01 | .64780+01 | .70667+00 | .60935+n2 | .43268+02 | .50000-02 |
| .70050+01 | .68925+01 | .13356+00 | .47970+02 | .44631+02 | .50000-02 |
| .80050+01 | .69033+01 | -.10684+00 | .41995+02 | .44666+02 | .5₀000-02 |
| .90050+01 | .67840+01 | -.13134+00 | .40995+n2 | .44278+02 | .5₀000-02 |
| .10005+02 | .66418+01 | -.15248+00 | .40000+n2 | .43812+02 | .50000-02 |

FIG. 3-17.  Output from Program in Figure 3-16. (First 6 columns only for DT = 0.005)

involving up to 500 or more differential equations are simulated today on large computers as standard procedure. Later chapters in this book indicate how these models are created and why they need to be solved.

The output from the general purpose print routine PRNTF is shown in Figure 3-17. Since it is designated as general purpose, it has to be in floating point as shown. However, there is no reason why a programmer cannot modify this format (line 4 PRNTF) to a more convenient fixed-point or integer format for specific cases.

## 3-13  ACCURACY  OF  INTEGRATION

A discussion of the effect of integration step size on the accuracy of integration for the first-order method was covered in Section 3-2. It was shown that the truncation errors were proportional to the step size used.

| | | | | $DT$ | | | |
|---|---|---|---|---|---|---|---|
| Time | 0.5 | 0.2 | 0.1 | 0.05 | 0.02 | 0.01 | 0.005 |
| 0 | 0.0 | 0.0 | 0.0 | 0.0 | 0.0 | 0.0 | 0.0 |
| 1 | 0.55019 | 0.50065 | 0.49560 | 0.49389 | 0.49340 | 0.49333 | 0.49331 |
| 2 | 1.5927 | 1.5515 | 1.5470 | 1.5456 | 1.5452 | 1.5451 | 1.5451 |
| 3 | 2.8848 | 2.8497 | 2.8458 | 2.8446 | 2.8442 | 2.8442 | 2.8442 |
| 4 | 4.2630 | 4.2328 | 4.2294 | 4.2283 | 4.2280 | 4.2280 | 4.2279 |
| 5 | 5.5542 | 5.5289 | 5.5261 | 5.5252 | 5.5249 | 5.5249 | 5.5249 |
| 6 | 6.4971 | 6.4772 | 6.4750 | 6.4743 | 6.4741 | 6.4741 | 6.4741 |
| 7 | 6.9073 | 6.8923 | 6.8908 | 6.8902 | 6.8901 | 6.8901 | 6.8901 |
| 8 | 6.9157 | 6.9036 | 6.9024 | 6.9019 | 6.9018 | 6.9018 | 6.9018 |
| 9 | 6.7948 | 6.7843 | 6.7833 | 6.7829 | 6.7828 | 6.7828 | 6.7828 |
| 10 | 6.6513 | 6.6422 | 6.6413 | 6.6410 | 6.6409 | 6.6409 | 6.6409 |

FIG. 3-18.  Numerical Solution of Tank Problem using Second order integration method

The second- and fourth-order integration methods can be shown to have errors proportional to the square and fourth power of the step size, respectively. This can be proved for linear systems by comparing the first term of the truncated series of the Taylor expansion for the derivative, as was done for the first-order method. A more practical approach for this discussion is to measure and compare the errors for various step sizes and integration methods on a typical example, such as the tank problem described earlier in this chapter.

Figure 3-18 shows the numerical result using the second-order method for step sizes ranging from 0.5 to 0.005 min. This table should be compared with Figure 3-7 for the first-order method in order to appreciate the faster convergence to a correct solution of the second-order method. A similar table for the fourth-order method would show an even faster convergence.

In order to compare the methods and the effect of step size, an error criterion is defined as the deviation from the correct solution at a specific point in the solution, say $t = 4$ min. Here the correct solution is $H = 4.2279$. The % error then will be:

$$\%e = \frac{100*(H - 4.2279)}{4.2279}$$

Figure 3-19 shows a plot on log–log scales of %e versus step size for each method of integration. The relative improvement of the fourth-order over second- and first-order integration methods at any step size is quite evident. Also, the rate of accuracy increase with a decrease in step size is superior for the higher-order methods; in fact it follows the relationship indicated by theory

$$\%e \propto (DT)^{IO}$$

where $IO$ = integration order. The fourth-order method seems to deviate from this relationship, probably because of the nonlinear characteristic of the equation being solved.

Since each method requires a different number of derivative evaluations per step, a more realistic comparison of the relative advantages would be to compare the number of derivative evaluations required per unit of time as a function of error. This is shown in Figure 3-20 for each integration method where, despite the increased number of derivative evaluations, there is a clear advantage of the higher-order methods in terms of the number of derivative evaluations or the equivalent computer running time required.

The conclusion from this analysis is that it appears to be more efficient to use a higher-order method, since for any specified accuracy (usually 1% is considered adequate) larger steps can be taken. However, if for any reason the maximum step size is limited by other considerations, it may be more

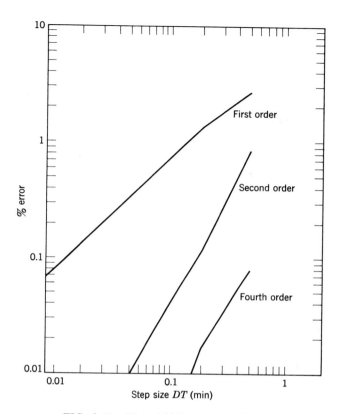

FIG. 3-19.  Plot of % Error versus Step Size

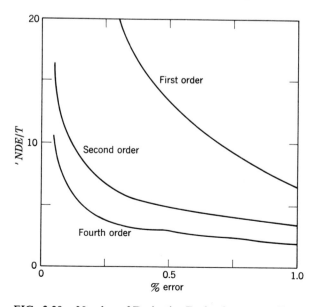

FIG. 3-20.  Number of Derivative Evaluations versus Error

efficient to use a lower-order method. For example, suppose that for the tank problem a print interval of 0.2 min had been specified and that an error criterion of 1.5% was considered adequate. Figure 3-19 shows that for these conditions, the first-order method with an integration step equal to the print interval would be satisfactory. The second- or fourth-order methods would provide unnecessary accuracy at the expense of twice and four times the computational effort. Another factor that more commonly limits step size is stability, which is explained in the next section.

### 3-14 STABILITY OF NUMERICAL INTEGRATION (Ref. 1)

A common difficulty that is inherent in numerical integration is stability. It arises in situations where there are several simultaneous differential equations to be simulated with a wide range of time constants. For a proper understanding of this phenomena, consider the following simple first-order linear equation:

$$\frac{dy}{dt} = \frac{1}{\tau} * (1. - y)$$

The analytical solution to this equation for the initial condition $y = 0$ at $t = 0$ is

$$y = 1. - e^{-\tau/t}$$

The time constant for this equation is the parameter $\tau$. If the differential equation is integrated numerically using the first-order method and a step size $DT = 2 * \tau$, the result will be critically stable, as shown in Figure 3-21, case $a$. If the step size is only slightly larger than the critical value $2 * \tau$ the result will oscillate with an ever-increasing amplitude. Such a situation is termed unstable. As the step size is progressively decreased, the numerical solution will approach the exact analytical solution.

For a simple first-order linear equation with a time constant $\tau$, the critical step size for each integration order is as follows:

First order $\quad DT_c = 2 * \tau$
Second order $\quad DT_c = 2 * \tau$
Fourth order $\quad DT_c = 3 * \tau$

When solving a single differential equation we are not concerned with this situation because in order to achieve any reasonable accuracy, the step size will be appreciably smaller than the characteristic time constant. The stability difficulty arises where there are several equations. To demonstrate

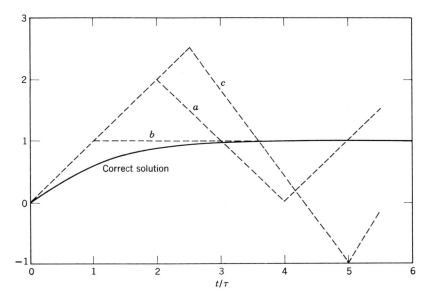

FIG. 3-21.   Stability of First-Order Differential Equation
Case *a* DT = 2*$\tau$
Case *b* DT = $\tau$
Case *c* DT > 2$\tau$

this, consider the process shown in Figure 3-22. Here, a large vessel is being heated by a constant temperature jacket. The two variables of concern are the fluid temperature $T$ and the temperature $TM$ sensed by a thermocouple inside a thermowell. A heat balance on the fluid contents of the vessel and on the thermowell will result in two differential equations which can be normalized to the following form:

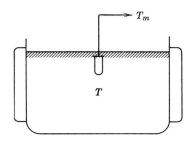

$$\frac{dT}{d\theta} = (1 - T) \qquad (3\text{-}14a)$$

$$\frac{d(TM)}{d\theta} = (T - TM) * \frac{1}{\tau} \quad (3\text{-}14b)$$

FIG.   3-22.   Large   vessel   and   Thermal Element

By changing the size of the thermowell, the value of $\tau$ can be changed. The ratio of time constants between equations 3-14a and 3-14b will be $\tau$ since the time constant for equation a is 1 min. This ratio is known as the "stiffness ratio."

Suppose in this case the thermowell time constant is 0.03 min, then, extending the findings described previously we can specify the maximum possible step size for numerical solution as 0.06 min for the first and second order and 0.09 min for the fourth order method.

The analytical solution to equations 3-14a and 3-14b is

$$TM = 1 - \frac{\tau}{\tau - 1} * e^{(-\theta/\tau)} + e^{-\theta}/(\tau - 1)$$

Comparing the numerical solution obtained with a first-order method at 90% critical step size, that is, DT = .054 min, produced results within 1% of the analytical solution in the region $TM = 0.5$. This measure of accuracy at almost the critical step size was made for this system for a range of stiffness values from 0.2 to 0.005. The results are shown summarized in Figure 3-23 for the three methods of integration, and it will be seen that for stiffness ratios of 0.03 or less, acceptable ($<1\%$) accuracies are obtained with the first-order method. For such situations then, the first-order method is more efficient than the second- or fourth-order methods by a ratio of the number of derivative evaluations required for each method, namely 2 and 4, respectively.

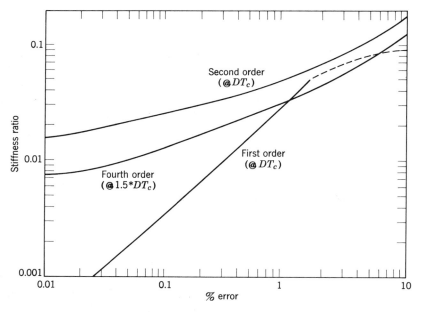

FIG. 3-23.   Error versus Stiffness Ratio

The above characteristic is accentuated when dealing with very large systems of nonlinear stiff equations. For this reason all three methods of integration are available in the integration subroutine INT described in this chapter. They provide the programmer with the opportunity to optimize computer time with accuracy and stability by experimenting with different methods of integration. The procedure should be as follows:

1. Starting with the fourth-order method, decrease the step size until stability is achieved. ($\rightarrow DT_c$)
2. Evaluate the accuracy by decreasing the step size to $\frac{1}{2} * DT_c$.
3. If accuracy at the critical step size is satisfactory, repeat the run using the second-order method.
4. If the second-order method gave satisfactory accuracy, repeat the run using the first-order method and evaluate the accuracy.

The above procedure should be adopted in cases where the computer running time for the solution is excessive.

Sophisticated methods have recently been developed (Ref. 2) to overcome the instability problem in stiff systems. A description of these methods, however, is beyond the scope of this text.

### 3-14  VARIABLE STEP-SIZE METHODS

A popular class of integration methods, called "variable step size methods," feature an ability for internally monitoring the integration error and automatically adjusting the step size so as to maintain the error within a specified tolerance. These methods find much use in situations involving "open" integrations, that is, systems without internal feedback or self correction, such as trajectory or navigational problems. The majority of chemical engineering problems are invariably "closed," containing sufficient internal feedback to make internal step size correction justifiable only in rare cases, since these methods require much greater computer time per solution than the direct methods described in this chapter. Two of the most common methods of automatic step size correction are described next.

### 3-14-1  Implicit Method

This method can be implemented with any integration order and consists of the following procedure (shown symbolically in Figure 3-24):

1. The integration is performed across the interval $DT$ and the result is stored ($a \rightarrow b$).
2. The result is recalculated by integrating twice across each half of the total interval. ($a \rightarrow c \rightarrow d$).

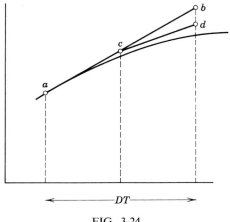

FIG. 3-24

3. The result of the single-step and the two half-step integrations are compared $(b - d)$.

4. If the difference is greater than a specified tolerance, the procedure (Steps 1, 2, and 3) is repeated using half the size of the original step.

5. The above cycle of operations is repeated until the error tolerance is satisfied by all of the integrated variables.

6. When a suitable step size is found, the calculation proceeds to the next interval, where a step twice the size will be attempted.

This method will try to increase the step size when possible, and will also decrease the step size in the regions where the error tolerance is violated. This method then has the following advantages:

1. It satisfies the error tolerance for all variables during the entire integration, assuring an accurate solution.

2. It attempts to increase the integration step size so as to reduce the calculation time.

These advantages recommend the method for beginners or occasional users who do not have the background for adopting the more direct methods described earlier in this chapter. For the assurance of accuracy provided by this method, there is a price to be paid in excessive computer time per solution. Whereas the single-step fourth-order method requires four derivative evaluations, this variable step method requires 11 evaluations. The procedure requires at least one extra trial step (twice the size) to be made at each interval; actually, a minimum of 22 derivative evaluations are required at each step. If a stiff set of differential equations is being solved, where at the

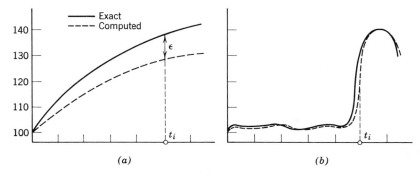

FIG. 3-25. Two Cases Having a Maximum 10% Error

critical step size the accuracy with a first-order method is satisfactory, it will be 22 times faster than the fourth-order variable-step method. Furthermore, by decreasing the step so that all the variables satisfy the error tolerance, the variable-step method may be reducing the step to an unnecessarily small level. This is because some of the intermediate variables whose derivative equations have small time constants can be severely in error without seriously affecting the accuracy of the primary variables of interest in the problem. Also, the concept of accuracy tolerance is certainly not clear cut, as discussed in Section 3-2. For example Figure 3-25 shows two cases, (a) unacceptable and (b) acceptable error, yet both results have the same 10% inaccuracy at $t_i$.

An arbitrarily set error tolerance, then, is not necessarily a good criterion of solution acceptability.

### 3-14-2 Explicit Methods (Runge-Kutta-Merson)

This method of variable step-size integration is more efficient than the implicit method discussed in the previous section. It is basically the fourth-order Runge-Kutta method having a truncation error proportional to the fourth power of the step size with a modification by Merson that provides an explicit estimate of the error with only one additional derivative evaluation. The procedure will be outlined by the following series of five derivative evaluations as applied to a generalized form of our tank differential equation described in Section 3-2.

$$\frac{dV}{dt} = f(V, t)$$

Evaluation 1    $dV_1 = f(V_n, t_n) * \dfrac{Dt}{3}$

Evaluation 2    $dV_2 = f\left(V_n + dV_1, t_n + \dfrac{Dt}{3}\right) * \dfrac{Dt}{3}$

Evaluation 3 $\quad dV_3 = f\left(V_n + \dfrac{dV_1}{2} + \dfrac{dV_2}{2}, t_n + \dfrac{Dt}{3}\right) * \dfrac{Dt}{3}$

Evaluation 4 $\quad dV_4 = f\left(V_n + \dfrac{3\,dV_1}{8} + \dfrac{9\,dV_3}{8}, t_n + \dfrac{Dt}{2}\right) * \dfrac{Dt}{3}$

Evaluation 5 $\quad dV_5 = f\left(V_n + \dfrac{3\,dV_1}{2} - \dfrac{9\,dV_3}{2} + 6\,dV_4, t_n + Dt\right) * \dfrac{Dt}{3}$

The final integration is made with the following weighted average derivative formula:

$$V_{n+1} = V_n + \tfrac{1}{6}(dV_1 + 4\,dV_4 + dV_5)$$

The estimated value for the truncation error is

$$\epsilon_t = \left(\tfrac{1}{5}\right)\left(dV_1 - \frac{9\,dV_3}{2} + 4\,dV_4 - \frac{dV_5}{2}\right)$$

In a general program, the above procedure is carried out for each variable, and the error $\epsilon_t$ is converted to a fractional or absolute error and compared with a specified tolerance. If any error is greater than the tolerance, the interval is halved and the integration repeated, and so on. On the succeeding step, following a legitimate step, a step twice the current size is attempted, as in the implicit method.

The advantage and disadvantages of the variable step methods discussed in the previous section also apply to the Runge-Kutta-Merson method, though the disadvantages are not as severe. It should be noticed, however, that the amount of arithmetic to be performed for the actual integration, apart from the number of derivative evaluations required, is significantly greater than in the implicit method, a factor that becomes even more pronounced with integration orders higher than four.

### 3-15  ALGEBRAIC SOLUTION OF DERIVATIVE EQUATIONS

Previous sections show that the critical factor that establishes the maximum step size possible for integrating a stiff system of differential equations. This factor is the ratio of the smallest time constant (establishing critical step size), to the largest time constant (establishing the range of the integration). If this ratio is very small, say less than $\frac{1}{20}$ or $\frac{1}{30}$, in some cases it is possible to increase the ratio by eliminating from the system those equations having the small time constants. This is done by converting them to algebraic equations. For example, suppose that our system of equations contains the following equation

$$\frac{dV}{dt} = \frac{1}{\tau}(f(t) - f(V))$$

where the time constant $\tau$ is relatively small compared with the dominant system time constant. If this is solved by integration, that is,

$$V = \int \frac{1}{\tau} (f(t) - f(V))\, dt$$

a small integration step is required. The alternative is to realize that since $\tau$ is small $\tau\, dV/dt$ will also be small compared with either $f(t)$ or $f(V)$. Thus as a first approximation the equation can be solved for $V$ by neglecting the derivative term, that is,

$$f(V) \approx f(t) \quad \text{or} \quad V \approx f^1(t)$$

This is the most common procedure adopted, usually during the initial formulation of a problem, where the analyst will ignore any mass and energy storages that are considered insignificant. He will then define his system as IN = OUT. For example, in the tank problem described earlier as

$$\frac{dH}{dt} = \frac{1}{A} (Q_i - Q_o)$$

if the crosssectional area is very small, the level $H$ rapidly comes to equilibrium so that

$$Q_o \approx Q_i$$

Neglecting the equilibrium, that is, the term $A(dH/dt)$, the equation is stated simply as $Q_o = Q_i$. An exact relationship can be solved by redefining the equation as

$$Q_o = Q_i - A \frac{dH}{dt}$$

This can be solved satisfactorily for $Q_o$ as long as the derivative term $A(dH/dt)$ is relatively small compared to either $Q_i$ or $Q_o$. If it is large, there is a danger of instability; in that case it should be solved by integration. The procedure for this exact solution is to obtain the variable in the differential ($H$), calculate a rate of change or differential of this variable, and substitute it in the equation, that is,

$$Q_o = Q_i - A * \frac{dH}{dt}$$

$$H = \frac{Q_o^2}{C_v^2}$$

The rate of change of $H$ has to be calculated as a backward derivative. This means that during the step $t_2 \to t_3$, the derivative for the previous step, that is,

$$\frac{H_2 - H_1}{t_2 - t_1}$$

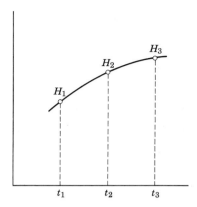

FIG. 3-26. Backward Derivative

(Figure 3-26) is used. The current derivative is unknown until the step is completed, when it will be used in the following step and so on. This procedure can be generalized by the subroutine DER described in the next section.

## 3-16 SUBROUTINE DER

This subroutine is used for generating the derivative of a variable Y with respect to any other problem variable X. The internal mechanics of the routine are keyed to the INT system for solving differential equations; thus it is not a "stand alone" subroutine that can be used with any program unless it is specially modified. The first item in the argument list (Figure 3-27) is the variable Y whose derivative is to be determined with respect to the variable X, listed as the second argument. The derivative DYDX will appear in the third location of the argument list, while the last item I specifies the CALL number. The derivative is recalculated only on completion of an integration step, that is, on legitimate cycle JS4 = 4 for the fourth-order method or JS = 2 for first- and second-order methods. On these cycles, then, the derivative is recalculated (line 10) by using the values of X and Y stored in the arrays AY(I), AX(X) on the previous legitimate cycle and the current values of X and Y in the argument list. The stored values X and Y are then replaced by the current X and Y (lines 7 and 8) before returning to the

```
 1 *            SUBROUTINE DER(Y,X,DYDX,I)
 2 *            COMMON/CINT/T,DT,J9,JN,DXA(500),XA(500),IO,JS4
 3 *            DIMENSION AY(10),AX(10)
 4 *            IF (T .LE. 0.) GO TO 6
 5 *            IF ((JS4 .EQ. 4) .OR. (JS .EQ. 2)) GO TO 5
 6 *            RETURN
 7 *          6 AY(I)=Y
 8 *            AX(I)=X
 9 *            RETURN
10 *          5 DYDX=(Y-AY(I))/(X-AX(I))
11 *            GO TO 6
12 *            END
```

FIG. 3-27. Listing for subroutine DER
Argument list:
Y = variable Y
X = variable X
DYDX = derivative Y versus X
I = call number

calling program. Since a previous history is required for establishing this derivative, the initial pass (T = 0) cannot calculate a derivative; thus it merely stores the initial values of X and Y in their respective arrays. The initial derivative can be estimated or calculated in the initiation section and entered into the argument list as an initial condition.

It should be remembered that the use of this subroutine in an equation implies an approximation because of the backward derivative. As the derivative term decreases, though, the inaccuracy will diminish accordingly.

## Problems

A batch reaction consists of the chemical sequence

$$A + B \underset{}{\overset{R_1}{\rightleftharpoons}} C$$

$$C + B \overset{R_2}{\longrightarrow} D$$

The differential equations defining the component mole balances for each specie are

$$\frac{dA}{d\theta} = -R_1$$

$$\frac{dB}{d\theta} = -R_1 - R_2$$

$$\frac{dC}{d\theta} = R_1 - R_2$$

$$\frac{dD}{d\theta} = R_2$$

where the reaction rates $R_i$ are

$$R_1 = k_1 AB - k_1' C$$
$$R_2 = k_2 BC$$

1. Obtain the compositions $A$, $B$, $C$, and $D$ as a function of time using the following information:

$k_1 = 0.001$     at $\theta = 0$, $A = 10$.
$k_1' = 0.015$     at $\theta = 0$, $B = 20$.
$k_2 = 0.001$     at $\theta = 0$, $C = D = 0$

(a) With a print interval of 10, continue the solution until $\theta = 200$, using the first-order method and a step size small enough to reduce the integration error below 1%. Define the % error in this case as $\epsilon = (B - B^*) 100/B^*$ for $B \approx 10.$, where $B$ is the calculated value and $B^*$ is the precise value.

(b) Repeat the solution using the second- and fourth-order methods. In each case determine the step size necessary to maintain $\epsilon)_{B\approx 10} < 1$. Compare the step sizes for all three methods and comment on the relative efficiency of the methods.

2. Repeat 1(a) and 1(b) with a stiff system by changing $k_1$ to 0.4 and $k_2$ to 0.01.

3. Repeat problem 2 by substituting for $R_1$ and $R_2$ in the differential equation for $C$ and solving for $C$ algebraically and neglecting the $dC/d\theta$ term.

4. Repeat problem 3 without neglecting the derivative term $dC/d\theta$. In problems 3 and 4 observe the relationship between accuracy, step size, and integration method.

5. Solve problem 1 by including the following Arrhenius coefficients.

$$k_1 = k_2 = 5 * 10^{18} * e^{(-15000/T)}$$

$$k_1' = 75 * 10^{18} * e^{(-15000/T)}$$

The temperature $T$ is obtained from a heat balance equation

$$\frac{dT}{d\theta} = (R_1 + R_2) * 10 + (300. - T)$$

6. Show how a variable-size integration step and print interval improves the calculation time and information density distribution in problem 5.

7. The equations $\dfrac{dY_1}{dt} = -0.04Y_1 + 10^4 Y_2 Y_3$

$$\frac{dY_2}{dt} = 0.04Y_1 - 10^4 Y_2 Y_3 - 3.10^7 Y_2^2$$

$$\frac{dY_3}{dt} = 3.10^7 Y_2^2$$

where $Y_1(0) = 1.$, $Y_2(0) = 0$, $Y_3(0) = 0.$ represent a system of reaction rate equations (Ref. 1) considered highly stiff (Ref. 3). Show that

(a) the solution using RK4 with $Dt = 0.001$ is unstable after $t = 18$.

(b) A stable solution can be obtained using a variable $Dt$ coupled to $Y_2$, that is, $Dt = f(Y_2)$.

(c) Almost the same solution can be obtained using the same $Dt$ as in (b) with the first-order integration method.

(d) By solving for $Y_2$ algebraically (i.e., $dY_2/dt = 0$) the same solution can be obtained with a step size several orders of magnitude larger.

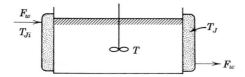

8. The exothermic reaction $A + B \rightarrow C + D$ is carried out in a jacketed reactor. A constant flow of cooling water $Fw$ passes through the jacket, entering at a temperature of $T_{Ji}(40°)$. Calculate the temperature and composition of the fluid mass, using the following data:

(a) Total volume of liquid = 30 ft³, no density change with reaction.

(b) Initial charge:

$$30 \text{ moles } A$$

$$24 \text{ moles } B$$

Initial temperature: 70°

(c) The reaction rate is second order, proportional to the concentration (moles/ft³) of each component. The rate coefficient is $k = 2.58.10^5 e^{-5000./T(°K)}$ ft³/(min mole).

(d) The heat of reaction is $10.8.10^4$ PCU/mole of $A$ or $B$ reacting, and the average heat capacity of the reaction mass is 300 PCU/mole °C.

(e) Assume that the average jacket temperature is also the outlet temperature and that the initial value is 40°C. The overall heat transfer coefficient between jacket and reactor contents is 40,000 PCU/°C min. The heat capacity of the water in the jacket is 2000 PCU/°C, and the cooling water flow rate $Fw = 6877$. lb/min.

## REFERENCES

1. "Review of Numerical Integration Techniques for Stiff Ordinary Differential Equations," J. Sienfeld, L. Lapidus, and H. Hwang, *Ind. Eng. Chem. Fund.* **9**, No. 2, 1970.
2. "The Automatic Integration of Ordinary Differential Equations," C. W. Gear, *Comm. of the ACM*, **14**, No. 3, March 1970.
3. "Solution of a set of Reaction Rate Equations," H. H. Robertson, in *Numerical Analysis*, J. Walsh, Ed., Thomson Book, Washington, D.C., 1967.

For a more advanced and detailed discussion of the topics in this chapter consult the following reference:

4. *Applied Numerical Methods*, B. Carnahan, H. A. Luther, and J. O. Wilkes, Wiley, New York, 1970.

# CHAPTER IV

# BASIC MODELING

Chapter 1 describes the procedure for the solution of problems using the analytical approach. It is pointed out that the difficulty of solving the large number of complex equations arising from the analysis has now been largely eliminated due to the general availability of powerful computers. Chapters 2 and 3 demonstrate this by developing a series of subroutines that can be used for the solution of complex algebraic/differential equations. These routines contain the procedure for numerical integration and iteration so that programming a series of equations is merely a matter of calling the appropriate subroutine. As a reminder, consider one differential equation from a set:

$$\frac{d}{dt}(NX_c) = F_R X_{CR} + R - V Y_c$$

The only programming effort required for this equation is the following FORTRAN equation in the derivative section

$$DNXC = FR * XCR + R - V * YC$$

and the integration subroutine

$$CALL \; INT(NXC, DNXC)$$

listed in the integration section, with the initial value of NXC provided. To obtain the variable XC (a composition) a division is made in the derivative section:

$$XC = NXC/N$$

The procedure for listing the equations and statements in the proper order is a rather elementary clerical procedure; thus in the remainder of this text, when developing mathematical models for physical systems, the procedure

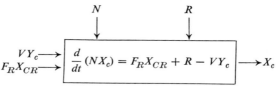

FIG. 4-1. Basic Information Flow Block

is implied by the use of information-flow block diagrams. For the equation shown above, block notation would appear as shown in Figure 4-1.

This picture represents the fact that given all the variables except one unknown ($X_c$), the equation can be readily solved for that unknown, using the appropriate computer procedure. The purpose of these information flow, equation block flow diagrams is to create a picture of the analytical equivalent of the physical system being analyzed. By this means it is hoped to eliminate much of the apparent sterility of mathematical equations and present them instead as elegant expressions of engineering relationships. Hopefully, this will encourage their use by engineers.

In a complex series of interrelated blocks, that is, a model, it is important for the analyst to distinguish between those equations to be solved algebraically with those to be solved by integration. The reason is that the analyst must recognize the presence of any implicit algebraic loops in his model in order to make suitable provision for this situation in his program. Any integration block is, therefore, denoted by a solid triangle on the output as in Figure 4-2.

This output can be considered as a point of origin when laying out the FORTRAN sequence in the derivative section, that is, the outputs of all the derivative equations are available at the start of the derivative section. For those models that do not have implicit algebraic loops, all the remaining variables, including the derivatives, can be obtained explicitly from these starting variables without ever using a variable that has not been defined in a prior statement. For those cases where this cannot be done, there is an implicit algebraic loop calling for the use of the CONV (see Chapter 2) subroutine. Examples of this situation are discussed in later chapters.

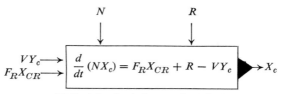

FIG. 4-2. Basic Integration Block

### Modeling Principles

To construct mathematical models with the preceding approach, the following simple rules must be observed:

1. There must be as many equations (nonredundant) as unknown values when dealing with physical systems.

2. Any equation can be solved for an unknown quantity, provided all the other unknown values are supplied from other equations.

3. Equations are assigned to solve for unknowns in such a way that each equation solves for one of its most significant quantities, based on an estimate of the physical aspects of the problem.

These rules, if applied correctly, should result in stable convergent mathematical models.

The remainder of this chapter develops the basic approach to mathematical modeling of physical systems by starting with a very simple system, common in many chemical processes, and gradually adding other typical complexities. The information flow block diagrams to be developed become larger as the physical system becomes more complex, it is shown, however, greater complexities merely require adding more equations to the model with a minimum of change to the basic model.

The physical system to be studied involves various aspects of a continuous flow, well-agitated vessel, with interacting variables such as flow, pressure, mixing, and chemical reaction. The first case is simply a tank with fluid flowing in at a known rate, $F_1$, and an outflow of $F_2$, also known. Note that these rates are time functions and need not generally be constant. It is required to determine the level $Z$ of the fluid in the tank at any time $t$ (Figure 4-3).

*Case 4-1*

The conservation equation required can be stated as

$$\text{rate of accumulation} = \text{inflow} - \text{outflow}$$

The rate of accumulation is the rate of change of fluid volume with respect to time, that is, $dV/dt$. If the cross-sectional area is $A$, the volume is $ZA$

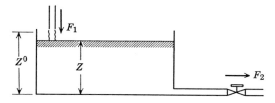

FIG. 4-3. Continuous Flow Tank

Inputs ——————————————→ System equations ——————————————→ Output

$$A \frac{dZ}{dt} = F_1 - F_2 \quad \xrightarrow{\frac{dZ}{dt}} \quad Z = \int \left(\frac{dZ}{dt}\right) dt \quad \xrightarrow{Z}$$

FIG. 4-4

and $dV/dt$ becomes $d(ZA)/dt = A\, dZ/dt$ since $A$ is constant. The equation for this system can be stated as

$$A \frac{dZ}{dt} = F_1 - F_2$$

This equation is shown in Figure 4-4, arranged as an information flow diagram.

The model expresses the fact that by supplying the values of the two flow rates $F_1$ and $F_2$ to the first equation as continuous functions of time the value of the derivative $dZ/dt$ can also be computed as a continuous function of time. This derivative can be continuously integrated (shown as the second block) to produce $Z$, the tank level, as a continuous time function. Usually it is convenient to include the second function, that of integrating the derivative, as part of the first block by assuming that if we can define a differential, $dZ/dt$, it can be integrated to give the variable $Z$. The model now simplifies to the diagram in Figure 4-5.

NOTE. Because $A$ is a constant, its value is assumed and does not, like $F_1$ and $F_2$, have to be supplied to the block.

If this equation is solved by a computer, supplying $F_1$ and $F_2$ and a starting value for the level in the vessel $Z^0$ (Figure 4-3), the value of $Z$ as a function of time will be obtained. The foregoing example was used to demonstrate the basic idea in representing an equation as an integral part of an information flow diagram. A more complex case would consider the situation in which both the inflow and outflow rates are influenced by the level in the tank, as is the next case.

Inputs ——→ System equation ——→ Output

$$A \frac{dZ}{dt} = F_1 - F_2 \quad \xrightarrow{Z}$$

FIG. 4-5.   Model of Tank in Figure 4-3

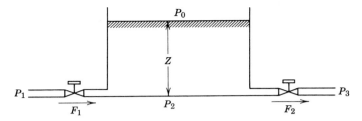

FIG. 4-6.   Tank System

*Case 4-2*

In this case (Figure 4-6) the inflow $F_1$ passes through a fixed inlet valve from a pressure source, $P_1$. The pressure on the downstream side of the inlet valve is $P_2$, that is, the hydrostatic pressure in the tank at the level of the valve. In a similar fashion the outflow passes through a fixed valve with the hydrostatic pressure $P_2$ on the upstream side discharging to pressure $P_3$. The flows $F_1$ and $F_2$ are influenced by a variable in the system, namely, the level $Z$, and as such they both become *dependent* variables, whereas the *independent* variables (besides time) are the pressures $P_0$, $P_1$, and $P_3$. Because there are four dependent variables in this system ($F_1$, $F_2$, $Z$, and $P_2$) four equations are required. The first is the same conservation equation used in the preceding example

$$\frac{dZ}{dt} = \frac{1}{A} \cdot (F_1 - F_2)$$

with the addition of two flow equations through the valves

$$F_1 = C_{V1}\sqrt{(P_1 - P_2)}$$
$$F_2 = C_{V2}\sqrt{(P_2 - P_3)}$$

where $C_V$ is the valve constant.

The fourth equation relates the pressure $P_2$ to the hydrostatic head $Z$

$$P_2 = P_0 + Z\phi$$

where $\phi$ is the density.

These four equations can be arranged in model form in three different ways, as shown in Figure 4-7. The difference between the models is the result of selecting a different equation to define each variable, and although each of these models is mathematically possible only the first model makes any sense from a physical viewpoint. In this arrangement each equation is used in its natural form: that is, the flows $F_1$ and $F_2$ are the result of exerting a pressure in a fluid across a valve. The unnatural use of this equation (models 2 and 3) would be to supply the flow and to derive from the equation

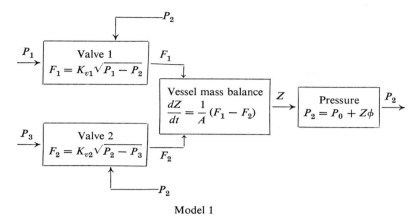

Model 1

FIG. 4-7. Alternative Models for Tank System shown in Figure 4-6

what the pressure would have to be to yield the flow. This is an example of abstract reasoning and does not represent the cause-and-effect relationship that occurs in nature. Likewise, the mass balance equation can be used mathematically to solve for either flow, $F_1$ (model 3) or $F_2$ (model 2), if both the level $Z$ as a time function and one of the flows are supplied. Here again, though, this is not consistent with cause and effect because the variations in flows $F_1$ and $F_2$ *cause* variations in level $Z$, and the mass balance equation should be used in this manner (model 1).

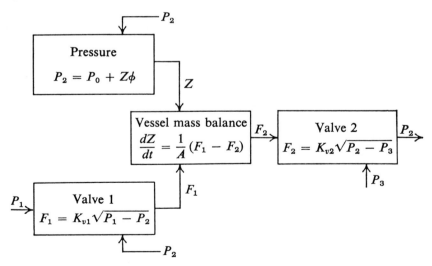

FIG. 4-7.   Model 2

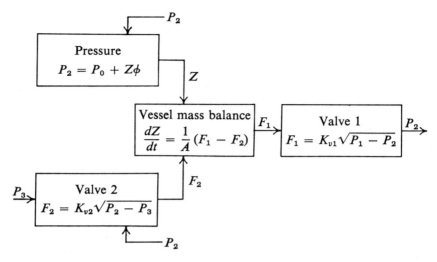

FIG. 4-7. Model 3.

Another example of abstract reasoning is illustrated by an alternative arrangement such as that shown in Figure 4-7 (model 2): "If one of the flows $(F_1)$ is supplied to the mass-balance equation, how must the other flow $(F_2)$ behave in order to result in the observed variation in vessel level $Z$." There are two important reasons for strongly favoring the "natural" model over the other mathematically possible arrangements. The first is that the natural model, being based strictly on cause-and-effect relationship, will provide insight to the analyst on the true mechanism of the system. Second, an unnatural model often leads to computational difficulties of instability and divergence, whereas a natural model is inherently computationally stable (excluding numerical instabilities as described in Chapter 2).

*Case 4-3*

The next case to be studied is similar to the preceding one except that the vessel is wholly enclosed (see Figure 4-8), and the pressure $P_0$ above the

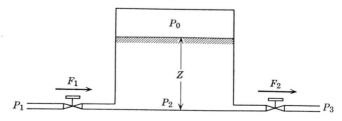

FIG. 4-8. Enclosed Vessel

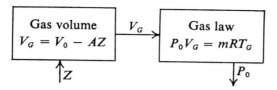

FIG. 4-9.  Model for Gas Volume

surface becomes a variable instead of a constant. It is now required to relate this pressure $P_0$ to the movement of the liquid surface. Clearly, movement of the surface up and down will compress and expand the trapped gas and cause the pressure to change. Assuming an ideal gas, we define the relationship between pressure and volume of a gas by the gas law equation:

$$P_0 V_G = mRT_G$$

where $P_0$ = gas pressure,
$V_G$ = gas volume,
$T_G$ = gas temperature,
$m$ = mass of gas,
$R$ = gas constant.

The assumption of isothermal expansion and compression will be made, that is the temperature $T_G$ remains constant and also that vaporization from the surface will be neglected; that is, the mass of gas $m$ remains constant.

If the cross section of the tank is $A$, the liquid volume is $AZ$, and if $V_0$ is the total tank volume the gas volume will be $V_G = V_0 - AZ$. These two equations are arranged in model form in Figure 4-9. Their addition to the model is shown in Figure 4-10.

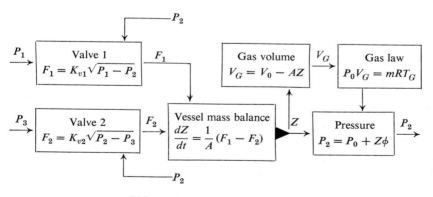

FIG. 4-10.  Model for Enclosed Vessel

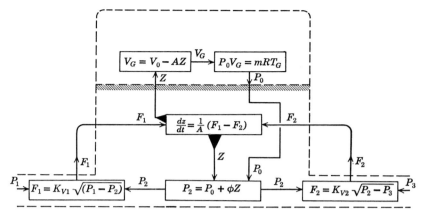

FIG. 4-11.  Pictorial Arrangement of Model

The solution of this set of equations, for example, would be the variation of the level $Z$ for a given regime of $P_1(t)$ and $P_3(t)$. Such solutions are difficult to obtain practically by mathematical techniques, but they can be readily programmed on a computer by any competent programmer; presenting the problem to the programmer as a natural model reduces the programmer's task to a simple procedure.

To visualize the relationship between the equations and the physical system, it sometimes helps to arrange the mathematical model to resemble a diagram of the physical layout of the system. For example, the model for Case 4-3 could be arranged to resemble the vessel shown in Figure 4-11. For very large problems an arrangement of the blocks in a pictorial sequence is useful in the construction and visualization of a "natural" model.

### Case 4-4

Case 4-3 assumed that gas volume compresses under isothermal conditions; now, the assumption of adiabatic compression is made, in which the gas temperature is no longer constant but varies with the compression. The same basic gas law holds ($P_0V_G = mRT_G$) but the temperature $T_G$, as well as $V_G$, must be supplied to determine the pressure $P_0$. Some other equation is needed to define the temperature; this is provided by the relationship between the work of compression on the gas and the sensible heat of the gas. In this adiabatic case all the heat equivalent of the work done on (or by) the gas will appear as sensible gas heat. From elementary thermodynamics the

$$V_G \quad \boxed{\dfrac{dT_G}{dt} = -\dfrac{P_0}{mC_vJ} \cdot \dfrac{dV_G}{dt}} \quad T_G$$

$$\uparrow P_0$$

FIG. 4-12.    Model for Adiabatic Gas
Volume

following equations can be stated:

$$\text{rate of work done on gas volume} = -P_0 \frac{dV_G}{dt}$$

$$\text{heat equivalent of this work} = -\frac{1}{J} \cdot P_0 \frac{dV_G}{dt}$$

where $J$ = joules heat equivalent of work.

The rate of change of sensible heat of a gas is $d/dt\,(mC_VT_G)$, and because $m$ and $C_V$ (specific heat) are constant this becomes $mC_V \cdot dT_G/dt$. When the work is equated to the sensible heat, the following equation can be defined:

$$\text{change in sensible heat} = \text{work done}$$

$$mC_V \frac{dT_G}{dt} = -\frac{P_0}{J}\frac{dV_G}{dt}$$

In model form it will be used as shown in Figure 4-12. Adding this effect to the basic model of Case 4-3 results in the new model shown in Figure 4-13.

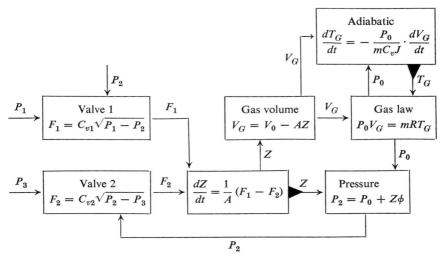

FIG. 4-13.    Model for Tank System Assuming Adiabatic Gas Volume

There is a temptation in the preceding example to use the well-known relationship $P_0 V^\gamma_G = $ constant for adiabatic compression. However, this suffers from the inflexibility of its limitation to *adiabatic* compression only. If there were some other heat loss in the gas volume, it would be included quite simply in the basic equation shown in the model, but it cannot be included in the adiabatic equation.

### 4-4-1  Computer Program

It is worthwhile at this point to demonstrate to the reader the simplicity of programming the model shown in Figure 4 13. It should be realized, though, that the derivative of $V_G$ can be obtained by differentiating the gas volume equation $dV_G/dt = -A dZ/dt$. The derivative of the level $Z$ is available from the mass balance equation. The essential statements in the program would be (omitting the housekeeping section)

```
C ** INITIATION SECTION **
      DATA CV1, CV2, M, CV, J, R, RO, A, VO/ Numerical values/
      T = 0.
      Z = 0.
      TG = 25.
      P1 = 20.
      P3 = 15.
C ** DERIVATIVE SECTION **
    5 VG = VO − A * Z
      PO = M * R * TG/VG
      P2 = PO + Z * RO
      F1 = CV1 * SQRT (P1 − P2)
      F2 = CV2 * SQRT (P2 − P3)
      DZ = (F1 − F2)/A
      DTG = −PO/(M * CV * J) * A * DZ
      CALL PRNTF (1., 20., NF, T, Z, F1, F2, PO, P2, VG, TG, DZ, DTG)
      GO TO (6,7), NF
C ** INTEGRATION SECTION **
    6 CALL INTI (T,. 1, 2)
      CALL INT (Z, DZ)
      CALL INT (TG, DTG)
      GO TO 5
    7 STOP
      END
```

This program follows the general procedure described in Chapter 3 and summarized as:

1. Specify the parameters and initial conditions.
2. Specify the derivative equations in proper order.
3. Call PRNTF routine, specifying print interval (1) and finish (20).
4. Call integration routine INT to integrate the derivatives generated in 2, starting with INTI.
5. Recycle to first line of derivative section.

*Case 4-5*

The basic elements of mixing and elementary kinetics are developed in the next examples which return to the case of a well-agitated vessel with known flows in and out, $F_1$ and $F_2$ (Figure 4-14).

The inlet and outlet flows consist of a solvent that contains two soluble components, $A$ and $B$. The inlet concentrations are $C_{A1}$ and $C_{B1}$. The equations relating the exit compositions to the inlet compositions are based on simple mass balances.

$$\text{rate of accumulation} = \text{inflow} - \text{outflow}$$

$$\frac{d}{dt}(VC_{A2}) = F_1 C_{A1} - F_2 C_{A2} \quad \text{(mass balance on component } A)$$

$$\frac{d}{dt}(VC_{B2}) = F_1 C_{B1} - F_2 C_{B2} \quad \text{(mass balance on component } B)$$

Note that because ideal mixing is assumed the composition leaving in the outflow is the same as in the vessel. As in Case 4-4, the volume $V$ in the tank is obtained by an overall mass balance

$$\frac{dV}{dt} = F_1 - F_2$$

These equations are arranged in model form in Figure 4-15. Here, again, several possible arrangements exist, but only the one illustrated makes sense from a physical standpoint. The same reasoning applies to all three equations;

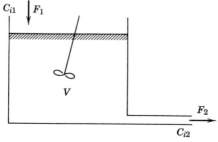

FIG 4-14.   Mixing Vessel

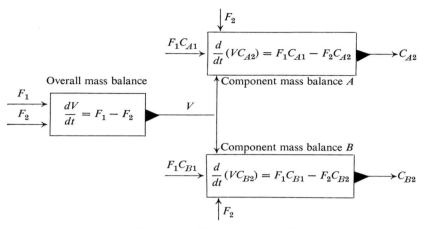

FIG. 4-15.   Model for Mixing Vessel

that is, supplying the input flows to the conservation equations establishes the quantity inside the tank. In the total mass balance this establishes the total volume $V$, whereas for the component equations supplying the component flows $FC_i$ provides the quantity in the tank $VC_i$ and, because $V$ is also supplied, the composition $C_i$ is obtained. It should also be observed that it is unnecessary to manipulate the equations in any way. The product terms $VC_i$, inside the differentials do not have to be differentiated by parts; they remain inside the differential.

*Case 4-6*

Continuing from the preceding example, suppose that components $A$ and $B$ react in the vessel, the reaction being defined by the stochiometric equation

$$A + B \xrightarrow{k_F} C + D$$

The effluent and vessel contents contain four components, $A$, $B$, $C$, and $D$. A rate expression defining the rate of reaction can be stated as

$$R = k_F \cdot V \cdot C_A \cdot C_B$$

where $R$ = moles/unit time and is the reaction rate in the volume $V$ and $k_F$ is the rate constant.

A convenient way of viewing this reaction is to consider $R$ as an outflow of components $A$ and $B$ and an inflow of components $C$ and $D$. The mass balance equations now become

$$\text{rate of accumulation} = \text{inflow} - (\text{outflow})$$

$$\frac{d}{dt}(VC_{A2}) = F_1C_{A1} - (F_2C_{A2} + R)$$

$$\frac{d}{dt}(VC_{B2}) = F_1C_{B1} - (F_2C_{B2} + R)$$

$$\frac{d}{dt}(VC_{C2}) = R - (F_2C_{C2})$$

$$\frac{d}{dt}(VC_{D2}) = R - (F_2C_{D2})$$

Starting with the model developed in Case 5, add the term $R$ to the $A$ and $B$ component material balances. The compositions $C_A$ and $C_B$ and the volume

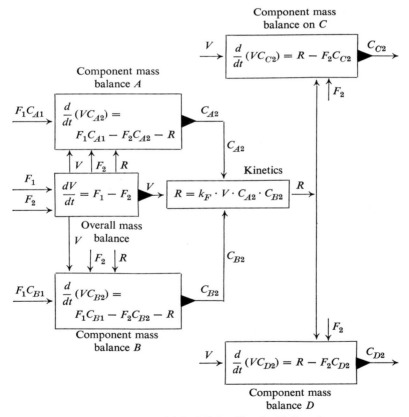

FIG. 4-16.  Model for Mixing Vessel with Reaction

$V$ are then supplied to the kinetic equation that produces $R$. This is fed back to the $A$ and $B$ component material balances and on to the component balances for $C$ and $D$. This sequence is shown diagrammatically in Figure 4-16.

*Case 4-7*

Continuing from Case 6, suppose the reaction is reversible, that is,

$$A + B \underset{k_R}{\overset{k_F}{\rightleftharpoons}} C + D$$

The reaction rate now becomes

$$R = k_F V C_{A2} C_{B2} - k_R V C_{C2} C_{D2}$$

For this case it is assumed that the reaction takes place in the enclosed vessel studied in Case 4-4, with a trapped gas space above the surface and the inlet flow supplied at the bottom of the vessel from a known pressure source $P_1$ through a fixed valve opening. Similarly, the outlet flow $F_2$ will flow to a known pressure, $P_3$, through a fixed valve. All the equations for such a system have already been developed in this chapter; the complete model for this final case is shown assembled in Figure 4-17.

## 4-8  SIMULTANEOUS MASS AND ENERGY BALANCES

The second part of this chapter develops the basic concepts of energy balances for simple flow systems. A chemical process invariably involves energy transfer simultaneously with mass transfer; it is therefore important that the factors involved in performing a correct heat balance be clearly understood. The same general approach used in Chapter 3 is adopted here, and the differential equations that define the dynamic situations are developed for each case.

*Case 4-8*

Figure 4-18 shows a vessel with a steam jacket, an inlet flow, $F_1$, and an outlet flow, $F_2$ (volume/time).

The holdup in the vessel, $V$ (volume), varies according to the equation

$$\frac{dV}{dt} = F_1 - F_2$$

The energy balance for the contents of the vessel is similar to the mass balance

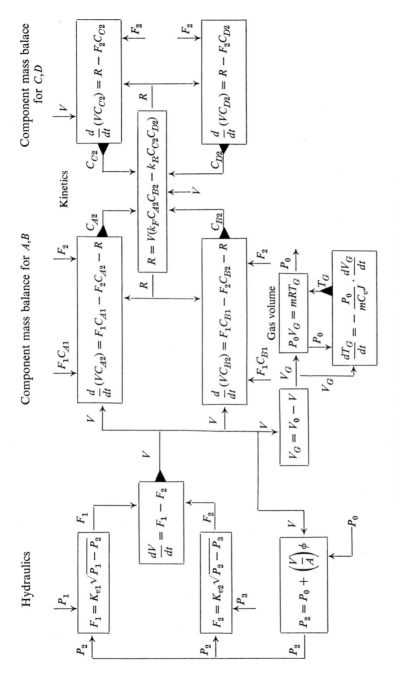

FIG. 4-17. Complete Model for Reaction in Enclosed Vessel

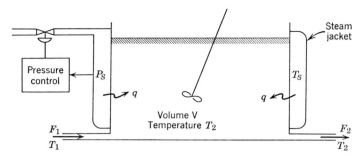

FIG. 4-18.  Vessel with Steam Jacket

used in Chapter 3; that is,

rate of change of heat energy in vessel = heat in − heat out
heat content of vessel = $Vc\phi T_2$
heat flowing into vessel = $F_1 c\phi T_1$
heat flowing out vessel = $F_2 c\phi T_2$
heat transferred from jacket $q = UA(T_2 - T_S)$

where $T_1$ = temperature of inlet flow,
$T_2$ = temperature of contents of vessel,
$U$ = overall heat transfer coefficient across jacket wall,
$A$ = jacket wall area, ft²,
$T_S$ = steam temperature in jacket,
$c$ = specific heat of fluid, PCU/lb°C,
$\phi$ = density, lb/ft³.

By substituting these terms in the energy balance statement we obtain the following equation:

$$\frac{d}{dt}(Vc\phi T_2) = F_1 c\phi T_1 - q - F_2 c\phi T_2$$

As in the example on mixing in Chapter 3, it is assumed here that the temperature $T_2$ is the same at all points inside the vessel and is therefore also the temperature of the outflow $F_2$. The temperature of the steam jacket $T_S$ is a function of the pressure, and it is assumed that this pressure $P_S$ is controlled to a known value. The steam temperature $T_S$ can be defined merely as a function of pressure; that is,

$$T_S = f(P_S)$$

This function, of course, is the well-known relationship between the boiling point of water and pressure. Note that is is unnecessary to approximate this function analytically, for, if a computer is to be used, it is more

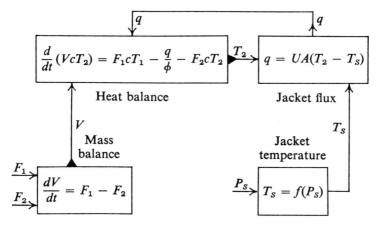

FIG. 4-19.   Model for Steam-Jacketed Vessel

convenient to set up this function in one of the computer's function generators. These equations can be assembled and are shown in model form in Figure 4-19.

The arrangement is quite logical; the mass balance is used to establish the holdup $V$; the heat transfer equation establishes $q$, and the heat balance equation defines the vessel temperature $T_2$. Again, the student is reminded that mathematically one other arrangement is possible, but this would not make sense from a natural "cause-and-effect" point of view.

*Case 4-9*

Two complications are added to the example in Case 4-8. The first is that instead of a single feed there are now two feed flows, $F_A$ and $F_B$, each with a different specific heat, $c_A$ and $c_B$. The second complication is to assume that the heat transfer area $A$ between the steam jacket and the contents of the vessel varies significantly because of the variation in level (Figure 4-20). Density variations are neglected.

The mass balance for this case becomes

$$\frac{dV}{dt} = F_A + F_B - F_2$$

The specific heat of the vessel contents is

$$c_2 = c_A \frac{C_A}{\phi_A} + c_B \frac{C_B}{\phi_B}$$

where $C$ = concentration lb moles/ft³
$\phi$ = density lb moles/ft³

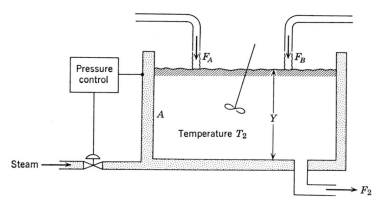

FIG. 4-20.   Steam-Jacketed Vessel with Mixing

Two more equations are required to establish the concentrations $C_A$ and $C_B$.

rate of accumulation = inflow − outflow

$$\frac{d}{dt}(C_A V) = F_A - F_2 C_A$$

$$\frac{d}{dt}(C_B V) = F_B - F_2 C_B$$

The variation in volume $V$ will cause the heat transfer area $A$ to vary according to the following equation:

$$A = \frac{D^2 \pi}{4} + \frac{4V}{D}$$

where $D$ = diameter of vessel.

Up to this point each example has been built up from a preceding, less complex, case. Experience has shown that those who are unfamiliar with systems engineering techniques or computer programming have difficulty in assembling the complete information flow model from its basic parts. There are, of course, several approaches to assembling a model, but the following guide lines should prove useful.

1. List the equations, defining all the symbols.

2. From physical considerations (i.e., natural cause and effect) decide how each equation is to be used (i.e., which unknown variable it will establish).

This example is used to illustrate the procedure.
(a) Component mass balance on $A \to C_A$
(b) Component mass balance on $B \to C_B$
(c) Overall mass balance   $\to V$
(d) Area   $\to A$
(e) Jacket heat transfer   $\to q$
(f) Specific heat   $\to c_2$
(g) Heat balance   $\to T_2$
(h) Boiling point function   $\to T_S$

3. Boundary inputs (and initial values) to the system are now listed. There should be as many initial condition values as the number of first-order differential equations.

| Boundary Inputs | Initial Values |
|---|---|
| 1. $F_A$ | for $V$ |
| 2. $F_B$ | for $C_A$ |
| 3. $P_S$ | for $C_B$ |
| 4. $F_2$ | for $T_2$ |
| 5. $T_A$ | |
| 6. $T_B$ | |

4. Arrange the equations pictorially in whatever form best suits personal preference, bearing in mind that if possible a main information flow sequence should be reflected, by connecting one equation block to the next (shown in Figure 4-21).

5. Finally, a check is made to ensure that all the variables required by each equation is produced by some other equation in the model or is a known boundary input.

The model may now be programmed directly for computation.


## 4-10   BOILING

Suppose a container of fluid is heated at a rate $q$ (PCU/time). A heat balance equation would state

rate of change of heat content = heat in − heat out

$$\frac{d}{dt}(VcT) = q - 0 \quad \text{(no heat loss)}$$

where $V$ = volume,
  $c$ = specific heat.

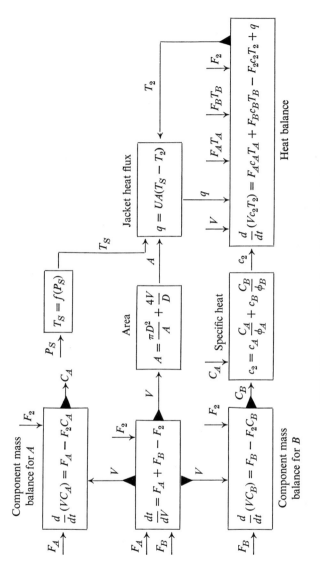

FIG. 4-21. Model for Jacketed Vessel with Mixing

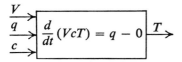

FIG. 4-22.   Heat Balance Model

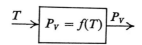

FIG. 4-23.   Vapor Pressure/Temperature Relationship

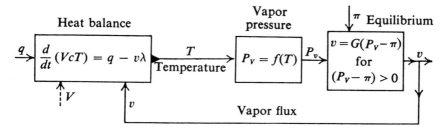

FIG. 4-24.   Model (Microscopic) for Equilibrium Balance

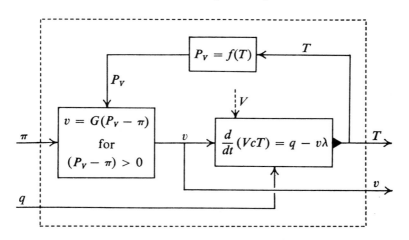

FIG. 4-25.   Input-Output Relationship for Microscopic Boiling Model

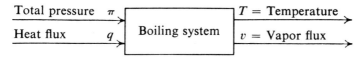

FIG. 4-26.   Macroscopic Input-Output Relationship

Since $V$, $q$, and $c$ are known, this equation can be used to establish the temperature (Figure 4-22) by supplying $V$, $q$, and $c$.

The vapor pressure exerted by the liquid varies with the temperature, as shown in Figure 4-23. Vaporization can be considered negligible until the temperature has reached the boiling point. At this temperature the vapor pressure $P_V$ tends to exceed the actual pressure $\pi$, that is, $P_V > \pi$, and this results in a stream of vapor to issue from the boiling liquid. Because there is no resistance to the departure of this vapor, a sufficient flow is emitted that (through the heat balance) automatically prevents the temperature from rising beyond the boiling point. The vapor pressure corresponding to the boiling temperature is infinitesimally greater than the total pressure, but this minute difference is sufficient to provide the vapor flow that maintains the status quo. The boiling heat balance can be shown as in Figure 4-24. The equilibrium equation that computes $v$ contains a gain factor $G$ that is large enough to keep $(P_V - \pi)$ very small.

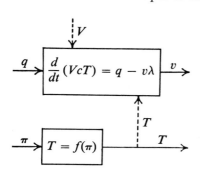

FIG. 4-27.   Macroscopic Model For Boiling

This model is the "natural" definition of the system. It can be regrouped as shown in Figure 4-25. The overall diagram in Figure 4-26 shows that the system has two inputs, $\pi$ and $q$, and two outputs, $T$ and $v$. The phenomena of boiling is such that these are virtually "noninteracting"; that is, the temperature responds *only* to the total pressure $\pi$ and the vapor flow *only* to the heat flux $q$. This leads to the more convenient macroscopic model shown in Figure 4-27. The heat balance is used to establish the vapor flux, whereas the system pressure $\pi$ dictates the temperature. In most cases the differential term $d/dt\ (VcT)$ is very small in comparison with $q$ and can be neglected. Because this latter scheme is the most convenient to implement on a computer, it is the one that is nearly always used. Although it does not parallel the microscopic natural model of Figure 4-25, it does duplicate the macroscopic model of Figure 4-26. Digitally, it avoids the time-consuming computing iteration loop (shown in Figure 4-24).

In summary, then, for a boiling single-component liquid the only way to change the temperature is to change the total pressure; changing the heating rate changes only the rate of evolution of vapor. The cause-and-effect relationships for a single-component boiling fluid can be stated as

Pressure ($P$) establishes the boiling temperature ($T$)
Heat flux ($q$) establishes the vapor rate ($v$)

*Case 4-10*

Returning to the first example of this chapter, the complexity of boiling will be assumed for the jacketed vessel. Figure 4-28 shows the feed flow supplied in liquid form and the exit flow withdrawn as vapor. The mathematical model for this boiler consists of simultaneous mass and energy balances. The mass balance on the liquid is

$$\frac{dV}{dt} = F_1 - v$$

where $F_1$ is the feed rate and $v$ is the boilup rate. The mass balance on the vapor is

$$\frac{dm}{dt} = v - v_E$$

where $v_E$ = flow of vapor through exit valve.

Because equilibrium is assumed to exist at all times between liquid and vapor, an energy balance for the vapor is not necessary, and the vapor temperature is assumed to be the same as the liquid temperature. The energy balance in the liquid is

change in sensible heat = heat in + jacket flux − heat content of vapor

$$\frac{d}{dt}(VcT) = F_1cT_1 + q - v(cT + \lambda)$$

where $(cT + \lambda)$ is an approximation of the vapor enthalpy. The pressure

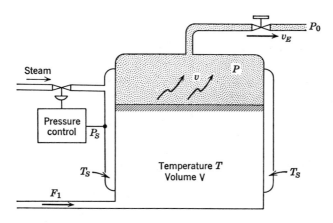

FIG. 4-28. Continuous Flow Boiling System

in the vapor space is obtained from the gas law relationship

$$PV_G = mRT \quad \text{and} \quad V_G = V_0 - \frac{V}{\phi} \quad \phi = \text{density, mass/unit vol}$$

where $V_0$ is the total volume of vessel. The temperature is obtained from the pressure/boiling temperature relationship

$$T = f(P) = c_2/(\text{Ln } P - c_1)$$

If the effluent valve is fixed, the flow rate of vapor through the valve $v_E$ will be

$$v_E = k_v\sqrt{P(P - P_0)}$$

The heat flux $q$ is the same as the first example in this chapter; that is,

$$q = UA(T_S - T)$$

With the equations for each part of the system defined, the procedure for assembling the model will now be followed.

A. Boundary values.
   1. Inlet flow $F_1$
   2. Inlet temperature $T_1$
   3. Jacket steam pressure $P_S$
   4. Exit pressure $P_0$

B. Equations.

1. Valve $\qquad\qquad\qquad\qquad v_E = k_v\sqrt{P(P - P_0)} \rightarrow v_E$

2. Gas law $\qquad\qquad\qquad\quad PV_G = mRT \rightarrow P$

3. Vapor mass balance $\qquad\quad \dfrac{dm}{dt} = v - v_E \rightarrow m$

4. Boiling point $\qquad\qquad\quad T = f(P) \rightarrow T$

5. Jacket heat $\qquad\qquad\qquad q = UA(T_S - T) \rightarrow q$

6. Heat balance $\qquad\qquad\quad \dfrac{d}{dt}(VcT) = F_1 cT_1 + q - (cT + \lambda)v \rightarrow v$

7. Mass balance on liquid $\qquad \dfrac{dV}{dt} = F_1 - v \rightarrow V$

8. Gas volume $\qquad\qquad\qquad V_G = V_0 - \dfrac{V}{\phi} \rightarrow V_G$

These equations are assembled in information flow in Figure 4-29.

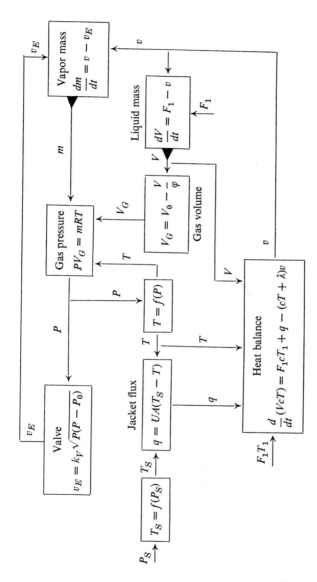

FIG. 4-29. Model for Continuous Flow, Boiling Jacketed Vessel

### 4-10-1 Computer Program for Continuous Flow Boiling

A computer program is assembled in this section to demonstrate the relative simplicity of simulation for solution of a modestly complex problem. The following conditions will apply to the system described in the preceding Case 4-10.

1. The liquid level is maintained at a fixed position by a level controller. This makes $V$ and $V_G$ constant and also the feed flow $F_1 = V_E$.
2. The liquid is initially cold and is heated up to its boiling point. After boiling starts, the pressure rises to its equilibrium level, raising the temperature to a higher value.

The first assumption will simplify the model to the first six equations listed in Case 4-10 Section B. The second assumption complicates the procedure for solution because the heat balance equation (6) has to be integrated until boiling starts, after which it must be solved algebraically. This is accomplished by the program shown in Figure 4-30 and 4-31. The first part of the program contains the input data for the various constants and the initial pressure and feed temperature T1. The initial temperature of the batch is assumed to be equal to the feed temperature (line 7). The initial moles of gas in the gas space is calculated from the pressure (line 6). The temperature level where the liquid is expected to boil (TB) is also calculated on line 5. This initiation section is followed by the equations that describe the rate of change of temperature prior to boiling. The derivative of the temperature DTC, generated on line 10 is integrated on line 14, following the INTI subroutine. The integration section (lines 13 and 14) is followed by a check on whether the temperature TC has reached the boiling temperature TB. If this is satisfied

```
1*          COMMON/CINT/XT,DT,JS,JN,DXA(500),XA(500),IO,JS4
2*          REAL M,LM,KV
3*          DATA TS,VC,VG,T1,KV,P/150.,1.E4,3.E4,15.,5.7.1./
4*          DATA C1,C2,LM,R,UA/13.96,-5210.6,9717.,1.98,1700./
5*          TB = C2/(LOG(P)-C1)-273.
6*          M = P*VG/(R*(TB+273))
7*          TC = T1
8*    C **  HEAT UP SECTION **
9*        8 Q = UA * (TS-TC)
10*         DTC = Q/VC
11*         CALL PRNTF(.5,20.,,NF,TIM,TC,VB,P,VE,Q,M,DTC,TB,0.)
12*         GO TO (5,6),NF
13*       5 CALL INTI(TIM,,1,4)
14*         CALL INT(TC,DTC)
15*         IF((TC.GE.TB).AND.(JS4.EQ.4)) GO TO 9
16*         GO TO 8
17*       9 CONTINUE
```

FIG. 4-30. Heatup Section of Continuous Flow Boiling Program

```
18*    C **   BOILING SECTION **
19*            P = M*R*(TC+273.)/VG
20*            TC1 = C2/(LOG(P)-C1)-273.
21*            CALL CONV(TC,TC1,1,NC)
22*            GO TO (3,9),NC
23*          3 Q = UA*(TS-TC)
24*            VB = Q/(TC-T1+LM)
25*            VE = KV*SQRT(P*(P-1,0))
26*            CALL PRNTF(.5,20.,NF,TIM,TC,VB,P,VE,Q,M,DTC,TB,0.)
27*            GO TO (4,6),NF
28*          4 CALL INTI(TIM,.1,4)
29*            CALL INT(M,VB-VE)
30*            GO TO 9
31*          6 STOP
32*            END
```

FIG. 4-31.   Boiling Section

and a legitimate pass also has been accomplished (JS4 from COMMON is 4), the calculation proceeds to the boiling-section equations.

In the boiling section the temperature is established from the pressure via the Antoine equation (line 20), and the pressure is established from the moles M and the temperature TC via the ideal gas law (line 19). Since these two equations contain TC implicitly, a call on CONV has to be made on line 21. The remaining equations are self explanatory. It should be noted that the derivative for M is calculated (VB − VE) in the argument for the INT routine. This is quite legal and sometimes convenient, as it is here, but this convenience should be exercised with care because of the greater possibility of sequencing errors. It should also be noted that both the PRNTF and INTI subroutine have been called twice in this same program. Perhaps the program could be shortened by using an IF statement, and it would be an interesting exercise to rearrange the program along these lines.

Part of the numerical results are shown in Figure 4-32. It will be seen that boiling starts between 5.5 and 6 min at a temperature of approximately 100°C. The pressure then rises from 1 atm to 1.668 atm, raising the temperature to 114.4°C.

## Exercise Problems

1. Part of a process consists of two vessels as shown in Figure 4-33. Components $A$ and $B$ flow into the first vessel at a rate $Q_{a1}$ and $Q_{b1}$ (ft³/min), respectively. The outlet flow is $Q_0$ (ft³/min) and flows into the second vessel in which a third inlet flow of component $B$ is added ($Q_{b2}$). The outlet flow of this vessel is $Q_1$. Assume that each vessel is well agitated and construct a model that defines the composition in $Q_1$ as a function of time if the imput flows vary with time.

| TIM | TC | VB | P | VE |
|---|---|---|---|---|
| .00000 | .15000+02 | .00000 | .10000+01 | .00000 |
| .50000+00 | .26001+02 | .00000 | .10000+01 | .00000 |
| .10000+01 | .36105+02 | .00000 | .10000+01 | .00000 |
| .15000+01 | .45386+02 | .00000 | .10000+01 | .00000 |
| .20000+01 | .53911+02 | .00000 | .10000+01 | .00000 |
| .25000+01 | .61741+02 | .00000 | .10000+01 | .00000 |
| .30000+01 | .68933+02 | .00000 | .10000+01 | .00000 |
| .35000+01 | .75539+02 | .00000 | .10000+01 | .00000 |
| .40000+01 | .81607+02 | .00000 | .10000+01 | .00000 |
| .45000+01 | .87180+02 | .00000 | .10000+01 | .00000 |
| .50000+01 | .92299+02 | .00000 | .10000+01 | .00000 |
| .55000+01 | .97001+02 | .00000 | .10000+01 | .00000 |
| .60000+01 | .10082+03 | .85293+01 | .10213+01 | .84103+00 |
| .65000+01 | .10311+03 | .81289+01 | .11121+01 | .20125+01 |
| .70000+01 | .10490+03 | .78185+01 | .11872+01 | .26870+01 |
| .75000+01 | .10634+03 | .75676+01 | .12510+01 | .31943+01 |
| .80000+01 | .10753+03 | .73606+01 | .13059+01 | .36025+01 |
| .85000+01 | .10852+03 | .71877+01 | .13533+01 | .39413+01 |
| .90000+01 | .10936+03 | .70418+01 | .13944+01 | .42273+01 |
| .95000+01 | .11007+03 | .69179+01 | .14302+01 | .44712+01 |
| .10000+02 | .11068+03 | .68120+01 | .14614+01 | .46808+01 |
| .10500+02 | .11120+03 | .67211+01 | .14887+01 | .48617+01 |
| .11000+02 | .11165+03 | .66428+01 | .15125+01 | .50184+01 |
| .11500+02 | .11204+03 | .65751+01 | .15333+01 | .51546+01 |
| .12000+02 | .11238+03 | .65165+01 | .15516+01 | .52732+01 |
| .12500+02 | .11267+03 | .64657+01 | .15676+01 | .53766+01 |
| .13000+02 | .11293+03 | .64214+01 | .15816+01 | .54669+01 |
| .13500+02 | .11315+03 | .63829+01 | .15939+01 | .55460+01 |
| .14000+02 | .11334+03 | .63492+01 | .16047+01 | .56151+01 |
| .14500+02 | .11351+03 | .63199+01 | .16142+01 | .56757+01 |
| .15000+02 | .11366+03 | .62942+01 | .16225+01 | .57287+01 |
| .15500+02 | .11379+03 | .62718+01 | .16299+01 | .57752+01 |
| .16000+02 | .11390+03 | .62522+01 | .16363+01 | .58161+01 |
| .16500+02 | .11400+03 | .62350+01 | .16419+01 | .58519+01 |
| .17000+02 | .11409+03 | .62199+01 | .16469+01 | .58833+01 |
| .17500+02 | .11416+03 | .62067+01 | .16512+01 | .59109+01 |
| .18000+02 | .11423+03 | .61951+01 | .16551+01 | .59351+01 |
| .18500+02 | .11429+03 | .61849+01 | .16584+01 | .59564+01 |
| .19000+02 | .11434+03 | .61760+01 | .16614+01 | .59751+01 |
| .19500+02 | .11438+03 | .61681+01 | .16640+01 | .59916+01 |
| .20000+02 | .11442+03 | .61613+01 | .16663+01 | .60060+01 |

FIG. 4-32.  Numerical Results of Continuous Flow Boiling

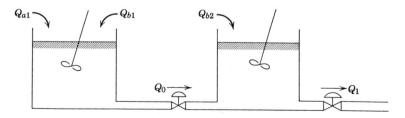

FIG. 4-33.  Series C.S.T.R

Using the following data calculate the composition in $Q_1$ during a startup.

(a) $Q_{a1} = 10$ ft$^3$/min
(b) $Q_{b1} = 0.$ for $t < 10$ min, 5 ft$^3$/min for $t > 10$ min
(c) $Q_{b2} = 0.$ for $t < 15$ min, 7 ft$^3$/min for $t > 15$ min
(d) $Q_0 = 15$ ft$^3$/min
(e) $Q_1 = 22$ ft$^3$/min
(f) Holdup in vessel 1 at time $t_0 = 100$ ft$^3$ pure $B$
(g) Holdup in vessel 2 at time $t_0 = 180$ ft$^3$ pure $B$

2. *Batch kinetics.* Two flow rates, $N_A$ and $N_B$ (number moles/min), each consisting of component $A$ and $B$, respectively, flow into a well-agitated vessel that has an outflow rate $F_0$ (ft$^3$/min). Inside the vessel the holdup is $H$ moles, and the following reaction occurs:

$$A + B \underset{k_2}{\overset{k_1}{\rightleftharpoons}} C + D$$

$$A + C \underset{k_4}{\overset{k_3}{\rightleftharpoons}} E$$

The reaction rate coefficients $k_n$ are known, and the general form of the reaction rate equation is

$$R_n = k_n \cdot H \cdot X_i X_j$$

where $X$'s are mole fractions and $k = $ min$^{-1}$.
The outflow through the valve

$$F_0 = k_v \sqrt{P - P_0}$$

where $P = $ psi, the pressure at the bottom of the vessel, and $P_0$ is known.
Define the mathematical model for the system using the symbols

$$A = \text{cross-sectional area of vessel (ft}^2)$$
$$\phi_i = \text{density of component } i \text{ lb/ft}^3$$
$$M_i = \text{molecular weight of component } i$$

Using the following data, calculate the compositions in the effluent stream during a startup period.

(a) Initial holdup $= 10$ moles of pure component $B$
(b) Feed flow $N_A = 10$ moles/min of $A$
    $N_B = 5$ moles/min $B$
(c) $k_v = 2.7$ ft$^3$/psi$^{1/2}$: $A = 10$ ft$^2$: $P_0 = 15$ psi
(d) Rate coefficients $k_1 = 1.5$, $k_2 = 0.2$, $k_3 = 2.1$, $k_4 = 0.05$

(e)

| Component | Mol. Weight | Density lb/ft³ |
|-----------|-------------|----------------|
| A | 24 | 75. |
| B | 36 | 65. |
| C | 40 | 80. |
| D | 20 | 60. |
| E | 64 | 80. |

3. A single component fluid, flowing at a rate $F$ lb/min, passes through a heat exchanger where heat is added at a rate $q$ (PCU/min). The inlet temperature to the heat exchanger is $T_1$, and after the exchanger the fluid flows through a restriction at which it partly flashes into a vessel of volume $V$ (Figure 4-34). The liquid fraction collects in the bottom of the vessel and flows out through a valve to a pressure $P_0$, whereas the vapor at a pressure $P$ flows out through a second valve to the same downstream pressure $P_0$. Construct a model that defines the variation in outlet vapor and liquid flow as a function of input flow $F$ and heat flux $q$.

Using the following data, simulate the model on a computer and calculate the vapor flow for an impulse flow $F = 10$ moles/min lasting 1 min.

(a) Vapor pressure P (atm) = exp (13.45 − 5040./$TA$) where $TA$ = °K
(b) $P$ at $t = 0$. is $P_o$ and $P_o = 1$ atm
(c) Heat capacity of liquid = 30 PCU/(mole °C)
(d) Latent heat = 1000 PCU/mole
(e) $T_1 = 80$°C
(f) $q = 8000$ PCU/min
(g) Volume of vessel = 5000 ft³, area = 200 ft²
(h) Vapor valve coefficient $k_v = 0.1$ mole/psi
(i) Liquid valve coefficient $k = 2$ mole/psi$^{\frac{1}{2}}$

4. Modify the computer program for Case 4-10 so as to include the change in sensible heat of the liquid in the heat balance.

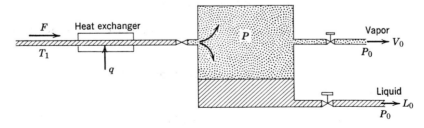

FIG. 4-34.  Flash Chamber Problem 3

## NOMENCLATURE

$A$    area or chemical component
$B$    chemical component
$C_i$    concentration of component $i$ moles/unit vol
$C_v$    gas specific heat
$c$    liquid specific heat
$D$    diameter
$F$    flow rate
$H$    holdup
$J$    joules constant
$C_v$    valve constant
$k$    reaction rate constant
$M$    molecular weight
$m$    mass or moles
$N$    mole flow rate
$P$    pressure
$q$    heat flux
$R$    reaction rate or gas constant (Case 4-3)
$T$    temperature
$t$    time
$U$    overall heat transfer
$V$    mass or volume
$v$    vapor flow rate
$Z$    level
$\phi$    density
$\lambda$    latent heat
$\pi$    total pressure or 3.1416

## REFERENCES

These two published examples are comparable in complexity to the examples discussed in this chapter.

1. "Analog Computer Study of a Semi-Batch Reactor," A. Carlson, *Inst. Control Systems*, April 1965.
2. "Process Control Problems Yield to the Analog Computer," C. W. Worley, R. G. Franks, and J. Pink, *Control Eng.*, June 1957.

The following texts provide more advanced treatment of the concepts developed in this chapter.

3. *Transport Phenomena*, R. B. Bird, W. E. Stewart, and E. N. Lightfoot, Wiley, New York, 1970.
4. *Process Analysis and Simulation—Deterministic Systems*, D. M. Himmelblau and K. B. Bischoff, Wiley, New York, 1968.

CHAPTER V

# MULTICOMPONENT VAPOR LIQUID EQUILIBRIUM

## 5-1 GENERAL STRUCTURE FOR DYFLO

The previous chapters develop a set of simple subroutines for performing common mathematical procedures, such as integrating, generating functions, and converging algebraic equations. We now proceed to a higher level and develop a set of subroutines that simulate common process operations, such as boiling, condensing, and accumulating. Such a set of routines will greatly facilitate programming complex process loops and will provide a convenient library for simulating the dynamic behavior of a process. The system evolves throughout the remainder of this book, and a complete summary is provided in the appendix. This library of routines requires a framework in which to function. This framework provides a common communication network that is internal to the system and requires little attention from the programmer.

### 5-1-1  Stream Array STRM (IS,IP)

In process simulation, we are generally dealing with streams or nodes. A stream can be either liquid or vapor, rarely solid, while a node represents a quantity of material that is not flowing. For example, Figure 5-1 shows a typical situation of a fluid stream IL, flowing to a vessel node IH, boiling off stream IV.

These streams and nodes have certain extensive properties, such as flow rate (streams) or quantity (nodes), compositions, temperature, and so on. It is convenient, then, to assign a number to each stream or node which, in turn, specifies a set of properties listed in an array. This will be a two-dimensional array called STRM (IS,IP). The first index of this array (IS) is the stream or node number, while the second index (IP) specifies a particular property of the stream according to the following schedule.

| Location IP | Property |
|---|---|
| $1 \to 20$ | Compositions $1 \to 20$, generally mole fractions $X_1 \to X_{20}$ |
| 21 | Flow rate (moles/min) or holdup for a node (moles) |
| 22 | Temperature (°C) |
| 23 | Enthalpy (PCU/mole) |
| 24 | Pressure (atm) |

Examples of the use of this system are STRM (5,21), which is the molar flow rate of stream 5, or STRM (7,8), which is composition number 8 in stream 7. The dimensions of this stream array will be nominally set to (300,24). It can, of course, be changed if necessary.

### 5-1-2 Data Array DATA (IC,ID)

The second array called DATA will contain the basic properties required for each component. The first index IC specifies the component number, and since the STRM array accommodates 20 components, the DATA dimension for IC will also be 20. The index ID specifies the location for a particular property according to the following schedule:

1. Antoine coefficient $C_1$ (see Section 5-4-3)
2. Antoine coefficient $C_2$ (see Section 5-4-3)
3. Antoine coefficient $C_3$ (see Section 5-4-3)
4. Vapor enthalpy coefficient $A_v$ (see Section 5-2)
5. Vapor enthalpy coefficient $B_v$ (see Section 5-2)
6. Latent heat at $0°$
7. Liquid enthalpy coefficient $A_L$ (see Section 5-2)
8. Liquid enthalpy coefficient $B_L$ (see Section 5-2)
9. Activity $\gamma$ (see Section 5-4-2)

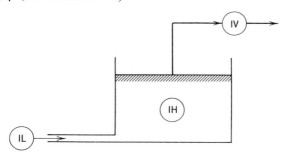

FIG. 5-1

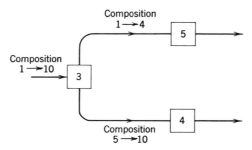

FIG. 5-2. Example of Use of NCF and NCL

The dimension ID will be 10, allowing one more space for special situations requiring an additional property.

For the sake of efficiency, two more constants will be used, NCF and NCL. These allow specifying the first (NCF) and last (NCL) component to be dealt with in any part of the process. For example, Figure 5-2 shows how node 3 of a process separates components 1 to 10 into group 1 to 4 going to node 5 and components 5 to 10 going to node 4.

When simulating node 5, NCF = 1 and NCL = 4, while for node 4, NCF = 5 and NCL = 10. As is seen later, the purpose is to save computation effort. (See Appendix B-6.)

The first two subroutines described are used to calculate the molal enthalpy for a stream whose temperature and composition are specified.

## 5-2 SUBROUTINES ENTHL (I) AND ENTHV (I)

The heat capacity $C_p$ of a pure component usually varies slightly with temperature and is expressed as $C_p = a + bT$, where $a$ and $b$ are constants and $T$ is the temperature. The enthalpy of the component at any temperature $T$ is

$$H = \int_o^T C_p \, dT = \int_o^T (a + bT) \, dT$$

$$H = aT + \frac{b}{2} T^2 + c$$

The integration constant $c$ is zero for liquids and will be the latent heat at $0°$ for vapor. In their final form, the vapor and liquid enthalpy of a component at temperature T are

vapor $H_v = \lambda + (A_v + B_v T)T$ PCU/mole
liquid $h_L = (A_L + B_L T)T$ PCU/mole
where $\lambda$ = latent heat at $0°$,
$A$ = heat capacity at $0°$,
$B = \frac{1}{2} \times$ temperature coefficient of heat capacity.

```
1 *        SUBROUTINE ENTHL(I)
2 *        COMMON/CD/STRM(300,24),DATA(20,10),RCT(22),NCF,NCL,LSTR
3 *        HV = 0.
4 *        DO 5 N = NCF,NCL
5 *     5  HV = HV + STRM(I,N)*(DATA(N,7)+DATA(N,8)*STRM(I,22))
6 *        STRM(I,23) = HV*STRM(I,22)
7 *        RETURN
8 *        END
```

FIG. 5-3.   Listing for Subroutine ENTHL
I = steam number

These coefficients $A_v$, $B_v$, $\lambda$, $A_L$, $B_L$ occupy positions 4 to 8, respectively in the DATA array as explained in the previous section. Since we will be dealing with compositions expressed as mole fractions, PCU/mole will be the units for the enthalpy.

A stream I having a temperature STRM(I,22) will have a molal enthalpy STRM(I,23). In order to calculate this enthalpy, given the composition and temperature, the following ideal mixing relation is used for liquids

$$\text{Enthalpy } (L) = \sum X_i(A_{Li} + B_{Li}T)T$$

This is programmed in the subroutine ENTHL, Figure 5-3 as HL = $T \sum X_i(A_{Li} + B_{Li}T)$.

The item I in the argument of subroutine ENTHL is the stream number for which the enthalpy is required. The name COMMON/CD/ contains the STRM and DATA arrays and the first and last component indexes NCF, NCL. The other items in COMMON are used and explained later. The enthalpy summation is done successively on line 5, and the final result, after doing 5 for all the components present in the stream (NCF → NCL), is transferred to the appropriate location in the stream array STRM(I,23) on line 6. A similar subroutine for calculating the enthalpy of a vapor stream (ENTHV) is shown in Figure 5-4, and should be self-evident.

```
1 *        SUBROUTINE ENTHV(I)
2 *        COMMON/CD/STRM(300,24),DATA(20,10),RCT(22),NCF,NCL,LSTR
3 *        HV = 0.
4 *        DO 5 N = NCF,NCL
5 *     5  HV=HV+STRM(I,N)*((DATA(N,4)+DATA(N,5)*STRM(I,22))*
6 *       1STRM(I,22)+DATA(N,6))
7 *        STRM(I,23) = HV
8 *        RETURN
9 *        END
```

FIG. 5-4.   Listing for Subroutine ENTHV
I = stream number

## 5-3   SUBROUTINE TEMP(I,L)

The inverse procedure of finding the enthalpy for a stream or node, given the temperature and composition, is to determine the temperature given the

```
 1 *              SUBROUTINE TEMP(I,L)
 2 *              COMMON/CD/STRM(300,24),DATA(20,10),PCT(22),NCF,NCL,LSTR
 3 *              J4=4+L
 4 *              J5=5+L
 5 *              SAX=0.
 6 *              SBX=0.
 7 *              SLM=0.
 8 *              DO 5 N=NCF,NCL
 9 *              SAX=SAX+DATA(N,J4)*STRM(I,N)
10 *              SBX=SBX+DATA(N,J5)*STRM(I,N)
11 *            5 IF(L.NE.3)SLM=SLM+DATA(N,6)*STRM(I,N)
12 *              T = STRM(I,22)
13 *            7 TD=(STRM(I,23)-SLM)/(SAX+T*SBX)
14 *              CALL CONV(T,TD,1,NC)
15 *              GO TO (6,7),NC
16 *            6 STRM(I,22)=T
17 *              RETURN
18 *              END
```

FIG. 5-5.   Listing for Subroutine
Argument list:
I = stream number
L = phase: liquid = 3, vapor = 0

enthalpy and composition. Subroutine TEMP(I,L) shown in Figure 5-5 performs this function. It serves both vapor and liquid streams, distinguished by specifying 3 for the second argument (symbol L) in the case of liquids, and 0 if the stream is a vapor. The stream number is I, which is the first item in the argument list. The listing shows that the numeral L (0 or 3) is used to specify the location of the proper enthalpy coefficients in the DATA array.

The calculation is made by the implicit method using the CONV subroutine (Chapter 2), with the assumption that on the initial pass a reasonable estimate for the temperature is available in STRM(I,22). It would appear that since the temperature T is a second-order term in the enthalpy expression, that is,

$$H_v = \lambda + A_v T + B_v T^2$$

or

$$h_L = A_L T + B_L T^2$$

that it could be determined explicitly. Unfortunately, there are some cases where the B coefficient is zero, which would make the second-order expression indeterminate. Of course, the situation could be circumvented by an IF statement, however, since the second order term is "weak," the implicit method shown in the subroutine converges rapidly. In fact, during process loop iterations, a single pass through the routine should be sufficient to update the temperature to a value within the tolerance limit internally specified in CONV. The implicit equation solved in line 13 is:

$$T = \frac{H_v - \sum \lambda_i X_i}{\sum A_i X_i + T \sum B_i X_i}$$

where the terms $\sum \lambda_i X_i (\mathrm{SLM})$, $\sum A_i X_i (\mathrm{SAX})$, $\sum B_i X_i (\mathrm{SBX})$ are accumulated on lines 11, 9, 10, respectively.

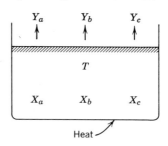

FIG. 5-6. Three-Component Boiling Mixture

## 5-4  VAPOR LIQUID EQUILIBRIUM

Before developing the subroutine that performs this calculation, a few words to explain the theory are in order. Consider, for example, a vessel containing, say, three miscible fluid components (Figure 5-6) $a$, $b$, and $c$. The liquid composition is $X_a$, $X_b$, and $X_c$ where $X_i$ is the mole fraction for component $i$. When this liquid is boiling at temperature $T$, a vapor stream will evolve having the composition $Y_a$, $Y_b$, $Y_c$. This vapor composition is generally different than the liquid compositions $X_i$, providing us with a convenient way of separating miscible components. When the vapor composition is the same as the liquid, the system is termed a constant boiling azeotropic mixture.

Since the vapor is issuing from the liquid, it is essentially in equilibrium with the liquid; hence the chemical potentials are equal. This can be expressed as

$$\varphi_i Y_i P = X_i F_i \gamma_i$$

where $X_i$ = liquid mole fraction of component $i$
$\quad\ Y_i$ = vapor mole fraction of component $i$
$\quad\ F_i$ = fugacity of pure component at the boiling temperature $T$
$\quad\ \varphi_i$ = fugacity of component $i$ in the mixture
$\quad\ p$ = total pressure
$\quad\ \gamma_i$ = activity of component $i$ in the liquid mixture

At moderate pressures and for ideal mixtures, this relationship can be simplified to Raoult's law:

$$PY_i = X_i P_i(T)$$

that is, the partial pressure in the vapor phase is equal to the liquid mole fraction times the vapor pressure of the pure component $P_i(T)$ at the boiling temperature $T$. Since nonideal mixtures are common, an activity coefficient $\gamma_i$ will be included in the equilibrium relationship for the sake of generality, that is,

$$Y_i = X_i \frac{P_i(T)}{P} \gamma_i \tag{5-1}$$

### 5-4-1  Activity

The activity $(\gamma_i)$ of a component $i$ in a liquid mixture is a function of composition and temperature. There are several methods for expressing these functional relationships, such as Van Laar or Wilson equations, and the reader is referred to the appropriate texts (Ref. 1, 2) for a more extensive coverage of this subject. Most sophisticated computer program libraries offer alternate methods for calculating activities; the programmer merely selects the method he desires. In order to demonstrate how these methods are used, a fairly elementary ternary Margules relationship is programmed in a subroutine ACTY. Other routines using more complex relationships could be assembled in a similar manner. In a two-component, nonideal mixture the log of the terminal activity $\gamma$ of component 1 out of component 2 (i.e., with $X_1 \to 0$.) is designated as $A_{12}$. Similarly, the activity of component 2 out of 1 will be $A_{21}$. Generally these binary terminal coefficients are expressed as functions of temperature, that is,

$$\ln \gamma_1 = A_{12} = a_{12} + \frac{b_{12}}{TK}$$

where $TK$ = temperature $°K$

For a three-component mixture, there will be six coefficients, namely

$$A_{12}, A_{21}$$
$$A_{13}, A_{31}$$
$$A_{23}, A_{32}$$

They represent the log of the terminal activities for each pair of the three-component mixture. In order to calculate the activity of each component out of the three components, the following Margules correlation is often satisfactory

$$\text{Ln } \gamma_1 = A_{12}X_2{}^2 + A_{13}X_3{}^2 + X_2X_3(A_{12} + A_{13} - A_{23})$$

By rotating the subscripts $1 \to 2 \to 3 \to 1$, the activities of the other two components will be

$$\text{Ln } \gamma_2 = A_{23}X_3{}^2 + A_{21}X_1{}^2 + X_3X_1(A_{23} + A_{21} - A_{31})$$
$$\text{Ln } \gamma_3 = A_{31}X_1{}^2 + A_{32}X_2{}^2 + X_1X_2(A_{31} + A_{32} - A_{12})$$

If one component is missing, the expressions simplify to binary equations, that is,

$$\text{Ln } \gamma_i = A_{ij}X_j{}^2$$

If more than three nonideal components are present in the mixture, expressions involving many more terms relating to the binary coefficients are required (Ref. 3).

## 5-4-2  Subroutine ACTY

Suppose we are about to perform an equilibrium calculation for a liquid mixture containing N components numbered from NCF to NCL. To perform this calculation (described in the next section) the activity coefficient of each component is required. For this example, it is known that three of these components are nonideal; thus it will be necessary to calculate their activities by the expressions developed above. This chore can be accomplished by the subroutine ACTY shown in Figure 5-7. The routine is suitable only for a three-component system treated in this manner. A more general routine could be written for systems with more than three components; in fact, most computer installations dealing with chemical engineering problems provide such extensive and sophisticated routines for activity calculations.

The argument N for the subroutine ACTY specifies the stream or node for which the activity of each component is to be determined. At loading time, an ideal activity coefficient of 1. is placed in number and location for each component of the DATA array by the DATA statement on line 3. The temperature coefficients are specified on line 4 and the temperature is extracted from location 22 of the N stream array (line 6). The A coefficients (i.e., the Ln of the terminal binary activities) for components 6, 7, and 8 are calculated on lines 7–12 and the compositions extracted from the stream array on lines 13–15. The activities are finally calculated on lines 16–18 and placed in their proper location in the DATA array.

```
 1*        SUBROUTINE ACTY(N)
 2*        COMMON/CD/STRM(300,24),DATA(20,10),RLT(22),NCF,NCL,LSTR
 3*        DATA(DATA(J,9),J=1,20)/20*1./
 4*        DATA SA67,SA68,SA78,B67,B68,B78/-.62,-.8,-.3,370.,380.,170./
 5*        DATA SA76,SA86,SA87,B76,B86,B87/-.6,-.7,-.4,365.,375.,185./
 6*        TK=STRM(N,22)+273.
 7*        A67 = SA67+B67/TK
 8*        A68 = SA68+B68/TK
 9*        A78 = SA78+B78/TK
10*        A76 = SA76+B76/TK
11*        A86 = SA86+B86/TK
12*        A87 = SA87+B87/TK
13*        X6 = STRM(N,6)
14*        X7 = STRM(N,7)
15*        X8 = STRM(N,8)
16*        DATA(6,9)=EXP(A67*X7**2+A68*X8**2+X7*X8*(A67+A68-A78),
17*        DATA(7,9)=EXP(A78*X8**2+A76*X6**2+X6*X8*(A78+A76-A86))
18*        DATA(8,9)=EXP(A86*X6**2+A87*X7**2+X6*X7*(A86+A87-A67))
19*        RETURN
20*        END
```

FIG. 5-7.  Listing for Subroutine ACTY
Argument list N = stream or node number

## 5-4-3  Equilibrium Vapor Composition

In order to calculate the vapor composition Y that is in equilibrium with the liquid composition X, it will be assumed that the ambient pressure is

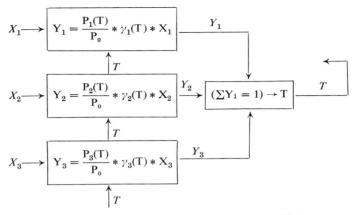

FIG. 5-8a. General Model for Vapor/Liquid Equilibrium

known. It will also be necessary to calculate the boiling temperature. Conceptually, the calculation scheme is shown in Figure 5-8a. Where the vapor compositions for each component is calculated by the equilibrium equation, initially using an estimated temperature T. The Y vapor compositions are then summed and compared to unity. If not within tolerance, a new value of temperature is assumed and the cycle repeated. There are several schemes for performing the recycle convergence, and one could use the CONV routine developed in Chapter 2 as follows:

$$E = 1 - \sum Y_i$$
$$T1 = T + G * E$$
$$\text{CALL CONV (T, T1, 1, NC)}$$

The disadvantage of this scheme is that is requires a custom-selected gain factor G, which will vary from system to system. An alternative method that is just as efficient, but does not require a customized gain factor, is to use the analytical Newton-Raphson technique (see Chapter 2). This approach is readily implemented here since we are using an analytical form for the component ($i$) vapor pressure $p_i(T)$, namely, the Antoine equation, that is,

$$P_i(T) = \text{EXP} \left( C_1 + \frac{C_2}{T + C_3} \right) \tag{5-1}$$

where $T = °C$

It will be recalled that the constants $C_1$, $C_2$, and $C_3$ occupy the first three locations in the DATA array for each component. For example, $C_2$ for component 5 will be found in DATA (5, 2). The rate of change of vapor composition $Y_i$ with temperature for each component $i$ is obtained from the derivative of the equilibrium equation (5-1), that is,

$$\frac{d}{dT}(Y_i) = \frac{d}{dT}\left\{ \text{EXP}\left( C_{1i} + \frac{C_{2i}}{T + C_{3i}} \right) * \gamma_i * \frac{X_i}{P} \right\} \qquad (5\text{-}2)$$

If we neglect the change in activity with temperature, this differentiation simplifies to the following expression:

$$\frac{dY_i}{dT} = Y_i * \frac{C_{2i}}{(T + C_{3i})^2}$$

For $N$ components, the rate of change of the sum of the $Y_i^s$ with temperature will be $SDY = \sum (dY_i/dT)$. Thus if we obtain an error $YER = 1. - \sum Y_i$, the next trial value for the temperature will be $T = T + YER/SDY$. The cycle is repeated to reduce $YER$ below a nominal tolerance of 0.001, which is adequate for most cases.

The subroutine EQUIL is shown in Figure 5-8b. The two items in the argument list are IL, the liquid stream or node number, and IV, the vapor stream in equilibrium with IL. On initial entry into this routine, it is assumed that a reasonable estimate for the temperature already exists in the stream array IL, which also contains the liquid mole fractions $X_i$ and the pressure (atm). Prior to calculating the vapor compositions, the activities are determined for all components by calling ACTY (line 5). The equilibrium $Y^s$ is calculated on the lines 7 and 8 and derivatives DY on line 10 for each component. The $Y^s$ and the derivatives are summed on lines 9 and 11, respectively. A new temperature is calculated on line 13, as explained previously, and a tolerance test is done on line 14. The reader's attention is again drawn to the feature

```
 1 *          SUBROUTINE EQUIL(IL,IV)
 2 *          COMMON/CD/STRM(300,24),DATA(20,10),RCT(22),NCF,NCL,LSTR
 3 *        7 SUM = 0.
 4 *          SDY = 0.
 5 *          CALL ACTY(IL)
 6 *          DO 5 N = NCF,NCL
 7 *          STRM(IV,N)=EXP(DATA(N,1)+DATA(N,2)/(STRM(IL,22)+DATA(N,3)))*
 8 *         1STRM(IL,N)/STRM(IL,24)*DATA(N,9)
 9 *          SUM = SUM + STRM(IV,N)
10 *          DY = -STRM(IV,N)*DATA(N,2)/(STRM(IL,22)+DATA(N,3))**2
11 *        5 SDY = SDY + DY
12 *          YER = 1. - SUM
13 *          STRM(IL,22) = STRM(IL,22) + YER/SDY
14 *          IF (ABS(YER) .LT. .001 ) GO TO 6
15 *          GO TO 7
16 *        6 STRM(IV,22)=STRM(IL,22)
17 *          RETURN
18 *          END
```

FIG. 5-8b.   Listing for Subroutine EQUIL  
Argument list:  
IL = liquid stream or node number  
IV = vapor stream or node number

whereby the DO loop is only done for those components specified between NCF and NCL (line 6).

To summarize, when the call EQUIL (IL,IV) is made, the vapor composition of IV is calculated and the equilibrium temperature for both streams IL and IV is determined by the subroutine and inserted in the proper location in the stream arrays.

## 5-5 DEW-POINT CALCULATION

It is sometimes necessary, as is seen later, to determine the dew point of a stream. This is the condition that exists when the temperature of a vapor is lowered to the point where the very first drop condenses. The dew-point temperature is this temperature, and the dew-point composition is the composition of the liquid drop in equilibrium with the vapor at the dew-point temperature. These values are calculated by the inverse procedure followed in EQUIL; that is, given the vapor composition $Y_i{}^s$, calculate the equilibrium liquid composition $X_i{}^s$ and the temperature. For this procedure the equilibrium equation is expressed as

$$X_i = \frac{PY_i}{P_i(T) \cdot \gamma_i}$$

The derivative $dX_i/dT$ required for the Newton-Raphson convergence will be

$$\frac{dX_i}{dT} = X_i \frac{C_2}{(T + C_3)^2}$$

The subroutine DEWPT (IV,IL) is shown in Figure 5-9, and if the EQUIL

```
 1 *          SUBROUTINE DEWPT(IV,IL)
 2 *          COMMON/CD/STRM(300,24),DATA(20,10),RCT(22),NCF,NCL,LSTR
 3 *        7 SUM=0.
 4 *          SDX=0.
 5 *          CALL ACTY(IL)
 6 *          DO 5 N=NCF,NCL
 7 *          PN=EXP(DATA(N,1)+DATA(N,2)/(STRM(IL,22)+DATA(N,3)))
 8 *          STRM(IL,N)=STRM(IV,N)*STRM(IL,24)/DATA(N,9)/PN
 9 *          SUM=SUM+STRM(IL,N)
10 *          DX=STRM(IL,N)*DATA(N,2)/(STRM(IL,22)+DATA(N,3))**2
11 *        5 SDX = SDX+DX
12 *          XER=1.-SUM
13 *          STRM(IL,22)=STRM(IL,22)+XER/SDX
14 *          IF(ABS(XER).LT..001)RETURN
15 *          GO TO 7
16 *          END
```

FIG. 5-9.   Listing for Subroutine DEWPT
          Argument list:
          IV = vapor flow
          IL = liquid node

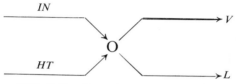

FIG. 5-10.  Generalized Flash operation

subroutine explanation in the previous section has been understood, the procedure in DEWPT should be selfevident. The first item in the argument list IV is the vapor node or stream for which the dew-point conditions are to be found. The dew-point temperature and liquid compositions are entered into IL, which usually is a dummy node.

## 5-6  GENERALIZED PHASE TRANSFORMATION

Processing operations often require changing the enthalpy of a particular stream by adding or withdrawing heat. In certain unit operations this change of enthalpy will result in changing the phase of the stream. A simplified generalized diagram for these transformations is shown in Figure 5-10 where heat (HT), positive or negative, is added to an input stream IN, liquid or vapor. This results in either a vapor stream, a liquid stream, or both. Table 5-1 shows nine possible combinations of these factors performed by familiar unit operations. A common approximation for those operations resulting in simultaneous vapor and liquid exit streams is to assume that both streams are in equilibrium with each other.

For those cases where this approximation is inadequate, special methods must be used to calculate the nature of the output streams. For example, the

**Table 5-1.  Classification of Enthalpy Changing Unit Operations**

| Input Stream | Heat Added | Output Streams | Unit Operation |
|---|---|---|---|
| 1. Liquid | Positive | Liquid only | Heater |
| 2. Liquid | Negative | Liquid only | Cooler |
| 3. Vapor | Positive | Vapor only | Superheater |
| 4. Vapor | Negative | Vapor only | Desuperheater |
| 5. Liquid | None | Vapor and liquid | Adiabatic flash |
| 6. Liquid | Positive | Vapor only | Boiler/vaporizer |
| 7. Vapor | Negative | Vapor and liquid | Partial condenser |
| 8. Vapor | Negative | Liquid only | Total condenser |
| 9. Liquid | Positive | Vapor and liquid | Flashing heat exchanger |

partial condenser can be treated as a first approximation by the assumption of exit equilibrium. A more refined approach involving diffusion resistance is sometimes required, which is treated in Chapter 9. In this chapter we restrict the discussion to equilibrium situations.

It would appear from Table 5-1 that merely specifying the sign and magnitude of the heat flux added to a given stream automatically specifies the type of unit operation involved. By and large this is true; however, in some cases it is not convenient to view the operation in this manner. A useful subroutine can, therefore, be developed that will encompass all nine unit operations listed in Table 5-1. The philosophy, however, will be to develop an auxiliary routine (HTEXCH) that can be used directly for those cases where there is only one exit stream. It will also be used as a nested routine in the general routine FLASH to be described later.

### 5-6-1 Subroutine Heat Exchange (HTEXCH)

This routine can be used directly for the following unit operations:

1. Heater
2. Cooler
3. Superheater
4. Desuperheater
5. Boiler-superheater
6. Total condenser

Since there is only one exit stream (see Figure 5-11), the total flow and composition of the exit stream JO will be identical to the inlet stream I (there is also no holdup). All that is necessary, then, is to determine the enthalpy and temperature of the exit stream JO for a given heat flux HT. This procedure is listed in subroutine HTEXCH(I,JO,HT,L) shown in Figure 5-12. The four items in the argument list are as follows:

$$I = \text{input stream number}$$
$$JO = \text{output stream number}$$
$$HT = \text{heat load (positive or negative)}$$
$$L = \text{phase of exit stream, liquid } L = 3, \text{ vapor} = 0$$

The routine transfers all the compositions from stream I to JO in the DO loop, calculates the molal enthalpy of the exit stream JO (line 6), transfers

FIG. 5-11. Flow
Streams for Sub-
routine HTEXCH

```
 1 *          SUBROUTINE HTEXCH(I,JO,HT,L)
 2 *          COMMON/CD/STRM(300,24),DATA(20,10),RCT(22),NCF,NCL,LSTR
 3 *          QF = HT/STRM(I,21)
 4 *          DO 5 N = NCF,NCL
 5 *        5 STRM(JO,N)=STRM(I,N)
 6 *          STRM(JO,23)=STRM(I,23)+QF
 7 *          STRM(JO,21)=STRM(I,21)
 8 *          CALL TEMP(JO,L)
 9 *          RETURN
10 *          END
```

FIG. 5-12.   Listing for Subroutine HTEXCH
Argument list:
I = input stream number
JO = output stream number
HT = heat load (positive or negative)
L = exit stream phase, liquid = 3, vapor = 0

the total flow (line 7), and concludes by calculating the temperature of stream JO for the appropriate phase L (line 8). It should be noted that since both the input stream flow and the heat flux HT are specified, the routine is not suitable for calculating the performance of a boiler producing saturated vapor. Since the heat flux HT will, in general, be more than sufficient to boil all the input stream, a superheated vapor will result. For the case of a saturated boiler or vaporizer, the subroutines CVBOIL and VVBOIL must be used. These are described in Section 5-9.

## 5-7   SUBROUTINE FLASH

For those unit operations listed in Table 5-1 where the input stream is converted to vapor and liquid exit streams that are in equilibrium, it is necessary to determine the total and component vapor/liquid split. If a heat flux is specified, it is also necessary to determine the exit equilibrium temperature. The special case of a partial condenser where the exit temperature is specified is treated in Section 5-10 (PCON).

The situation that is to be analyzed is shown in Figure 5-13. An input stream $F$, plus a heat flux $HT$, yields a vapor stream $V$ and a liquid stream $L$. If the input stream is vapor and the heat flux is negative, we have a

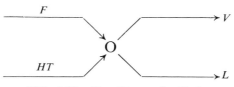

FIG. 5-13.   Flow Streams for Flash

partial or total condenser. If the input stream is liquid and the heat flux is positive, we have a flashing boiler. A single routine FLASH will serve both situations.

### 5-7-1   Heat Balance

The heat balance, as recommended in Chapter 4, will be used to calculate the magnitude of the exit vapor stream. The relationship is

$$\text{heat in} = \text{heat out}$$

$$Fh_F + HT = VH_v + Lh_L$$

where $F$, $V$, and $L$ are flow rates (moles/min)

$$HT = \text{heat flux}$$

$H_v$, $h_L$, $h_F$ are molal enthalpies

Defining $R = V/F$, the heat balance reduces to

$$R = \frac{\left(h_F + \dfrac{HT}{F} - h_L\right)}{(H_v - h_L)}$$

### 5-7-2   Component Balance

In general form the component balance is

$$FX_{Fi} = VY_i + LX_i$$

If we define $H_i = Y_i/X_i$, then the vapor composition can be established from the component balance, also using $R$ obtained from the heat balance, that is,

$$Y_i = X_{Fi} \frac{H_i}{1. + R(H_i - 1.)}$$

This calculation is performed for each component $i$.

### 5-7-3   Equilibrium

The vapor/liquid composition ratio $H_i$ defined above is obtained from the equilibrium equation:

$$H_i = \frac{P_i(T)\gamma_i}{P}$$

where

$$P_i(T) = \exp{(C_1 + C_2/(T + C_3))}$$

$$\gamma_i = \text{activity coefficient (DATA(i, 9))},$$

$$P = \text{total pressure}\quad (\text{STRM(L, 24))}.$$

### 5-7-4 Liquid Composition

Having obtained $H_i$ and $Y_i$, the liquid composition is $X_i = Y_i/H_i$.

### 5-7-5 Temperature

The Newton-Raphson procedure programmed in the subroutine EQUIL (see Section 5-4-3) is also used here. The derivative $dY_i/dT$ is calculated from the following equation:

$$\frac{dY_i}{dT} = Y_i \frac{C_2}{(T + C_3)^2}$$

and

$$T = T + \sum \frac{(dY_i/dT)}{YER}$$

where $YER = 1 - \sum Y_i$.

The overall scheme is shown in Figure 5-14 as a simplified information flow model. The numbers in the boxes refer to the line numbers of the listing for the subroutine FLASH shown in Figure 5-15. The argument list for the subroutine FLASH(I,JV,JL,HT) is as follows:

$I$ = input stream number
$JV$ = output vapor stream number
$JL$ = output liquid stream number
$HT$ = input heat flux

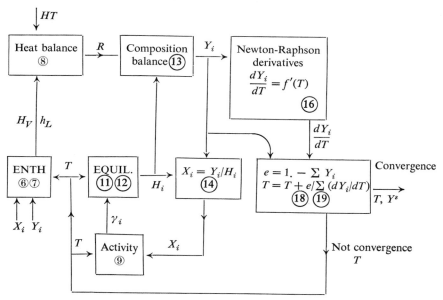

FIG. 5-14. Information Flow for Flash

```
 1 *                    SUBROUTINE FLASH(I,JV,JL,HT)
 2 *                    COMMON/CD/STRM(300,24),DATA(20,10),RCT(22),NCF,NCL,LSTR
 3 *                    QF=HT/STRM(I,21)
 4 *                    R=STRM(JV,21)/STRM(I,21)
 5 *                    CALL ACTY(JL)
 6 *                  7 SUM=0.
 7 *                    SDY=0.
 8 *                    DO 5 N=NCF,NCL
 9 *                    H = EXP(DATA(N,1) + DATA(N,2)/(STRM(JL,22)+DATA(N,3)))*DATA(N,9)/
10 *                  1STRM(JL,24)
11 *                    STRM(JV,N) = STRM(I,N)*H/(1.+R*(H-1.))
12 *                    STRM(JL,N) = STRM(JV,N)/H
13 *                    SUM = SUM + STRM(JV,N)
14 *                    DY = -STRM(JV,N)*DATA(N,2)/(STRM(JL,22)+DATA(N,3))**2
15 *                  5 SDY = SDY + DY
16 *                    YER = 1. - SUM
17 *                    STRM(JL,22) = STRM(JL,22) + YER/SDY
18 *                    IF (ABS(YER) .LT. .001) GO TO 6
19 *                    GO TO 7
20 *                  6 STRM(JV,22)=STRM(JL,22)
21 *                    CALL ENTHV(JV)
22 *                    CALL ENTHL(JL)
23 *                    R1=(STRM(I,23)+QF-STRM(JL,23))/(STRM(JV,23)-STRM(JL,23))
24 *                    IF(R1.LE.0.) GO TO 9
25 *                    IF(R1.GE.1.) GO TO 10
26 *                    CALL CONV(R,R1,1,NC)
27 *                    GO TO (8,7),NC
28 *                  8 STRM(JV,21)=R*STRM(I,21)
29 *                    STRM(JL,21)=(1.-R)*STRM(I,21)
30 *                    RETURN
31 *                  9 CALL HTEXCH(I,JL,HT,3)
32 *                    STRM(JV,21)=0.
33 *                    RETURN
34 *                 10 CALL HTEXCH(I,JV,HT,0)
35 *                    STRM(JL,21)=0.
36 *                    RETURN
37 *                    END
```

FIG. 5-15.   Listing for Subroutine FLASH
Argument list:
I = input stream number
JV = output vapor stream number
JL = putput liquid stream number
HT = input heat load

The sequence of operations shown pictorially in Figure 5-14 is programmed from line 6 to line 22. When convergence on temperature (STRM(JL,22)) has been achieved (line 21) the calculation proceeds to the lower section (lines 23 to 27), where a check is made on R. If R is greater than 1 or less than 0., it means that the operation is either a superheater or total condenser, respectively. If this is the case, the subroutine HTEXCH is called with the appropriate value of L (3 or 0) on lines 28 or 31. This feature allows using FLASH for situations where for some reason the operation has temporarily eliminated one of the two exit streams. For cases where there is always only one exit stream, HTEXCH should be used directly.

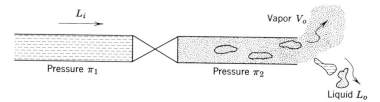

FIG. 5-16

## 5-8    ADIABATIC FLASH

A common situation to be found in chemical processes is where a liquid stream passes through a restriction, such as an orifice plate or control valve, and drops to a lower pressure (Figure 5-16). The enthalpy of both inlet and outlet streams is essentially the same (neglecting expansion work); however, the downstream pressure $\pi_2$ is sometimes sufficiently lower to cause a vapor phase to form. Or, if the inlet stream is already in two phases, an increase in the vapor phase will occur. It should be evident that this situation can be readily handled by the FLASH program with two exit streams $L$ and $V$ and the heat flux $HT = 0$.

## 5-9    BOILING OPERATIONS

Several processing units, such as boilers and evaporators, operate by appropriate controls on the principle that the heat flux supplied determines the vapor flow out of the boiler. For the case where the holdup can be neglected, the inflow will be equal to the outflow. The subroutine CVBOIL (for Constant Volume Boiler) has been developed to simulate this situation. The routine is listed in Figure 5-17, and the three items in the argument list

```
 1 *        SUBROUTINE CVBOIL(I,JV,HT)
 2 *        COMMON/CD/STRM(300,24),DATA(20,10),RCT(22),NCF,NCL,LSTR
 3 *        CALL DEWPT(I,JV)
 4 *        DO 5 N=NCF,NCL
 5 *      5 STRM(JV,N)=STRM(I,N)
 6 *        CALL ENTHV(JV)
 7 *        STRM(JV,21)=HT/(STRM(JV,23)-STRM(I,23))
 8 *        STRM(I,21)=STRM(JV,21)
 9 *        RETURN
10 *        END
```

FIG. 5-17.   Listing for CVBOIL
I = input stream number
JV = output vapor stream number
HT = heat flux

are:

> I = input stream number
> JV = output vapor stream number
> HT = heat flux

Since the composition of the exit stream is the same as the inlet stream, the boiling temperature will be the dew point of the inlet composition. The first operation is to determine this temperature by calling the subroutine DEWPT(I,JV). The boiling temperature will be entered in STRM(JV,22) location, and after transferring all the feed compositions to this JV stream (line 5), the enthalpy of JV is established (line 6). The heat balance (line 7) then determines the flow rate of JV, and the feed flow STRM(I,21) is also made equal to JV (line 8).

For the case where a boiler has a significant holdup, a different routine is required, called VVBOIL. This is described in Section 5-12.

## 5-10 PARTIAL CONDENSER (PCON)

In the previous sections, it was pointed out that the generalized routine FLASH could be used to establish a vapor/liquid split given a specific heat flux *HT*. In cases where a partial condenser operation is described as having a specific cooling load, the FLASH routine will perform the necessary calculations, even in the event that the cooling load causes total condensation with subcooling (via HTEXCH). Another, perhaps more typical, situation is to specify the condensate temperature and hence the vapor temperature leaving the condenser. This is shown symbolically in Figure 5-18. The model for this situation is shown in Figure 5-19. It is similar to the model for the FLASH subprogram except that here, since the equilibrium temperature *T* is specified, the convergence is performed for *R*, the vapor/feed split.

The derivatives $dY_i/dR$ required for the Newton-Raphson convergence is obtained by differentiating the component balance equation, that is,

$$Y_i = X_{Fi} \frac{H_i}{1 + R(H_i - 1)}$$

FIG. 5-18

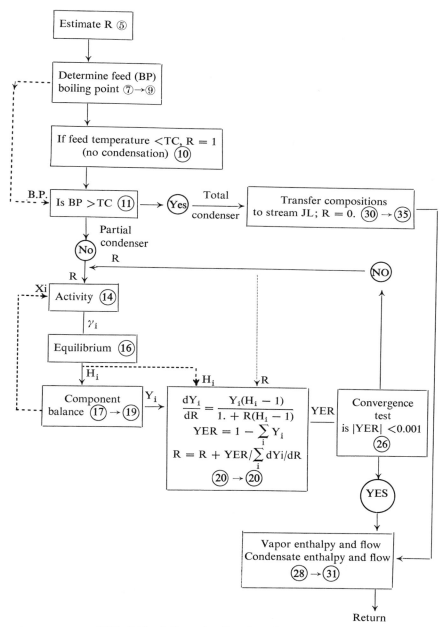

FIG. 5-19. Information flow for subroutine PCON

133

where $Y_i$ = exit vapor composition,

$$H_i = Y_i/X_i, \qquad R = V/F, \qquad X_{Fi} = \text{feed composition.}$$

Differentiating

$$\frac{dY_i}{dR} = -\frac{Y_i(H_i - 1)}{1 + R(H_i - 1)}$$

The iteration on $R$ will be

$$R = R + \frac{(1 - \sum Y_i)}{\sum \left(\dfrac{dY_i}{dR}\right)}$$

The listing for the subroutine PCON(I,JV,JL,TC) is shown in Figure 5-20. The items in the argument list are:

I = input stream number      JV = output vapor stream number
JL = output condensate stream number   TC = exit condensate temperature

```
 1 *        SUBROUTINE PCON(I,JV,JL,TC)
 2 *        COMMON/CD/STRM(300,24),DATA(20,10),RCT(22),NCF,NCL,LSTR
 3 *        STRM(JV,22)=TC
 4 *        STRM(JL,22)=TC
 5 *        R=STRM(JV,21)/STRM(I,21)
 6 *        IF(R.GT..95) R=.95
 7 *        TMP=STRM(I,22)
 8 *        CALL EQUIL(I,200)
 9 *        STRM(I,22)=TMP
10 *        IF(TMP.LE.TC)R = 1.
11 *        IF(STRM(200,22).GT.TC) GO TO 8
12 *      7 SDY=0.
13 *        SUM=0.
14 *        CALL ACTY(JL)
15 *        DO 5 N=NCF,NCL
16 *        H=EXP(DATA(N,1)+DATA(N,2)/(TC+DATA(N,3)))*DATA(N,9)/STRM(JL,24)
17 *        DEN=1.+R*(H-1.)
18 *        STRM(JV,N)=STRM(I,N)*H/DEN
19 *        STRM(JL,N)=STRM(JV,N)/H
20 *        DY=-STRM(JV,N)*(H-1.)/DEN
21 *        SUM=SUM+STRM(JV,N)
22 *      5 SDY=SDY+DY
23 *        IF(SDY.GE.0.) SDY=-1.
24 *        YER=1.-SUM
25 *        R=R+YER/SDY
26 *        IF(ABS(YER).LT..001 ) GO TO 10
27 *        GO TO 7
28 *     10 CALL ENTHV(JV)
29 *        CALL ENTHL(JL)
30 *        STRM(JV,21)=STRM(I,21)*R
31 *        STRM(JL,21)=STRM(I,21)-STRM(JV,21)
32 *        RETURN
33 *      8 DO 9 N=NCF,NCL
34 *      9 STRM(JL,N)=STRM(I,N)
35 *        R=0.
36 *        GO TO 10
37 *        END
```

FIG. 5-20.   Listing for Subroutine PCON
Argument list:
I = input stream number
JV = output vapor stream number
JL = condensate stream number
TL = specified condensate temperature

A preliminary estimate for R is made on line 5, and the condensate temperature TC is entered in streams JV and JL (lines 3 and 4), after which the sequence of calculations follows the information flow diagram shown in Figure 5-19. Initially, however, the temperature of the inlet stream is stored temporarily as TMP (line 7) prior to bringing the inlet stream to equilibrium with a dummy stream 300 (line 8). This changes the temperature of stream I to its boiling temperature, a value also stored in the dummy stream 300 (see EQUIL 5-4-3). The actual temperature TMP is replaced in stream I (line 9), followed by the following two tests:

1. (Line 10) If the condenser temperature TC is greater than the inlet temperature of stream I, presumably a vapor stream, then no condensation will occur. R, the ratio of exit vapor to feed flow, is made 1. The calculations proceed through the subroutine, but will make only one pass, and, essentially, the entire extensive properties of the input stream I are transferred to the output vapor stream JV (see lines 17 and 18 with R = 1.0).

2. (Line 11) If TC is less than the theoretical boiling temperature of I, now stored in dummy stream 300, the partial condenser is now a total condenser, and the entire feed stream I is transferred to the exit condensate stream JL (lines 33 to 35).

Because of a possible reversal of slope of the function $\sum Y(R)$, a nominal slope of $-1$ is assigned to $\sum dY/dR$ if it is $> 0$. (line 23), forcing the convergence to the proper value of R. On completion of the convergence, the enthalpies of the two exit streams are established, (lines 28, 29) and the flow rates are calculated using the converged value of R.

## 5-11  SINGLE-PHASE HOLDUP

All the previous programs described in this chapter perform entirely algebraic calculations, because it was assumed that there was no energy or material storage in the unit operations simulated. In order to extend the library of subroutines being developed in this book to be able to simulate dynamic situations, some of the routines that follow incorporate dynamic holdup, which is capable of storing material or energy. The first example in this class is a storage vessel, or simply a volume with a single-phase input stream (*I*) and an output stream (*IO*) having the same phase (Figure 5-21). The storage volume contains a total of *HL* moles of holdup, and assuming efficient mixing, the exit-stream extensive properties (composition, temperature, etc.) will be identical with those of the holdup. It is not necessary, then, to assign a node number to the holdup, and *HL* merely symbolizes the total moles in the holdup.

An additional flow term will be added to the component and total mole balances, representing the internal loss or gain due to any reactions which may be occurring. If there are no reactions, this term will be zero. These

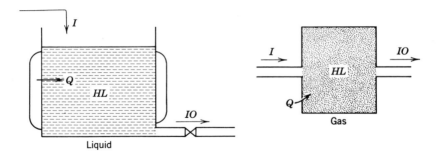

FIG. 5-21.   Single-Phase Holdup

quantities are stored in the reaction array RCT available from, COMMON/CD/ and will be generated by an appropriate routine defining the reaction rates. This feature is used in some examples in the following chapters, but for the moment assume that locations $1 \to 20$ in the RCT array represent the net reaction flow, if any, for components $1 \to 20$. Location RCT(21) is the total net gain or loss of moles due to all the reactions, while RCT(22) is the total net reaction heat generated. It is also convenient to include in the heat balance an external flux (Q), representing heating or cooling of the vessel contents. The formulation of the component and energy balance is demonstrated in Chapter 4, but the following development offers a convenient modification based on the fundamental relationship:

$$\text{accumulation} = \text{IN} - \text{OUT}$$

The mole balance for component n becomes

$$\frac{d}{dt} X_n = \frac{F_{in}X_{in} - F_{out}X_n + RCT(n) - X_n \dfrac{dHL}{dt}}{HL} \qquad (5\text{-}11\text{-}1)$$

Now the total mole balance $dHL/dt = F_{in} - F_{out} + RCT(21)$. Substituting this expression in equation 5-11-1 we obtain for the component balance

$$\frac{dX_n}{dt} = \frac{F_{in} * (X_{in} - X_n) + RCT(n) - RCT(21) * X_n}{HL}$$

Similarly, the heat balance is

$$\frac{d(EN_0)}{dt} = \frac{F_{in}(EN_i - EN_0) + RCT(22) + Q}{HL}$$

where $EN_0$ = molal enthalpy.

```
 1 *          SUBROUTINE HLDP(I,IO,L,HL,Q)
 2 *          COMMON/CD/STRM(300,24),DATA(20,10),RCT(22),NCF,NCL,LSTR
 3 *          LOGICAL LSTR
 4 *          IF (LSTR) GO TO 7
 5 *          DHL = STRM(I,21) - STRM(IO,21)+RCT(21)
 6 *          HIN=STRM(I,21)*STRM(I,23)+Q+RCT(22)
 7 *          DEN=(HIN-STRM(IO,23)*(STRM(IO,21)+DHL))/HL
 8 *          DO 6 N = NCF,NCL
 9 *          DFX=STRM(I,21)*(STRM(I,N)-STRM(IO,N))
10 *          DX=(DFX+RCT(N)-STRM(IO,N)*RCT(21))/HL
11 *        6 CALL INT(STRM(IO,N),DX)
12 *          CALL INT(HL,DHL)
13 *          CALL INT(STRM(IO,23),DEN)
14 *          CALL TEMP(IO,L)
15 *          RETURN
16 *        7 IF(L.EQ.3) CALL ENTHL(IO)
17 *          IF(L.EQ.0) CALL ENTHV(IO)
18 *          RETURN
19 *          END
```

FIG. 5-22.   Listing for Subroutine HLDP
Argument list:
I = input stream number
IO = output stream number
L = phase LIQ. = 3, VAP = 0
HL = holdup
Q = heat flux

These equations are programmed in the subroutine HLDP $(I,IO,L,HL,Q)$ shown in Figure 5.22, which requires some explanation. First, the arguments are:

I = input stream number
IO = output stream number
L = phase of output stream; liquid = 3, vapor = 0
HL = holdup (moles)
Q = heat flux

The equations discussed above and programmed on lines $5 \to 10$ define the derivatives of the holdup and enthalpy and composition of the exit stream IO. These derivatives are integrated within the subroutine by calling on the integration procedure INT (lines $11 \to 13$) described in Chapter 3. The inclusion of the integration within the subroutine is a great convenience for the programmer: however, the price to be paid for this convenience is that he must sequence the call statements in the main program in a direction opposite to that of the process flow stream. This is demonstrated in Chapter 7 dealing with staged operations.

The last item in COMMON/CD/ is the logical variable LSTR. It is required for all the routines containing internal integration, and its purpose is to divert the sequence back to the main program on the *first* pass through the routine. It will be recalled from Chapter 3 that the organization of the INT system requires the independent variable routine INTI to be the first routine called before any dependent variable routines INT are called. This enables INTI to reset the calling counter JN and the pass counters JS4 or JS to their proper values. When the INT subroutines are embedded within the unit operation subroutines, as in HLDP, it becomes necessary to skip over the integrations on the first pass, since the subroutines are listed in the derivative section. The logical variable LSTR (Logical Start) is "true," only on the first pass, thus accomplishing this function. The organization of a main calling program involving the variable LSTR is demonstrated in Example 5-14.

To complete the description of the subroutine HLDP, the final step before returning is to call TEMP, which will establish the temperature of the exit stream IO from its enthalpy and composition.

### 5-12 BOILER WITH HOLDUP

Section 5-9 covered the situation of a constant holdup boiler (CVBOIL) where the inflow was instantaneously flashed into vapor. We will now develop a subroutine for a variable volume boiler where the vapor is in equilibrium with the liquid holdup, which is different from the feed composition. Figure 5-23 shows the system to be simulated, consisting of a liquid volume, node *IHL*, an input stream *I*, and an output vapor stream *JV*. The boiler receives a heat flux $Q$ and may also be hosting a reaction, pertinent data for which will be located in the array RCT, as explained in Section 5-11. The basic equations for this system are similar to those for HLDP, described in the previous section. They are

Overall mole balance HL:

$$\frac{d(HL)}{dt} = F_I - F_{JV} + RCT(21)$$

where HL = moles holdup, F = flow rates

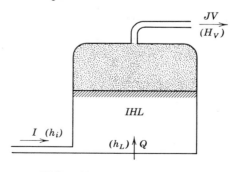

FIG. 5-23. Boiler with Holdup

Heat balance:

total input enthalpy = heat flux + reaction and input enthalpy

$$QP = Q + RCT(22) + F_I \cdot H_I$$

Boilup:

$$F_{JV} = \frac{QP - h_L \cdot \dfrac{dHL}{dt} - HL \cdot \dfrac{dh_L}{dt}}{H_V}$$

where $h_L$ and $H_V$ = molal enthalpies.

The last term in the numerator $dh_L/dt$ represents the accumulation of sensible heat in the holdup, usually small, but will be accounted for here.

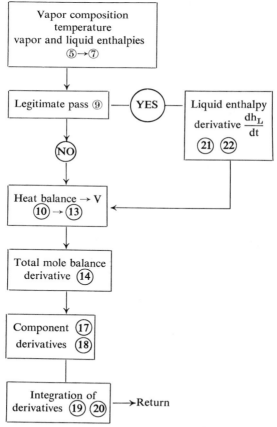

FIG. 5-24   Information Flow for VVBOIL

The generation of this term requires obtaining the derivative of the liquid enthalpy $h_L$, accomplished in a manner similar to the procedure used in subroutine DER (Chapter 3); that is, a "backward" derivative is calculated.

Component (N) balance:

input flow $FNI = F_I X_{NI} + RCT(N)$

accumulation = input − output

$$\frac{d}{dt}(HLX_N) = FNI - F_{JV}Y_N$$

The above equation reduces to the following, more convenient form, by partial differentiation:

$$DERN = \frac{dX_N}{dt} = \frac{FNI - F_{JV}Y_N - X_N \dfrac{dHL}{dt}}{HL}$$

A simplified information flow diagram of the above equations is shown in Figure 5-24, and the subroutine VVBOIL (I, JV, IHL, Q, IC) is listed in Figure 5-25. The arguments for VVBOIL are

I = input stream number       IHL = holdup node number

JV = output vapor stream number       Q = heat flux

```
 1 *        SUBROUTINE VVBOIL(I,JV,IHL,Q)
 2 *        COMMON/CD/STRM(300,24),DATA(20,10),RCT(22),NCF,NCL,LSTR
 3 *        COMMON/CINT/T,DT,JS,JN,DXA(500),XA(500),IO,JS4
 4 *        LOGICAL LSTR
 5 *        CALL EQUIL(IHL,JV)
 6 *        CALL ENTHV(JV)
 7 *        CALL ENTHL(IHL)
 8 *        IF(LSTR) STRM(I,20)=STRM(IHL,23)
 9 *        IF((JS4.EQ.4).OR.(JS.EQ.2))GO TO 7
10 *     10 QP=Q+RCT(22)+STRM(I,21)*STRM(I,23)
11 *        FI=STRM(I,21)+RCT(21)
12 *        BH=STRM(JV,23)-STRM(IHL,23)
13 *        STRM(JV,21)=(QP-STRM(IHL,23)*FI-STRM(JV,20))/BH
14 *        DHL=STRM(I,21)-STRM(JV,21)+RCT(21)
15 *        IF(LSTR) RETURN
16 *        DO 9 N=NCF,NCL
17 *        FNI = STRM(I,21)*STRM(I,N)+RCT(N)
18 *        DERN=(FNI-STRM(JV,21)*STRM(JV,N)-STRM(IHL,N)*DHL)/STRM(IHL,21)
19 *      9 CALL INT(STRM(IHL,N),DERN)
20 *        CALL INT(STRM(IHL,21),DHL)
21 *        RETURN
22 *      7 STRM(JV,20)=(STRM(IHL,23)-STRM(I,20))*STRM(IHL,21)/DT
23 *        STRM(I,20)=STRM(IHL,23)
24 *        GO TO 10
25 *        END
```

FIG. 5-25. Listing for Subroutine VVBOIL
Argument list:
I = input stream number
JV = output vapor stream number
IHL = holdup node number
Q = heat flux to boiler

As for the HLDP routine, the integrations for the derivatives generated are performed internally; thus the logical variable LSTR (lines 8, 15) again comes into play for the initial pass. The information flow diagram indicates by line numbers where in the subroutine each equation is calculated, and generally it is quite similar to previous routines. The derivative of the enthalpy, $dh_L/dt$ required for the heat balance is calculated in address 7 by subtracting the current enthalpy of the holdup (STRM(IHL,23)) from the enthalpy on the previous legitimate (line 22) integration pass, stored in location 20 of the input stream array (line 23). This change in enthalpy is divided by the integration step size DT to obtain the derivative and is multiplied by the holdup volume STRM(IHL,21) to form the total derivative term stored in SRTM(JV,20), which is required by the heat balance (line 13).

It will also be observed that since the holdup accumulates compositions, the $X^s$ are specified on each pass, and the vapor compositions and temperature are established directly (EQUIL) without an algebraic convergence.

## 5-13 OUTPUT EDITING: SUBROUTINES PRL AND RPRL

Chapter 3 explains the internal structure of a PRNTF subroutine that would print a line of specified variables at selected intervals along the independent variable. A similar routine is available for printing the state of a stream or node in a suitable format. A typical example is shown in Figure 5-34. The routine is called PRL and is listed in Appendix B-1. Since it merely contains a procedure for organizing and displaying numerical values in appropriate form and is of little engineering interest, no description of its internal structure will be given here other than the following:

1. The arguments for the routine PRL (PRI, FNR, NF, L1, L2, L3, L4, L5, L6, L7, L8, L9, L10, L11, L12) are as follows:
    PRI = print interval, referred to the independent variable.
    FNR = finish run, that is, time to stop.
    NF = logical finish index, NF = TRUE when finished.
L1 → L12 = numbers of streams specified for output display.
2. The subroutine incorporates the following automatic features:
(a) It will only print the compositions specified between NCF and NCL.
(b) Any stream numbers specified as 0 in the argument list will be ignored.
(c) It switches the logical variable LSTR to FALSE.

The capacity of PRL is 12 streams, but if more are required the companion routine RPRL (Repeat print) can be used (listed in Appendix B-2) which is triggered automatically by PRL and offers the same features (a) and (b) as PRL. Both routines are used in the main program in a similar manner to PRNTF and PRNTR (see Chapter 3).

The subroutines described in this chapter deal mainly with two-phase equilibrium processes. More programs are added to this library in later chapters, but at this point, several examples are developed that demonstrate the use of some of these programs.

### 5-14 CASE 5-1—BATCH DISTILLATION EXAMPLE

A three-component fluid mixture is to be partially separated by batch distillation in a still, as shown in Figure 5-26. The still is heated by a jacket maintained at a temperature $T_J = 130°C$ by a jacket pressure controller.

The heat flux from the jacket to the batch will be

$$Q = 1400(T_J - T)$$

where $Q = PCU/min$

$T$ = batch temperature °C

A computer program is required that will calculate the temperature and composition of the batch and distillate vapor during the first hour of operation. The following basic data and initial conditions apply:

| Component | Antoine Coefficients | | | | | | Enthalpy Coefficients | | |
|---|---|---|---|---|---|---|---|---|---|
|  | $C_1$ | $C_2$ | $C_3$ | $A_v$ | $B_v$ | $\lambda$ | $A_L$ | $B_L$ |
| 1 | 13.96 | −5210. | 273. | 8. | 0.01 | 9020. | 20. | 0.02 |
| 2 | 15.2 | −6050. | 273. | 12.2 | 0.02 | 11500. | 32. | 0.01 |
| 3 | 15.4 | −4957. | 273. | 6.5 | 0.01 | 7500. | 16. | 0.03 |

Total moles in the initial batch is 350, and the initial composition is $X_1 = 0.43$, $X_2 = 0.31$, $X_3 = 0.26$. The still is operating at atmospheric pressure, and it will be assumed that all three components are ideal and that boiling starts at time = 0.

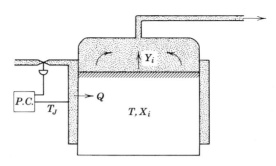

FIG. 5-26. Batch Still

## System Analysis

Although the eventual program for this process will make use of one of the subroutines developed in this chapter, it is worthwhile to formulate all the equations involved to obtain an understanding of the system and to appreciate the programming simplifications provided by the use of subroutines.

1. Heat balance:

$$\text{heat flux } Q = 1400(130 - T) \text{ PCU/min}$$

$$\text{boilup } V = \left[ \frac{Q - \dfrac{d}{dt}(HL \cdot h_L)}{H_v} \right]$$

where $H_v = \sum Y_i[\lambda_i + (A_{vi} + B_{vi}T)T]$ vapor enthalpy/mole,
$h_L = [X_i(A_{Li} + B_{Li}T)T]$ liquid enthalpy/mole,
$HL$ = total moles in batch.

2. Total mole balance:

$$\frac{d(HL)}{dt} = -V$$

3. Component mole balance:

$$\frac{d}{dt}(HL \cdot X_1) = -VY_1$$

$$\frac{d}{dt}(HL \cdot X_2) = -VY_2$$

$$\frac{d}{dt}(HL \cdot X_3) = -VY_3$$

4. Equilibrium:

vapor pressure for each component $i$

$$P_i = \text{EXP}\left(C_{1i} + \frac{C_{2i}}{T + C_{3i}}\right)$$

where $C_{Ji}$ = Antoine coefficients

vapor composition $Y_i = P_i X_i / 1.$ (pressure = 1. atm)

5. Temperature:

$$Y_1 + Y_2 + Y_3 = 1. \rightarrow T$$

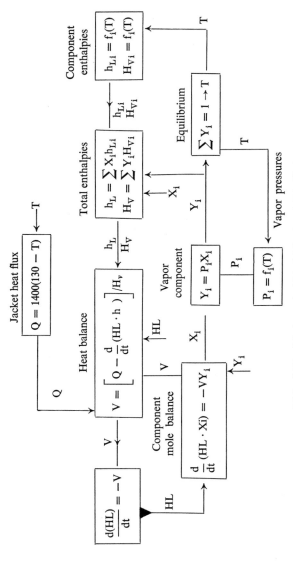

FIG. 5-27. Mathematical Model for Batch Distillation

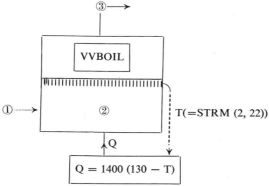

FIG. 5-28.   Stream Numbers for VVBOIL Simulation
of Batch Distillation

This last equation summarizes the procedure that will iterate the batch temperature $T$ so as to maintain $\sum_i Y_i = 1$ to within a suitable tolerance.

An information flow diagram specifying how the equations are to be used to establish the dependent variables is shown in Figure 5-27. All the equations could be programmed individually and structured in a form suitable for integration using the INT system described in Chapter 3. However, this process can be simulated with much less effort using the subroutine VVBOIL. Although this encompasses the more general case of including a liquid feed, such a feed can be assigned a dummy stream number (N°· 1) having zero flow (Figure 5-28), and the routine will then be duplicating a simple batch distillation. The batch will be assigned the STRM node number 2 and the vapor flow the STRM number 3.

The program calling VVBOIL is shown in Figure 5-29. The structure follows the general grouping required for the INT system, although since VVBOIL contains internal integration, the title "derivative section" does not really apply here. The following features about the program should be noted.

1. Component numbers NCF and NCL are specified in the data section (line 8).

2. The logical variable for the initial dummy pass LSTR is set to TRUE in the initiation section (line 10) and set to FALSE in the print subroutine PRL (line 15).

3. Since the heat flux Q is required for VVBOIL, it is the first line of the derivative section. To establish Q on the first pass the boiling temperature (STRM (2, 22)) must be available; thus EQUIL is called on the previous line to accomplish this task.

4. There are only two programming statements required to simulate the

```
 1*     C ** DATA SECTION **
 2*           COMMON/CD/STRM(300,24),DATA(20,10),RCT(22),NCF,NCL,LSTR
 3*           LOGICAL LSTR,NF
 4*           DATA(DATA(1,N),N=1,8)/13.46,-5210.,273.,8.,.01,9020.,20.,.02/
 5*           DATA(DATA(2,N),N=1,8)/15.2,-6050.,273.,12.2,.02,11500.,32.,.01/
 6*           DATA(DATA(3,N),N=1,8)/15.4,-5312.,273.,6.5,.01,7500.,16.,.03/
 7*     C ** INITIATION SECTION **
 8*           DATA(STRM(2,N),N=1,3)/.43,.31,.26/NCF,NCL/1,3/
 9*           DATA(STRM(2,N),N=21,24)/350.,95.35,0.,1./
10*           LSTR=.TRUE.
11*           CALL EQUIL(2,3)
12*     C ** DERIVATIVE SECTION **
13*         7 Q=1400.*(130.-STRM(2,22))
14*           CALL VVBOIL(1,3,2,Q)
15*           CALL PRL(10.,60.,NF,2,3,0,0,0,0,0,0,0,0,0,0)
16*           IF(NF)GO TO 10
17*     C ** INTEGRATION SECTION **
18*         5 CALL INTI(TIM,1,,4)
19*           GO TO 7
20*        10 STOP
21*           END
```

FIG. 5-29. Listing for Program Simulating Batch Distillation

entire process, namely, the heat flux Q and the call VVBOIL subroutine. All the other statements are data input and output or procedural details.

4. The activity subroutine for these ideal components is shown in Figure 5-30. The output results are shown in Figure 5-31 as the state of the batch (2) and vapor stream (3) at 5 min intervals. This output was produced from the PRL subroutine called on line 15 of the main program and discussed in Section 5-13.

### 5-15 CASE 5-2—TWO-STAGE BATCH DISTILLATION

A two-stage separation can be achieved for the batch distillation in the previous example by adding a partial condenser on the vapor line and refluxing the condensate via a hold tank back to the batch distillation. The system is shown in Figure 5-32 and, evidently, its purpose is to concentrate the high boiling component (#2) to a specified purity. The addition of the partial condenser and the reflux hold tank adds another 10 or 20 equations to those specified for the original system (Figure 5-27). However, it will only require an additional four statements added to the DYFLO program. There are now five streams/nodes, which are shown in Figure 5-33. The hold tank

```
1 *           SUBROUTINE ACTY(N)
2 *           COMMON/CD/STRM(300,24),DATA(20,10),RCT(22),NCF,NCL,LSTR
3 *           DATA (DATA(I,9),I=1,20)/20*1./
4 *           RETURN
5 *           END
```

FIG. 5-30. Activity Subroutine (Ideal) for Batch Distillation Examples

```
TIME =   .0000
STRM NO        2              3
FLOW        .3500+03      .7291+01
TEMP        .9535+02      .9535+02
ENTHAL      .2340+04      .8994+04
PRESS       .1000+01      .0000
COMP 1      .4300+00      .2169+00
COMP 2      .3100+00      .9110-01
COMP 3      .2600+00      .6920+00

TIME =   .1000+02
STRM NO        2              3
FLOW        .2912+03      .5291+01
TEMP        .1003+03      .1003+03
ENTHAL      .2539+04      .9326+04
PRESS       .1000+01      .0000
COMP 1      .4668+00      .2840+00
COMP 2      .3508+00      .1282+00
COMP 3      .1824+00      .5878+00

TIME =   .2000+02
STRM NO        2              3
FLOW        .2440+03      .4216+01
TEMP        .1056+03      .1056+03
ENTHAL      .2748+04      .9752+04
PRESS       .1000+01      .0000
COMP 1      .4947+00      .3657+00
COMP 2      .3894+00      .1783+00
COMP 3      .1160+00      .4560+00

TIME =   .3000+02
STRM NO        2              3
FLOW        .2069+03      .3320+01
TEMP        .1104+03      .1104+03
ENTHAL      .2938+04      .1022+05
PRESS       .1000+01      .0000
COMP 1      .5105+00      .4487+00
COMP 2      .4223+00      .2364+00
COMP 3      .6716-01      .3149+00
```

FIG. 5-31. Output results for the Batch Distillation Program (Figure 5-29)

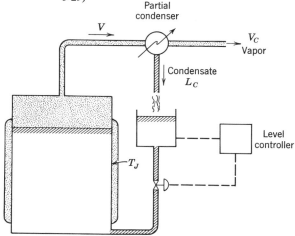

FIG. 5-32. Two-Stage Batch Distillation

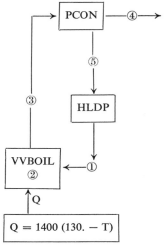

FIG. 5-33.   A Signal Flow Showing Stream Numbers

molar volume is controlled to a constant value of 50 moles, and the initial concentration is the same as the distillation batch. The condensate temperature TC = 87°C.

The complete program for this enlarged system is shown in Figure 5-34. The extra statements for the additional units are marked in column 73 to 80. The numerical results for the first 50 min are shown in Figure 5-35.

**Exercise Problems**

1. It is not unusual for the coefficients $C_1$, $C_2$, $C_3$ of the Antoine vapor pressure equation to be unavailable. Construct a program that will read in three pairs of coordinates of vapor pressure $P_i$ versus temperature $T$, and proceed to calculate the Antoine coefficients in the equation

$$P_i = \text{EXP} \left( C_1 + \frac{C_2}{T + C_3} \right)$$

and place these coefficients in their proper location in the data array.

2. How would the batch distillation program described in this chapter be modified to accommodate the following:

(a) decreasing the heat transfer area and therefore the coefficient as the batch volume decreases.

(b) the inclusion of a fourth component, virtually nonboiling, but miscible, such as a soluble polymer.

```
1*     C ** DATA SECTION **
2*           COMMON/CD/STRM(300,24),DATA(20,10),RCT(22),NCF,NCL,LSTR
3*           LOGICAL LSTR,NF
4*           DATA(DATA(1,N),N=1,8)/13.,46.,-5210.,273.,8.,.01,9020.,20.,.02/
5*           DATA(DATA(2,N),N=1,8)/15.2,-6050.,273.,12.,.02,11500.,32.,.01/
6*           DATA(DATA(3,N),N=1,8)/15.4,-5312.,273.,6.5,.01,7500.,16.,.03/
7*     C ** INITIATION SECTION **
8*           DATA(STRM(2,N),N=1,3)/.43,.31,.26/NCF,NCL/1,3/
9*           DATA(STRM(2,N),N=21,24)/350.,95.35,0.,.1./
10*          DATA(STRM(1,N),N=1,3)/.43,.31,.26/
11*          DATA(STRM(1,N),N=21,23)/1.835,95.35,2340./
12*          DATA (STRM(J,24),J=1,5)/5*1./
13*          LSTR=.TRUE.
14*          CALL EQUIL(2,3)
15*    C ** DERIVATIVE SECTION **
16*        7 Q=1400.*(130.-STRM(2,22))
17*          CALL VVBOIL(1,3,2,Q)
18*          CALL PCON(3,4,5,87.)
19*          STRM(1,21)=STRM(5,21)
20*          CALL HLDP(5,1,3,50.,0.)
21*          CALL PRL(10,.60.,NF,2,3,4,5,1,0,0,0,0,0,0)
22*          IF(NF)GO TO 10
23*    C ** INTEGRATION SECTION **
24*        5 CALL INTI(TIM,1.,4)
25*          GO TO 7
26*       10 STOP
27*          END
```

```
TIME =  .0000
STRM NO       2          3          4          5          1
FLOW      .3500+03   .7291+01   .4810+01   .2481+01   .2481+01
TEMP      .9535+02   .9535+02   .8700+02   .8700+02   .9535+02
ENTHAL    .2340+04   .8994+04   .8543+04   .1960+04   .2340+04
PRESS     .1000+01   .1000+01   .1000+01   .1000+01   .1000+01
COMP 1    .4300+00   .2169+00   .1359+00   .3739+00   .4300+00
COMP 2    .3100+00   .9110-01   .3868-01   .1926+00   .3100+00
COMP 3    .2600+00   .6920+00   .8255+00   .4333+00   .2600+00

TIME =  .1000+02
STRM NO       2          3          4          5          1
FLOW      .3174+03   .5425+01   .2336+01   .3089+01   .3089+01
TEMP      .9969+02   .9969+02   .8700+02   .8700+02   .9204+02
ENTHAL    .2515+04   .9282+04   .8541+04   .1955+04   .2181+04
PRESS     .1000+01   .1000+01   .1000+01   .1000+01   .1000+01
COMP 1    .4634+00   .2757+00   .1380+00   .3797+00   .4084+00
COMP 2    .3457+00   .1230+00   .3765-01   .1875+00   .2601+00
COMP 3    .1909+00   .6014+00   .8244+00   .4328+00   .3315+00

TIME =  .2000+02
STRM NO       2          3          4          5          1
FLOW      .3006+03   .4795+01   .1131+01   .3664+01   .3664+01
TEMP      .1029+03   .1029+03   .8700+02   .8700+02   .8961+02
ENTHAL    .2641+04   .9526+04   .8541+04   .1954+04   .2070+04
PRESS     .1000+01   .1000+01   .1000+01   .1000+01   .1000+01
COMP 1    .4837+00   .3243+00   .1387+00   .3816+00   .3950+00
COMP 2    .3688+00   .1508+00   .3731-01   .1858+00   .2239+00
COMP 3    .1475+00   .5249+00   .8240+00   .4326+00   .3812+00

TIME =  .3000+02
STRM NO       2          3          4          5          1
FLOW      .2928+03   .4463+01   .5015+00   .3962+01   .3962+01
TEMP      .1048+03   .1048+03   .8700+02   .8700+02   .8823+02
ENTHAL    .2714+04   .9680+04   .8541+04   .1953+04   .2008+04
PRESS     .1000+01   .1000+01   .1000+01   .1000+01   .1000+01
COMP 1    .4940+00   .3547+00   .1388+00   .3821+00   .3881+00
COMP 2    .3811+00   .1688+00   .3723-01   .1854+00   .2035+00
COMP 3    .1249+00   .4765+00   .8239+00   .4325+00   .4085+00

TIME =  .4000+02
STRM NO       2          3          4          5          1
FLOW      .2894+03   .4312+01   .2116+00   .4100+01   .4100+01
TEMP      .1057+03   .1057+03   .8700+02   .8700+02   .8755+02
ENTHAL    .2749+04   .9760+04   .8541+04   .1953+04   .1977+04
PRESS     .1000+01   .1000+01   .1000+01   .1000+01   .1000+01
COMP 1    .4987+00   .3702+00   .1389+00   .3822+00   .3848+00
COMP 2    .3869+00   .1780+00   .3721-01   .1853+00   .1934+00
COMP 3    .1145+00   .4517+00   .8239+00   .4325+00   .4218+00

TIME =  .5000+02
STRM NO       2          3          4          5          1
FLOW      .2880+03   .4250+01   .8451-01   .4166+01   .4166+01
TEMP      .1061+03   .1061+03   .8700+02   .8700+02   .8724+02
ENTHAL    .2765+04   .9791+04   .8533+04   .1952+04   .1963+04
PRESS     .1000+01   .1000+01   .1000+01   .1000+01   .1000+01
COMP 1    .5007+00   .3770+00   .1389+00   .3822+00   .3831+00
```

FIG. 5-35.  Numerical Output for Two-stage Distillation

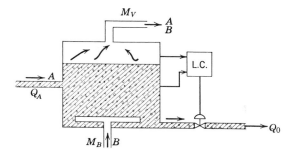

FIG. 5-36

(c) nonideality of components 1 and 2, that is, the terminal activity of 1 out of 2 is 1.6 and 2 out of 1 is 1.3. Both of these activities are invariant with temperature.

3. Compare the operation of the two-stage distillation system using:
(a) a fixed condensate temperature, or,
(b) a fixed condenser cooling load.

4. Suppose the flow stream F in Problem 4-6 (Chapter 4) consists of the three ideal components that were used in Case 5-1 (Chapter 5) with the composition $X_1 = 0.43$, $X_2 = 0.31$, $X_3 = 0.26$. Assuming equilibrium only at the flash point, modify the program of problems 4-6 accordingly.

5. A reactor (Figure 5-36) consists of a vessel with a gas sparger in the bottom. Reagent gas $B$ is sparged into the reaction mixture at a rate $M_b$ (moles/min). The reaction

$$A + B \rightarrow C$$

takes place in the liquid phase ($B$ is soluble in the liquid). Reagent $A$ flows into the reactor at a rate $Q_a$ (ft³/min). The reaction is exothermic, and the vapor effluent ($M_v$ moles/min) contains only components $A$ and $B$ ($C$ is negligible). A level controller maintains the level constant by regulating the outlet flow $Q_0$. Assuming that the reactor contents are well agitated, construct a model that establishes the relation between the outlet compositions and the flow rates of reagents $A$ and $B$.

## REFERENCES

1. *Vapor Liquid Equilibrium*, E. Hale, J. Pick, V. Fried, and D. Villim, Pergamon Press, 1958.
2. "Multi-Component Equilibria," R. V. Orye and J. M. Prausnitz, *Ind. Eng. Chem.*, **57**, No. 5, 1965.
3. "Some Methods of Handling Non-Ideal Vapor-Liquid Equilibria in Digital Computer Systems," N. G. O'Brien and R. L. Turner, *Chem. Eng. Progr. Symp. Ser.*, **56**, No. 31, 1960.

# CHAPTER VI

# REACTION KINETICS

This chapter demonstrates the basic methods that are used for simulating many different classes of chemical reactions typical in industrial situations. Unlike unit processes dealing with separation such as distillation columns or phase change such as condensers, it is most difficult to generalize reaction kinetics, and virtually useless to create general unit operations subroutines as is done in previous chapters for other common operations. The reason, of course, is that almost every reaction is unique, characterized by a specific chemical model, involving one or more phases, and invariably taking place in one of a great variety of reactor configurations. The influence of catalysts, heterogeneous mechanisms, multiphase energy transfer within the reaction environment, all greatly expand the different varieties of reaction mechanisms and process geometries. There is, however, a basic framework for simulating reaction equations by numerical integration that is similar to the procedures described previously for systems of ordinary differential equations and can be characterized as shown in the next section.

## 6-1 GENERAL MODELING SCHEME

In general, reagent flows are brought together (Figure 6-1) in a reactor environment where reaction rates are formed as a function of compositions, temperature, and possibly other variables. These rates will partially eliminate the original components and form new components. For example, for the simple reaction

$$A + B \xrightarrow{R} C + D$$

the reaction rate $R$ is the rate of disappearance of components $A$ and $B$, and is also the rate of appearance of components $C$ and $D$. $R$ will have the same

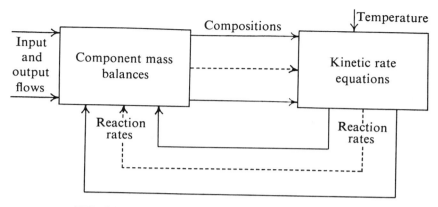

FIG. 6-1. General Modeling Scheme for Reaction Kinetics

units as a physical flow rate, most commonly moles/min, and for a complete definition should be coupled to one of the components. In this case, of course, $R$ is defined as the rate of disappearance of $A$ and will also be the rate of disappearance of $B$ and the rate of appearance of $C$ and $D$. In the reaction

$$A + 2B \xrightarrow{R} C + D$$

if $R$ is the rate of disappearance of $A$, the rate of disappearance of $B$ will be $2R$ since 2 moles of $B$ react with each mole of $A$.

The reaction rates are usually proportional to the reagent compositions; thus the compositions can be regarded as the "potential" creating the reaction flow rates. In turn, these reaction flow rates are included in the component balances together with other physical flows which are solved for the compositions. The direction of information flow is shown in Figure 6-1, where the equations describing a reaction situation are divided into two groups, the component mass balances that are solved for the compositions and the kinetic rate equations that define the reaction rates. Following this procedure reduces the solution of the most complex kinetic mechanisms to a matter of simple bookkeeping, elegantly handled by a suitable computer.

The solution sequence and information transfer is similar to that used for fluid flow, electrical energy, heat transients, and so on which can be stated as the universal axiom.

Flux accumulation produces potentials, and differences in potentials produce fluxes.

This is demonstrated many times in the previous chapters and is used as a guide for the computer simulation of the reaction systems selected as examples for this chapter.

## 6-2   LIQUID PHASE CSTR

The Continuous Stirred Tank Reaction (CSTR) is the most common system in industry although it does have numerous variations regarding the introduction and extraction of energy and materials. As an example, consider the reactions

$$A + B \underset{k_1^-}{\overset{k_1}{\rightleftarrows}} C + D$$

$$C + B \underset{k_2^-}{\overset{k_2}{\rightleftarrows}} E$$

$$A + E \underset{k_3^-}{\overset{k_3}{\rightleftarrows}} F$$

The original reagents $A$ and $B$ are to produce the product $E$ which in turn produces a byproduct $F$ by further reaction with reagent $A$. The most convenient way of expressing the reaction rates is in the following form using a rate coefficient $k_i$ and an equilibrium coefficient $K_i$.

$$R_1 = k_1 \left( \frac{C_A C_B - C_C C_D}{K_1} \right) \cdot V \tag{6-1}$$

$$R_2 = k_2 \left( \frac{C_C C_B - C_E}{K_2} \right) \cdot V \tag{6-2}$$

$$R_3 = k_3 \left( \frac{C_A C_E - C_F}{K_3} \right) \cdot V \tag{6-3}$$

Various systems of units can be used in these equations, though the concentration terms $C_i$ are correctly defined as mass/unit volume. A consistent unit system would be:

$V$ = reaction volume (ft³),
$C_i$ = moles (i/ft³),
$K_1$ = equilibrium constant, dimensionless,
$K_2, K_3$ = ft³/mole,
$k_i$ = moles/(min, ft³, (moles)²/(ft³)²) = ft³/(min, mole),
$R$ = reaction rate moles/min.

The equilibrium constant is the ratio between the forward and reverse rate coefficient, that is,

$$K_i = \frac{k_i}{k_i^-}$$

Unfortunately, many CSTRs involve simultaneous evaporation where the equilibrium relationships are expressed in terms of liquid and vapor mole fractions $X_i$ and $Y_i$. It is sometimes convenient in such a simulation to unify the units by expressing the reaction equation in terms of mole fractions instead of composition units (moles/ft³) as stated originally.

A transformation of the reaction equations can be made by considering the units of the basic reaction equation

$$R = k * C_A * C_B * V$$

The units are:

$$\frac{\text{moles}}{\text{min}} = \frac{\text{ft}^3}{\text{min mole}} * \frac{\text{moles}_A}{\text{ft}^3} * \frac{\text{moles}_B}{\text{ft}^3} * \text{ft}^3$$

By substituting mole fraction $X$ for molar concentration $C$ the units become

$$\frac{\text{moles}}{\text{min}} = \frac{\text{ft}^3}{\text{min mole}} * \frac{\text{moles}_A}{\text{moles}} * \frac{\text{moles}_B}{\text{moles}} * \left(\frac{\text{moles}}{\text{ft}^3}\right)^2 * \text{moles} * \left(\frac{\text{ft}^3}{\text{mole}}\right)$$

In symbol form the last equation can be written

$$R = kX_A X_B \frac{M}{V} M$$

The term $M/V$ is the molar density of the reaction mixture; if the fluctuations in this term resulting from changes in composition are small it can be combined with the reaction coefficient $k$. The reaction equation then becomes

$$R = kX_A X_B M$$

$$\frac{\text{mole}}{\text{min}} = \frac{1}{\text{min mole}} \frac{\text{mole}_A}{\text{mole}} \frac{\text{mole}_B}{\text{mole}} \text{mole}$$

where $M$ is the total molar volume of the reaction mixture. Applying this transformation to the reactor in this example, the reaction equations now become:

$$R_1 = k_1 M \left(X_A X_B - \frac{X_C X_D}{K_1}\right) \tag{6-4}$$

$$R_2 = k_2 M \left(X_C X_B - \frac{X_E}{K_2}\right) \tag{6-5}$$

$$R_3 = k_3 M \left(X_A X_E - \frac{X_F}{K_3}\right) \tag{6-6}$$

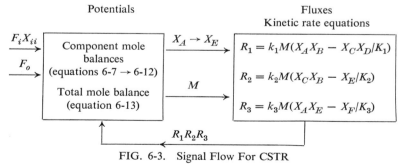

FIG. 6-2.  CSTR

Figure 6-2 shows an elementary diagram of the CSTR with an input fluid feed, whose rate is $F_i$ moles/min and an output flow having a rate $F_o$. Now the underlying assumption for a CSTR is that the outlet flow has the same temperature and composition as the reactor contents. (This is also referred to as a "backmix" reactor.) The component mass balance equations now become

$$\frac{d}{db}(MX_A) = F_iX_{iA} - F_oX_A - R_1 - R_3 \tag{6-7}$$

$$\frac{d}{dt}(MX_B) = F_iX_{iB} - F_oX_B - R_1 - R_2 \tag{6-8}$$

$$\frac{d}{dt}(MX_C) = R_1 - R_2 - F_oX_C \tag{6-9}$$

$$\frac{d}{dt}(MX_D) = R_1 - F_oX_D \tag{6-10}$$

$$\frac{d}{dt}(MX_E) = R_2 - R_3 - F_oX_E \tag{6-11}$$

$$\frac{d}{dt}(MX_F) = R_3 - F_oX_F \tag{6-12}$$

The sum of these component balances will give the total mole balance, that is,

$$\frac{d}{dt}M = F_i - F_o - R_2 - R_3 \tag{6-13}$$

This can be readily derived by realizing that $\sum_i X_i = 1$.

If the assumption is made that the reaction takes place in isothermal conditions, then the model is complete and is shown in Figure 6-3 in signal flow form.

FIG. 6-3.  Signal Flow For CSTR

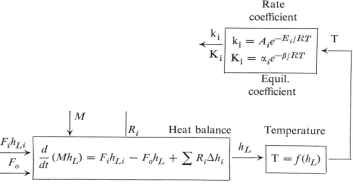

FIG. 6-4.   Signal Flow for Enthalpy Balance for CSTR

If the reactor is operating under adiabatic conditions, then an energy balance is required to establish the temperature. The most elementary form for such an energy balance is as follows:

$$\frac{d}{dt}(Mh_L) = F_i h_{Li} - F_o h_L + \sum R_i \Delta h_i \qquad (6\text{-}14)$$

where $h_L$ is the liquid enthalpy and $\Delta h_i$ is the heat of reaction for each individual reaction. Having obtained the enthalpy in the reactor $h_L$, the temperature is derived as a function of $h_L$ based on the component enthalpy characteristics in the reactor. This sequence is shown in Figure 6-4 as an addition to the model in Figure 6-3.

The influence of temperature on the reaction is mostly via the rate coefficient $k_i$. This is usually expressed as an Arrhenius function $k = Ae^{-ER/T}$ where $A$ is the frequency factor, $E$ the activation energy, $R$ the gas constant, and $T$ the absolute temperature. The equilibrium constant $K$ has the same exponential form as the rate coefficient.

### 6-2-1   Computer Program

It will be recalled that the DYFLO program (Chapter 5) was deliberately structured to accommodate any reactions that may be occurring. This is achieved through the array RCT(J) where the first 20 locations carry the net rate of change for each component due to any reaction. Location 21 is the net rate of change of total moles and location 22 is the net rate of heat generation due to all the reactions. The unit operation required for this reactor is the HLDP subroutine which calculates the exit compositions and temperature. If the inlet and outlet streams are numbered 1 and 2 as shown in Figure 6-2, the coding required for the reactor is simply to call HLDP

```
            SUBROUTINE REACT(I, HLD)
1           COMMON/CD/STRM(300, 24),DATA(20, 10),RCT(22),NCF,NCL,LSTR
2           RTA=(STRM(I, 22)+273.)*R
3           DATA A/...../E/...../EE/...../DH/...../R/../
4           DO 5 J=1,3
5           AK(J)=A(J)*EXP(−E(J)/RTA)
6       5   EK(J)=AE(J)*EXP(−EE(J)/RTA)
7           R1=AK(1)*HLD*(STRM(I, 1)*STRM(I, 2)−STRM(I, 3)*STRM(I, 4)/EK(1))
8           R2=AK(2)*HLD*(STRM(I, 3)*STRM(I, 2)−STRM(I, 5)/EK(2))
9           R3=AK(3)*HLD*(STRM(I, 1)*STRM(I, 5)−STRM(I, 6)/EK(3))
10          RCT(1)=−R1−R3
11          RCT(2)=−R1−R2
12          RCT(3)=R1−R2
13          RCT(4)=R1
14          RCT(5)=R2−R3
15          RCT(6)=R3
16          RCT(21)=−R2−R3
17          RCT(22)=R1*DH(1)+R2*DH(2)+R3*DH(3)
            RETURN
```

FIG. 6-5.  Subroutine REACT

with the appropriate entries in the argument list. Coupled with the HLDP routine, an auxiliary routine is required that calculates all the reaction values that are entered in the RCT array. A listing for this routine REACT is shown in Figure 6-5. There are six components in this reactor, A → F and they will occupy locations 1 → 6 in the STRM array. All the data required for the rate (k) and equilibrium coefficient (K) are subscripted (lines 5, 6) allowing these coefficients to be calculated in a DO loop (line 4). The inputs to the routine are the line number for the exit stream I and the reaction holdup HLD. The temperature and composition of stream I will be the same as that of the reactor contents. The three reaction rates are calculated on lines 7 to 9 and the net reaction flow rates for each component as established by equations 6-7 → 6-12 are entered in the RCT array on lines 10 to 17. (Note. If there is more than one holdup in the process this array must be cleared after use, see Appendix B5.) The total rate of moles and heat formation are calculated on lines 16 and 17, respectively. The main program is shown in Figure 6-6. It now consists merely of two lines of coding, the call on the REACT and

```
6   CALL REACT(2, VOL)
    CALL HLDP(1, 2, 3, VOL, 0.)
    CALL PRL(1., 10., LF, 1, 2, 0, 0, 0, 0, 0, 0, 0, 0, 0, 0)
    IF(LF) GO TO 5
    CALL INTI(TIM, .1, 4)
    GO TO 6
5   STOP
```

FIG. 6-6.  Main Program for CSTR

HLDP subroutines. The remaining statements are PRL and INTI routines that control the output printing and integration procedure. Data initiation for the feed line array and the initial conditions for the reactor effluent stream (2) are not shown in Figure 6-6.

## 6-2-2  Batch Reactor Simulation

The program discussed in the previous section can readily be converted to a batch reaction by simply assigning a zero flow to the input and output streams, that is,

$$STRM(1, 21) = 0$$
$$STRM(2, 21) = 0$$

If the reaction is not adiabatic, there is provision in the HLDP subroutine for a heat flux to an exterior source such as a cooling jacket. This heat flux is the last item in the argument list, so it could be calculated in the main program by a jacket heat flux statement such as

$$Q = UA*(TJ - STRM(2, 22)) \tag{6-15}$$

A typical application of computer simulation of batch reactions is to match laboratory data by solving the reaction equations using trial values for the rate coefficients. For situations where there is a final steady-state equilibrium, the analysis of a sample of the mixture yields the compositions that are in equilibrium. Substituting these values in the reaction equations will yield the equilibrium coefficients $K_i$. If samples of the reaction mixture are extracted and analyzed during the reaction phase before reaching equilibrium, the experimental composition versus time trajectory is determined for each component. A number of successive runs on the computer is now made with judicious values for the rate coefficients $k_i$ and the calculated values of the equilibrium coefficients $K_i$. By comparing the calculated and experimental composition/time functions adjustments can be made to the rate coefficients and thus arrive at optimum values that give the best match to the data. For most situations of industrial interest, this procedure is invariably successful because the reactions exhibit only minor coupling. This means that, for example in Case 6-2, adjustments to $k_1$ largely influence the compositions A, B, C, and D, and has a minor effect on the other compositions. The tuning procedure now consists of adjusting the coefficients for the major reactions, proceeding to the minor reactions and repeating this sequence until a reasonable match has been achieved. If it is impossible to obtain a satisfactory match to the experimental functions no matter how the coefficients are changed, then this indicates that the postulated kinetic model does not apply to the reaction and a new mechanism must be constructed. A good indication of this can be obtained when the shape of the experimental curves are not similar to the characteristic shapes obtained from the computer model.

It should also be evident that if the experiment is conducted under isothermal conditions, then repeating the experiment at two different temperature levels will yield two reaction coefficients for each reaction, one at each temperature. By proper manipulation these will yield both the frequency factor (A) and activation energy (E) required to specify the rate coefficient as an Arrhenius temperature function.

## 6-3 RADICAL KINETICS

Reaction models that do not involve radical kinetics are reasonably straightforward to simulate as demonstrated by the example in Section 6-2. Unfortunately, the majority of reaction models of industrial interest are large, complex, and invariably contain radical mechanisms requiring appropriate treatment within the model. These radical relationships when expressed mathematically can give rise to systems of differential equations so "stiff" that a numerical solution is virtually impossible by normal methods. If the reader is not familiar with the concept of "stiffness," an adequate explanation is provided in Chapter 3, Section 13. To overcome this problem, it is necessary to resort to algebraic reduction as demonstrated in the following section.

### 6-3-1 Elementary Reduction of Radical Mechanism

The reaction $A + 2B \rightleftharpoons C$ takes place in a single phase environment. As a first approximation, the reaction rate definition could be coupled with the stoichiometric relationship, that is,

$$R = k\left(AB^2 - \frac{C}{K_E}\right) \tag{6-16}$$

where $A$, $B$, $C$ symbolize concentrations.

This relationship might well be adequate, but a more precise relationship may be required and would be derived as follows. Since there are three molecules involved in the primary section, that is, $A$ and $2B$, an intermediate step is postulated based on bimolecular collision theory, that is,

$$A + B \underset{}{\overset{k_1}{\rightleftharpoons}} e \tag{6-17}$$

$$e + B \underset{}{\overset{k_2}{\rightleftharpoons}} C \tag{6-18}$$

where $e$ is an intermediate specie. The reaction rates for each of these steps would be

$$R_1 = k_1\left(AB - \frac{e}{K_1}\right) \tag{6-19}$$

$$R_2 = k_2\left(eB - \frac{C}{K_2}\right) \tag{6-20}$$

It will be assumed that the composition level of the postulated intermediate *e* is so small as to be virtually undetectable. Proceeding to the component balance equations, assuming a unit batch volume, we have for components *A* and *e*

$$\frac{d}{dt}A = -R_1 \tag{6-21}$$

$$\frac{d}{dt}e = R_1 - R_2 \tag{6-22}$$

Substituting for $R_1$ and $R_2$ and rearranging terms:

$$\frac{dA}{dt} = -k_1\left(AB - \frac{e}{K_1}\right) = k_1\frac{e}{K_1} - (k_1B)A \tag{6-23}$$

$$\frac{de}{dt} = \left(k_1AB + k_2\frac{C}{K_2}\right) - \left(\frac{k_1}{K_1} + k_2B\right)e \tag{6-24}$$

Now the stiffness ratio between these two differential equations is essentially the ratio of the coefficients of the negative feedback term, that is,

$$\text{ratio} = \frac{k_1/K_1 + k_2B}{k_1B} = \frac{1}{BK_1} + \frac{k_2}{k_1} \tag{6-25}$$

In order for component *e* to have a very low concentration, either $k_2$ is very large (i.e., $k_2 \gg k_1$) or $K_1$ is very small, or perhaps both cases apply. In any event, either or both terms in the ratio, that is, $1/BK_1$ or $k_2/k_1$ becomes a large number, implying a high stiffness ratio. Simulating equations 6-23 and 6-24 on a digital computer would then require a very small integration step, in order to maintain stability, leading to excessive running times to traverse a reasonable period of real time. In fact, there have been examples where the stiffness ratio was so large that computer running times of weeks or months could have been required for a complete solution. Clearly this is not a reasonable approach; thus we resort to the "steady-state theory."

### Steady-State Theory

Unfortunately steady-state theory is a misnomer because it implies that the composition of intermediate components are static during a reaction cycle; whereas they immediately change as the reaction proceeds. The theory is really the application of the concept that since the intermediate composition is so small, there is no significant accumulation of this component. Therefore, the rate of formation of the intermediate is practically identical to its rate of destruction; that is, the rates are in equilibrium or "steady state." A concomitant result is that since both reactions may be functions of the composition of the intermediate as well as other compositions, the intermediate composition, small as it is, will have to change in order to maintain this equilibrium.

Applying this approach to our example, the rigorous way to derive the steady-state equation, especially useful for more complex situations, is simply to set the derivative to zero in the intermediate component balance equation (6-22), that is, $0 = R_1 - R_2$, or

$$R_1 = R_2 \tag{6-26}$$

In this case this result should be self-evident from the theory; that is, the rate of formation of component $e(R_1)$ is the same as the rate of destruction ($R_2$). Substituting for $R_1$ and $R_2$ we obtain

$$k_1\left(AB - \frac{e}{K_1}\right) = k_2\left(eB - \frac{C}{K_2}\right) \tag{6-27}$$

which with several manipulative steps reduces to a definition of the composition of $e$:

$$e = \frac{k_1AB + k_2C/K_2}{k_2B + k_1/K_1} = \frac{ABK_1 + (k_2K_1/k_1K_2)C}{(K_1k_2/k_1)B + 1} \tag{6-28}$$

Substituting for $e$ in $R_1$ we obtain

$$R_1 = k_1\left(AB - \frac{k_1AB + k_2C/K_2}{(k_2B + k_1/K_1)K_1}\right) \tag{6-29}$$

which reduces to

$$R_1 = \frac{k_1k_2}{k_1/K_1 + k_2B}\left(AB^2 - \frac{C}{K_1K_2}\right) \tag{6-30}$$

In terms of basic coefficients this further reduces to

$$R_1 = \frac{k'}{1 + k_\alpha B}\left(AB^2 - \frac{C}{K_E}\right) \tag{6-31}$$

The model for the system now becomes Figure 6-7. Since the equation for

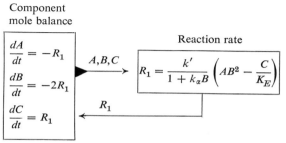

FIG. 6-7. Model for Reaction Involving Intermediate

the intermediate component $e$ has been eliminated, the stiffness ratio has been reduced to unity, permitting an efficient computer solution. If the ratio of the reaction rate coefficients $k_2/k_1$ is known or estimated and if the individual equilibrium constants ($K_1$ and $K_2$) are also known, then the composition of $e$ can be obtained directly by solving the algebraic equation 6-28. Generally, this is rarely the case; in fact, the level of $e$ is determined largely by the assumptions made for the values of these coefficients, so the $e$ composition is rather academic. In any event, it is important to observe that in the reduced model (Figure 6-7) for the reaction, not only has $e$ been eliminated, but also the separate coefficients $k_1$, $k_2$, and so on have been lumped into single overall coefficients whose values can be determined directly from a measure of the composition in the reaction phase as follows.

1. Equilibrium constant $K_E$: This is obtained quite simply by allowing the reaction mixture to come to equilibrium, at which time $R = 0$ or

$$AB^2 = \frac{C}{K_E}, \quad \text{that is,} \quad K_E = \frac{C}{AB^2} \tag{6-32}$$

Determining the composition $A$, $B$, and $C$ from a sample will provide the value for $K_E$ by substituting in equation 6-32.

2. Primary rate coefficient $k'$: If only component $C$ is placed in the reaction environment (i.e., $A$ and $B = 0$) and by sampling procedures the rate of change of $C$ is determined, then, by extrapolation, the initial reaction rate is determined. Since $A$ and $B = 0$, this rate is

$$R_1 = k'\left(\frac{-C}{K_E}\right) \tag{6-34}$$

and establishes $k'$.

3. If the procedure in 2 is repeated with an initial mixture of $A$ and $B$, then the rate will be

$$R_1 = \frac{k'}{1 + k_\alpha B}(AB^2) \tag{6-35}$$

which will establish $k_\alpha$.

The above sequence of tests is common practice in experimental kinetics and the only purpose in this description was to demonstrate the value of reducing the postulated model into a set of equations containing the minimum number of constants that can be established from measurable data.

## 6-3-2  Rate-Limiting Steps

Returning to equation 6-30, if the first reaction is nonreversible, then the rate reduces to the form

$$R_1 = k_1 AB$$

simply by substituting $K_1 = \infty$ in the rate equation. If the second reaction (equation 6-20) is nonreversible, that is, $K_2 = \infty$, then from equation 6-30 and 6-31 the rate simplifies to

$$R_1 = \frac{k'}{1 + k_\alpha B} AB^2 \qquad (6\text{-}36)$$

If the first reaction is rate limiting, that is, $k_2 \gg k_1$, then the rate becomes $R_1 = k_1(AB - C/BK_1K_2)$, and if the second reaction is rate limiting, that is, $k_1 \gg k_2$ then the reaction simplifies to

$$R_1 = k_1 K_1 \left( AB^2 - \frac{C}{K_1 K_C} \right) = k_1 \left( K_1 AB^2 - \frac{C}{K_2} \right) \qquad (6\text{-}37)$$

In conclusion, in situations where the model of a reaction scheme contains fast intermediates, typically radical species, it is recommended that the differential equation defining the mass balance on trace intermediate components be replaced by steady-state algebraic relationships or, where possible, be eliminated entirely by substitution, as demonstrated in this section. This allows efficient solution and reduces the number of constants to a minimum commensurate with the data.

The next two examples will demonstrate the application of the steady-state theory to typical situations.

### 6-4 HETEROGENEOUS KINETICS

The following example will demonstrate the applicability of modeling and simulation to a reaction system involving three phases, where several phenomena are occurring simultaneously, making manual calculations virtually intractable. This reaction takes place in a high-pressure autoclave containing an agitated liquid and gas phase (Figure 6-8) and consists of hydrogenating

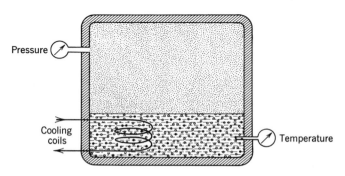

FIG. 6-8.   Batch Autoclave

an organic compound $A$ to compound $C$ via an intermediate $B$, that is,

$$A + H_2 \rightarrow B$$
$$B + H_2 \rightarrow C$$

This reaction occurs on the surface of a metal catalyst in the form of loose particles, the reagents diffusing from the liquid phase to the catalyst surface reacting and returning to the liquid phase. The three phases are enclosed in the autoclave. During the batch reaction a degree of control over the internal temperature can be achieved by a series of cooling coils contacting the liquid phase. The reaction is highly exothermic, which will cause a significant rise in temperature and pressure during the course of the reaction, monitored by suitable instrumentation. Several sets of test data are available, consisting of temperature and pressure transients for different starting and operating conditions. As part of a process improvement program, it is necessary to develop a computer model that matches the available data. The model will then be used to explore improvements in productivity and process design.

Because of the nature of the reaction, a significant exchange of material between the liquid and vapor phases occurs as the reaction proceeds with the simultaneous division of heat energy. However, all phases will have the same temperature since the agitation has sufficient violence to maintain equilibrium.

### 6-4-1 Reaction Sequence

The first step in assembling the model is to develop the reaction-rate equations. Since a catalyst is involved, a simple stoichiometric relationship is inadequate to describe the reaction, and so "Langmuir Isotherm" expressions are used to relate the reaction rates to the process variables. Theory postulates that a unit volume of catalyst contains a number $(S)$ of active sites on the surface of the catalyst particles. The reactants, contained in a solvent, diffuse to the catalyst surface, react or become attached to the active sites, and in a high energy state, react with each other. The products then diffuse back to the fluid surrounding the catalyst. Each step can be characterized as follows:

Step 1    $A + \sigma \rightleftarrows A_\sigma$    $A$ reacts with $\sigma$-free active site

Step 2    $G + \sigma \rightleftarrows H_\sigma$    $H_2(G)$ reacts with $\sigma$-free active site

Step 3    $A_\sigma + G_\sigma \rightarrow B_\sigma$    $A$ and $H_2$ on active sites react

Step 4    $B + \sigma \rightleftarrows B_\sigma$    $B$ reacts with $\sigma$-free active site

Step 5    $B_\sigma + G_\sigma \rightarrow C_\sigma$    $B$ and $H_2$ on active sites react

Step 6    $C + \sigma \rightleftarrows C_\sigma$    $C$ reacts with $\sigma$-free active site.

To develop the rate equations based on this mechanism, it is necessary to visualize the unit volume of catalyst with the total number of active sites $S$ as being partly occupied by the reacting species $I_\sigma$. The total sites $S$ consist of the unoccupied sites $\sigma$ and the sum of the occupied sites, that is,

$$S = \sigma + (A_\sigma + B_\sigma + C_\sigma + G_\sigma) \tag{6-33}$$

It is implied here that the solvent is essentially inert and does not occupy any active sites on the catalyst. Symbolizing $A$, $B$, $C$, and $G$ as the liquid mole fractions of the reacting species, where $G$ is the $H_2$, the following reaction rate equation can be stated:

Step 1 $$R_1 = k_1\left(A \cdot \sigma - \frac{A_\sigma}{K_A}\right) \tag{6-34a}$$

Step 2 $$R_2 = k_2\left(G \cdot \sigma - \frac{G_\sigma}{K_G}\right) \tag{6-34b}$$

Step 3 $$R_3 = k_3 A_\sigma G_\sigma \tag{6-34c}$$

Step 4 $$R_4 = k_4\left(B \cdot \sigma - \frac{B_\sigma}{K_B}\right) \tag{6-34d}$$

Step 5 $$R_5 = k_5 B_\sigma \cdot G_\sigma \tag{6-34e}$$

Step 6 $$R_6 = k_6\left(C \cdot \sigma - \frac{C_\sigma}{K_C}\right) \tag{6-34f}$$

In these equations, $k_i$ is the rate constant and $K_i$ is the equilibrium constant. Note that only Steps 3 and 5 are irreversible. If mass balance differential equations were to be based on the above reaction rates, we would have as an example for active site species $A_\sigma$.

$$\frac{d}{dt}(A_\sigma) = R_1 - R_3$$

Here, however, is another situation that would lead to impossibly stiff equations because the quantity $A_\sigma$ is very small compared with the flow in and out, $R_1$ and $R_3$. We are therefore forced to resort to essentially the steady-state theory where a rate-limiting step is specified for each of the two primary reactions $A \to B$ and $B \to C$. The other steps in the sequence are assumed to be in equilibrium. Experimental methods for determining which of the postulated steps is rate limiting are described in the literature (Ref. 14) and it will be assumed here that these details of the kinetic mechanism have already been established with the following conclusions:

1. The rate limitation for the first reaction $A \to B$ is the resistance to diffusion of the $A$ species migrating to the catalyst surface. This is symbolized

by Step 1 in the reaction sequence, that is,

$$R_1 = R_A = k_1\left(A \cdot \sigma - \frac{A_\sigma}{K_A}\right) \tag{6-35}$$

2. The rate limitation for the second reaction $B \rightarrow C$ is the surface reaction Step 5, that is,

$$R_B = R_5 = k_5(B_\sigma \cdot G_\sigma) \tag{6-36}$$

It follows from these two definitions that the remaining reversible steps are in equilibrium, leading to the following approximations:

$$B_\sigma = K_B B \cdot \sigma \tag{6-37}$$
$$G_\sigma = K_G G \cdot \sigma \tag{6-38}$$
$$C_\sigma = K_C C \cdot \sigma \tag{6-39}$$

Since for the first reaction the diffusion step is rate limiting, a further approximation can be made by neglecting $A_\sigma$. Substituting these expressions for the number of occupied active sites in the total site equation, we obtain

$$S = \sigma + K_B B \cdot \sigma + K_G G \cdot \sigma + K_C C \cdot \sigma \tag{6-40}$$

Rearranging:

$$\sigma = \frac{S}{1 + K_B B + K_G G + K_C C} \tag{6-41}$$

The expression for the first reaction rate, neglecting the reversible term is

$$R_A = k_1 A \cdot \sigma \tag{6-42}$$

Substituting for $\sigma$:

$$R_A = \frac{k_1 A S}{1 + K_B B + K_G G + K_C C} \tag{6-43}$$

For the second equation $B \rightarrow C$ the rate-limiting step is the surface reaction (equation 5). Substituting for $B_\sigma$ and $G_\sigma$ it can be stated as

$$R_B = k_5(K_B \cdot B \cdot \sigma)(K_G G \cdot \sigma) \tag{6-44}$$

Combining constants and substituting for the free active sites $\sigma$

$$R_B = \frac{k_5 S^2 \cdot B \cdot G}{(1 + K_B B + K_G G + K_C C)^2} \tag{6-45}$$

The total number of sites will be proportional to the mass of catalyst for a fixed particle size and geometry; thus if $W$ is the pounds of catalyst used, the proportionality factor can be combined with the rate constant, and the reaction rates are defined using $W$ in place of $S$.

### 6-4-2 Component Balances

Having defined the reaction rates in terms of the liquid compositions, the next logical step is to write the mole balance equations. This can be done readily by considering only the total moles of each component within the reactor and ignoring temporarily the vapor/liquid split. The equations are

Component A $$\frac{dM_A}{dt} = -R_A \tag{6-46a}$$

Component B $$\frac{dM_B}{dt} = R_A - R_B \tag{6-46b}$$

Component C $$\frac{dM_C}{dt} = R_B \tag{6-46c}$$

Component G $$\frac{dM_G}{dt} = -R_A - R_B \tag{6-46g}$$

Total moles $$\frac{dM}{dt} = -R_A - R_B \tag{6-46T}$$

The heat balance involves several terms which have to be accounted for as follows:

1. Reaction heat: The exothermic reaction heat is 14,700 and 15,600 PCU/mole reacted for $R_A$ and $R_B$, respectively.
2. Heat flux to reactor wall: Neglecting the heat loss from the exterior surface of the reactor, the heat absorbed by the wall is

$$HTW = (UA)_W * (T - T_W) \text{ PCU/min} \tag{6-47}$$

where $(UA)_W = 200$ PCU/min °C

$T_W =$ wall temperature °C

$T =$ temperature of reactor contents °C

The wall temperature is calculated from the heat balance:

$$\frac{dT_W}{dt} = \frac{HTW}{600} \tag{6-48}$$

where 600 PCU/°C is the total heat capacity of the wall.

3. Heat flux to cooling coils: A high flow rate of coolant at 100°C circulates through the coils providing the following heat flux:

$$HTC = (UA)_C(T\text{-}100) \tag{6-49}$$

The coil heat transfer coefficient is $(UA)_C = 100$ PCU/°C min. The overall heat balance for the reactor contents will be

$$\frac{dQ}{dt} = R_A * 14{,}700 * R_B \, 15{,}600 - HTW - HTC \qquad (6\text{-}50)$$

where $Q$ is total enthalpy in both phases.

### 6-4-3 Vapor Liquid Split

Since the reaction rates are established by the liquid composition $X_i$, it is necessary to determine continually the total vapor/liquid split in the reactor and the distribution of all components. In general, the total enthalpy determines the vapor/liquid split, and equilibrium relationships determine the liquid and vapor composition. The components will be assumed to be ideal, that is, all activity coefficients $= 1$. The pressure inside the vessel will be established by the total moles of gas forced into the vapor space by the heat energy. If the vapor moles are known, the pressure can be calculated from the ideal gas law as a first approximation by

$$\text{pressure (PIC)} = M_V(T°K) * \frac{R}{V}$$

The gas constant $R$ divided by the volume $V$ is 0.007. This neglects the change in volume due to the change in liquid mass, though this refinement could be incorporated if necessary. The pressure, in turn, determines the temperature level achieved by the component equilibrium relationships.

It will not be necessary at this point to write the equilibrium and split equations because the computer program that follows will use available subprograms that already incorporate these relationships. To help visualize

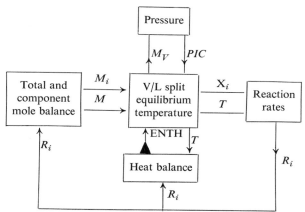

FIG. 6-9.   Information Flow for Heterogeneous Batch Reactor

the flow of information, Figure 6-9 shows the major groupings of the equations in the model and the primary variables interconnecting these blocks.

### 6-4-4 Computer Program

From the description of the model, it should be apparent that besides the reaction kinetics, the bulk of the model involves vapor/liquid equilibrium and enthalpy calculations. Since the DYFLO program (Chapter 5) was created specifically to handle energy transfer and mixed phases in equilibrium, it will be convenient to use some of the subroutines available from this program to simulate this batch-reactor system. Apparently, even though the unit-operation subroutines were designed for continuous-flow processes, they can be readily adapted for use in a nonflow or static-batch situation.

The subroutine HLDP, primarily intended to simulate a vessel with a variable holdup (moles) with an input and an output flow, also accounts for an external heat flux and internal reactions via the RCT array in COMMON. This subroutine HLDP can be used to simulate a batch process in several ways. The most convenient method in this case is to specify both input and output streams as the same stream (2 in Figure 6-10). By specifying the total mole holdup as the "flow" of stream 2, that is, STRM (2,21), this stream will automatically contain (via the composition) the total quantity of each component and the molar enthalpy (STRM (2,23)). If the contents are specified as liquid by inserting 3 in the third argument of HLDP, the temperature of stream 2 is that which it would have if it were all liquid with the calculated enthalpy. Therefore, it is a "dummy" temperature and is of no practical value.

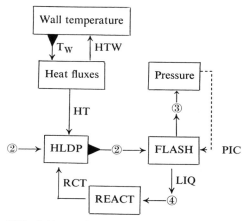

FIG. 6-10. Dyflo Arrangement for Heterogeneous Batch Reactor

```
 1*           COMMON/CD/STRM(300,24),DATA(20,10),RCT(22),NCF,NCL,LSTR
 2*     C ***INITIATION AND DATA SECTION***
 3*           DATA NCF/1/NCL/5/TW/75./TC/100./
 4*           LOGICAL NF,LSTR
 5*           CALL DAT
 6*           LSTR=.TRUE.
 7*     C ***DERIVATIVE SECTION***
 8*         5 CALL FLASH(2,3,4,0.)
 9*           PIC=STRM(3,21)*(STRM(3,22)+273.)*.007
10*           CALL CONV(STRM(4,24),PIC,2,NC)
11*           GO TO (4,5),NC
12*         4 CALL REACT(100.)
13*           HTW=200.*(STRM(4,22)-TW)
14*           HTC=100.*(STRM(4,22)-TC)
15*           HT=-(HTW+HTC)
16*           CALL HLDP(2,2,3,STRM(2,21),HT)
17*           DTW=HTW/600.
18*         6 CALL PRL(5.,40.,NF,2,3,4,0,0,0,0,0,0,0,0,0)
19*           IF(NF) STOP
20*     C ***INTEGRATION SECTION***
21*           CALL INTI(TIM,.1,1)
22*           CALL INT(TW,DTW)
23*           GO TO 5
24*           END
```

FIG. 6-11. Main Program

To obtain the actual vapor/liquid split "stream" 2, representing the total holdup, is fed to the FLASH subroutine that splits this stream into a vapor stream (3) and a liquid stream 4 (Figure 6-10). Since the "flow" of stream 2 is actually the total holdup moles, the "flows" of streams 3 and 4 will also be the vapor and liquid mole holdups, with their respective molar enthalpies, compositions, and temperature. Primarily, then, the total reactor system is programmed by three subroutines, namely, HLDP, FLASH, and REACT. The calls for these subroutines are on lines 16, 8, and 12 of the main program shown in Figure 6-11.

It will be recalled that in order for FLASH to calculate the split, it requires the system pressure for the equilibrium calculation. In this application, the pressure is established by the total moles in the vapor space (line 9); thus an iterative calculation is programmed using CONV as on line 10.

### 6-4-5  Subroutines REACT and DAT

Having converged on the split, the temperature and composition of the liquid phase are available for use in calculating the reaction rates. This is done (on line 12) in subroutine REACT shown in Figure 6-12. All the equilibrium and rate constants are calculated as exponential functions of the absolute temperature (lines $4 \rightarrow 9$), followed by the two reaction rates $R_A$ and $R_B$ (lines $10 \rightarrow 12$). The catalyst mass W required for the reaction rates is supplied via the argument. The net rate of change of moles for each component due to

```
1*          SUBROUTINE REACT(W)
2*          COMMON/CD/STRM(300,24),DATA(20,10),RCT(22),NCF,NCL,LS
3*          REAL K1,KG,KB,KC,K5
4*          TR=(STRM(4,22)+273.)*1.98
5*          K1=13.E4*EXP(-11130./TR)
6*          K5=2.E4*EXP(-12200./TR)
7*          KB=1.5E-3*EXP(4468./TR)
8*          KC=8.*EXP(-1930./TR)
9*          KG=.0144*EXP(-1130./TR)
10*         DEN=1.+KB*STRM(4,2)+KC*STRM(4,3)+KG*STRM(4,4)
11*         RA=W*K1*STRM(4,1)/DEN
12*         RB=W**2*K5*STRM(4,2)*STRM(4,4)/DEN**2
13*         RCT(1)=-RA
14*         RCT(2)=RA-RB
15*         RCT(3)=RB
16*         RCT(4)=-RA-RB
17*         RCT(21)=-RA-RB
18*         RCT(22)=RA*14700.+RB*15600.
19*         RETURN
20*         END
```

FIG. 6-12.  Subroutine React

these two reactions is entered in the RCT array on lines 13–16 where components A, B, C, G, and the solvent are numbered 1 to 5, respectively. The total moles and net reaction heat are entered in the appropriate location of the RCT array on lines 17 and 18, respectively.

The basic data for each of the five components are entered in the appropriate location of the DATA array in subroutine DAT (Figure 6-13). These consist of the Antoine coefficients for the vapor pressure and the enthalpy coefficients. The initial total quantities charged to the reaction in liquid form at 85° are as follows:

Moles A        = 5.         Component #1

Moles G        = 12.        Component #4

Moles solvent = 10.         Component #5

Total moles   = 27.

Initial temperature 75°

```
1*          SUBROUTINE DAT
2*          COMMON/CD/STRM(300,24),DATA(20,10),RCT(22),NCF,NCL,LSTR
3*          DATA(DATA(1,J),J=1,8)/10.5,-2955.,273.,16.,0.,5805.,24.,0./
4*          DATA(DATA(2,J),J=1,8)/10.5,-3010.,273.,19.,,1,8060.,30.,,.03/
5*          DATA(DATA(3,J),J=1,8)/11.2,-3400.,273.,22.,,2,9726.,33.,,.02/
6*          DATA(DATA(4,J),J=1,8)/10.5,-2750.,273.,6.9,.05,2600.,22.,,1/
7*          DATA(DATA(5,J),J=1,8)/13.,-4600.,273.,33.,,.06,12174.,41.,,0./
8*          DATA(STRM(4,J),J=1,24)/,18,0.,0.,,4,,42,15*0.,25.,75,2400.,7./
9*          DATA(STRM(2,J),J=1,24)/,185,0.,0.,,445,37,15*0.,27.,85,2700.,7./
10*         CALL ENTHL(2)
11*         RETURN
12*         END
```

FIG. 6-13.  Subroutine DAT

These quantities are transformed to mole fractions of the total 27 moles charged and entered as initial values of STRM 2 array on line 9, with an estimate of the quantity and quality of the liquid phase (STRM 4) on line 8. A call on ENTHL(2) for stream 2 calculates the total starting enthalpy. Other data, such as initial wall temperature $T_W$, are supplied in the main program (Figure 6-4) on line 3.

### 6-4-6  Operating Characteristics of the Computer Program

It was found during the initial trials that under certain conditions, the FLASH subroutine had difficulty converging on the split ratio R (Figure 5-15, Chapter 5) because of the inability of the CONV subroutine (line 26) to achieve stability. This effect was overcome by replacing CONV with the alternative partial substitution subroutine CPS (Figure 2-9, Chapter 2), with a substitution ratio of 0.5. Apart from this difficulty, the program sequence is quite straightforward, as shown in Figure 6-11, and follows the recommended structure. The results for one particular case corresponding to the data supplied in the main program and subroutine DAT are shown plotted by the computer (coding instructions for the plot routine are omitted from the main program) in Figure 6-15. Several spot values of the STRM data at 5-min intervals are shown in Figure 6-14. This program could now be used to adjust reaction coefficients to match plant data or to determine how process conditions, such as maximum temperature or pressure, will change with variations in the amounts of materials charged to the reactor initially, or it could be the basis of an investigation of the controllability of the system using an external controller that manipulated the coolant temperature TC (see Chapter 11, Problem 3).

### 6-5  SEMIBATCH SOLUTION COPOLYMERIZATION

Industrial chemical products, such as paints, films, plastics, and synthetic textiles, are based on the manufacture of polymers under widely varying conditions. The two most general types of polymers are homo- and co-polymers, and two most general conditions are either solution or emulsion polymerizations. Analytical definitions of the reaction mechanism invariably involve numerous sets of complex differential equations that are readily solved by computer simulation. The following case is an example of one class of system, which, although fairly elementary, demonstrates the procedure required for computer solution.

This example concerns a problem in the manufacture of copolymers. Many polymer compounds are made in batch quantities. In order to control the uniformity of the product, continuous additions of monomers are made

```
TIME =   .0000
STRM NO       2          3          4
FLOW      .2700+02   .3081+01   .2392+02
TEMP      .8500+02   .7743+02   .7743+02
ENTHAL    .2821+04   .4688+04   .2580+04
PRESS     .7000+01   .0000      .7557+01
COMP 1    .1850+00   .1925+00   .1840+00
COMP 2    .0000      .0000      .0000
COMP 3    .0000      .0000      .0000
COMP 4    .4450+00   .7595+00   .4045+00
COMP 5    .3700+00   .4797-01   .4115+00

TIME =   .5000+01
STRM NO       2          3          4
FLOW      .2441+02   .4804+01   .1970+02
TEMP      .1242+03   .1092+03   .1092+03
ENTHAL    .4450+04   .6359+04   .3915+04
PRESS     .7000+01   .0000      .1285+02
COMP 1    .1216+00   .1464+00   .1181+00
COMP 2    .5991-01   .6203-01   .5777-01
COMP 3    .2307-01   .1761-01   .2259-01
COMP 4    .3861+00   .6753+00   .3185+00
COMP 5    .4092+00   .9864-01   .4831+00

TIME =   .1000+02
STRM NO       2          3          4
FLOW      .1841+02   .6283+01   .1219+02
TEMP      .2108+03   .1536+03   .1536+03
ENTHAL    .8474+04   .1313+05   .6034+04
PRESS     .7000+01   .0000      .1876+02
COMP 1    .8700-02   .1383-01   .7279-02
COMP 2    .6043-01   .8407-01   .5033-01
COMP 3    .2017+00   .2394+00   .1776+00
COMP 4    .1876+00   .3428+00   .1116+00
COMP 5    .5416+00   .3199+00   .6532+00

TIME =   .1500+02
STRM NO       2          3          4
FLOW      .1731+02   .4996+01   .1232+02
TEMP      .2024+03   .1458+03   .1458+03
ENTHAL    .8096+04   .1397+05   .5738+04
PRESS     .7000+01   .0000      .1465+02
COMP 1    .1395-03   .2415-03   .1129-03
COMP 2    .1897-01   .2899-01   .1546-01
COMP 3    .2689+00   .3501+00   .2352+00
COMP 4    .1360+00   .2769+00   .7936-01
COMP 5    .5760+00   .3437+00   .6698+00
```

FIG. 6-14.   Computed Results at 5-Min Intervals

during the batch run; hence the term "semibatch." The kinetics involved in the reaction is based on a free-radical mechanism that is explained in the next section.

## 6-5-1   Radical Generation

The radicals are generated by the decomposition of an initiator. This initiator is added to a vessel containing a solvent that is miscible with all the

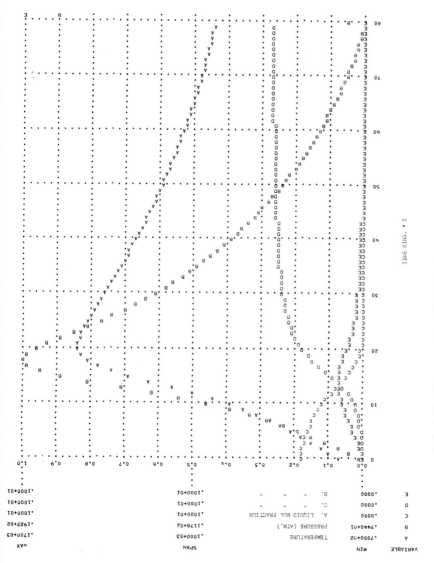

FIG. 6-15.  Computer Plot of Variables in Heterogeneous Batch Kinetics

175

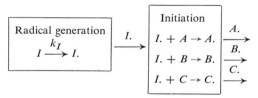

FIG. 6-16. Radical Generation and Chain Initiation

ingredients to be used in the reaction. The initiator decomposes as follows:

$$I \xrightarrow{k_1} I.$$ (6-51)

The $I.$ is a symbol for the initiator radical. The rate of decomposition is expressed by

$$R_I = k_I I \text{ moles/(min mole)}$$ (6-52)

The free radicals ($I.$) react with monomer molecules to form chain radicals, which at the initiation stage will consist of one monomer molecule with an attached radical. The system to be studied consists of three types of monomers, A, B, and C; hence the initiation stage can be symbolized, as in Figure 6-16.

The primary chain radicals $A.$, $B.$, and $C.$ can now react with more monomers and thereby create a growing chain. This growth phase is symbolized as in Figure 6-17, where $A.$, $B.$, and $C.$ now represent polymer chains ending with a radical attached to an $A$, $B$, or $C$ monomer. Since there are three monomers, there will be nine possible reactions as shown in Figure 6-17. The $k_{JK}$'s are the "propagation" rate constants.

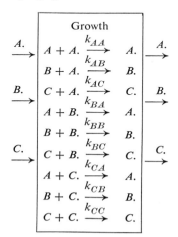

FIG. 6-17. Radical Chain Growth

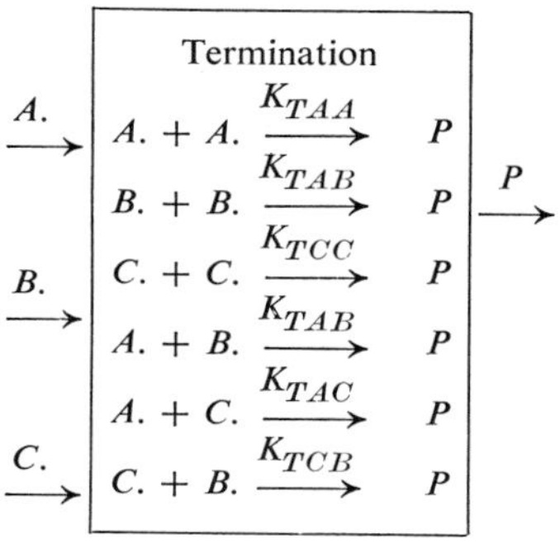

Termination

| A. | A. + A. | $\xrightarrow{K_{TAA}}$ | P | |
|---|---|---|---|---|
| $\xrightarrow{\quad}$ | B. + B. | $\xrightarrow{K_{TAB}}$ | P | P |
| B. | C. + C. | $\xrightarrow{K_{TCC}}$ | P | $\xrightarrow{\quad}$ |
| $\xrightarrow{\quad}$ | A. + B. | $\xrightarrow{K_{TAB}}$ | P | |
| | A. + C. | $\xrightarrow{K_{TAC}}$ | P | |
| C. | C. + B. | $\xrightarrow{K_{TCB}}$ | P | |
| $\xrightarrow{\quad}$ | | | | |

FIG. 6-18.  Termination Step

The fourth step in the sequence is the termination reaction where two chain radicals react to form a "dead" polymer molecule. There are six possible reactions in this last step as shown in Figure 6-18. The $K_{TJK}$'s are the "termination" rate constants. This sequence continues through these four steps until all the monomer present is converted to polymer, or the rate of radical formation from the initiator decreases to zero.

There is one key concept in this whole procedure that must be thoroughly understood in order to obtain a grasp of the entire model. This concept, called the steady-state theory described in Section 6-3-1, serves a useful purpose in simplifying the mathematical definition of the model. Its application to polymerization kinetics is covered in the next section.

### 6-5-2  Steady-State Theory in Polymer Kinetics

This theory is based on the fact that the rate of chain growth is extremely rapid, in fact, almost instantaneous. A mole balance on the total radical population can be summarized as

$$\text{accumulation} = \text{generation} - \text{termination}$$

$$\frac{d}{d_\theta}(R.) = k_I I - \sum K_{TJK}(R._J)(R._K) \qquad (6\text{-}53)$$

where $R._J$ and $R._K$ are the mole concentrations of radical chains ending in component $J$ and $K$. It follows from the steady-state theory that since the actual concentration of radicals at any time is very small compared with the rates of generation and termination, it can be neglected. The mole-balance equation then reduces to the algebraic form

$$\sum K_{TJK}(R._J)(R._K) = k_I I \qquad (6\text{-}54)$$

which says that the rate of radical generation equals the rate of termination. Applying this expression to our three-monomer example, we obtain the overall radical balance

$$k_I I = A.(K_{TAA}A. + K_{TAB}B. + K_{TAC}C.) + B.(K_{TBC}C. + K_{TBB}B.)$$
$$+ C.^2 K_{TCC} \tag{6-55}$$

The same concept is also applied to each of the component radical balances; that is, the rate of formation is equal to the rate of disappearance. The following expressions result:

rate of formation = rate of disappearance

$$A(k_{CA}(C.) + k_{BA}(B.)) = A.(k_{AB}B + k_{AC}C) - \text{balance on } A. \tag{6-56}$$

$$B[k_{CB}(C.) + k_{AB}(A.)] = B.(k_{BC}C + k_{BA}A) - \text{balance on } B. \tag{6-57}$$

$$C(k_{BC}(B.) + k_{AC}(A.)) = C.(k_{CB}B + k_{CA}A) - \text{balance on } C. \tag{6-58}$$

Since the rate of termination $K_{TJK}(R._J)(R._K)$ is much smaller than the rate of disappearance due to propagation, the termination step is neglected in the above equations

It should also be observed that any one of the above three equations is redundant, since adding any two will result in the remaining equation.

These equations will be used to establish the values of $B.$ and $C.$ (moles/liter). The overall radical balance ($OMB$) will establish the $A.$ concentration, since this component is the largest of the three, while the component balances ($CMB$) on $B.$ and $C.$ will solve for $B.$ and $C.$, respectively. An information flow diagram of this arrangement is shown in Figure 6-19, where there are several implicit loops which would require converging the minor loops $B. \rightarrow C. \rightarrow B.$ and $B. \rightarrow A. \rightarrow B.$ within the major loop $A. \rightarrow B. \rightarrow C. \rightarrow A.$ Since this would require inefficient use of computer time, the inner loop $B. \rightarrow C. \rightarrow B.$ is eliminated by substituting for $C.$ in the component balance for $B.$ equation. This results in an expression for $B.$ involving only radical $A.$ and monomer concentrations $A$, $B$, and $C$.

$$B. = A. \frac{B (k_{CB} k_{CA} C + k_{AB}(k_{CB}B + k_{CA}A))}{A (k_{AC} k_{CA} C + k_{BA}(k_{CB}B + k_{CA}A))} \tag{6-59}$$

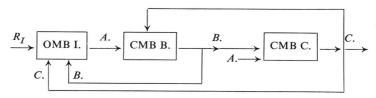

FIG. 6-19. Radial Balance Information Flow

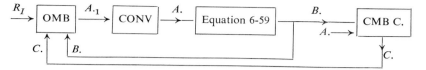

FIG. 6-20. Radial Balance Reduction of Implicit Loops

With this scheme, it is assumed that a value for $A.$ will provide explicitly a value for $B.$, then $C. A.$ can be recalculated and compared with the initial value assumed, and if not within a specified tolerance, a new value for $A.$ is assumed and the cycle repeated (Figure 6-20). The digital program for this section, "radical balance," is shown in the REACT Subroutine, Figure 6-21, lines 15 → 22, where the CONV subroutine is called to perform the necessary convergence. The initial value of $A.$ (AD) is supplied as input data (line 6).

### 6-5-3 Monomer Balances

The reaction rate of each monomer will depend on the monomer and radical concentrations. Considering only monomer $A$, it can react with chain

```
 1*            SUBROUTINE REACT(VA,VB,VC,VI,VS,RA,RB,RC,RI)
 2*            REAL KAA,KBB,KCC,KAB,KAC,KBC,KBA,KCA,KCB
 3*            REAL KTAA,KTBB,KTCC,KTAB,KTAC,KTBC,KI
 4*            DATA KAA/3.E5/KBB/6.E5/KCC/2.5E5/KAB/5.E5/
 5*            DATA KAC/4.E5/KBC/3.5E5/KBA/7.E5/KCA/5.5E5/KCB/4.5E5/
 6*            DATA KTAA/6.E7/KTBB/4.E7/KTCC/8.E7/AD/2.E-8/
 7*            DATA KTAB/5.E7/KTAC/4.E7/KTBC/7.E7/KI/6.E-3/
 8*    C       CATALYST REACTION RATE
 9*            RI = KI*VI
10*    C       MONOMER COMPOSITION
11*            V = VS + VA*.0416 + VB*.0776+VC*.1
12*            A = VA/V
13*            B = VB/V
14*            C = VC/V
15*    C       RADICAL BALANCE
16*          4 BDD = KAC*C*KCA+KBA*(KCB*B+KCA*A)
17*            BD = AD/A*B*(KCB*C*KAC+KAB*KCB*B+KAB*KCA*A)/BDD
18*            CD = C*(KBC*BD+KAC*AD)/(KCB*B+KCA*A)
19*            ADG = BD*(KTBC*CD+KTBB*BD)+CD**2*KTCC-RI/V
20*            ADB = KTAB*BD+KTAC*CD
21*            AD1 = (-ADB+SQRT(ADB**2-4.*KTAA*ADG))/(2.*KTAA)
22*            CALL CONV(AD,AD1,1,NC)
23*            GO TO (5,4),NC
24*    C       REACTION RATES
25*          5 RA = VA*(KAA*AD+KBA*BD+KCA*CD)
26*            RB = VB*(KBB*BD+KAB*AD+KCB*CD)
27*            RC = VC*(KCC*CD+KAC*AD+KBC*BD)
28*            RETURN
29*            END
```

FIG. 6-21. REACT Subroutine for Polymer Kinetics

radical $A.$, $B.$, or $C.$; thus its reaction rate $RA$ can be stated as

$$RA = V * A * (k_{AA}A. + k_{BA}B. + k_{CA}C.) \tag{6-60}$$

where $RA$ = moles A reacting/min,

$\quad V$ = volume of reactor liters,

$\quad A$ = monomer concentration moles/liter,

$A.$, $B.$, $C.$ = chain radical concentrations moles/liter,

$\quad k_{JK}$ = propagation constants.

Similarly for the other two monomers we have

$$RB = V * B * (k_{BB}B. + k_{AB}A. + k_{CB}C.) \tag{6-61}$$

$$RC = V * C * (k_{CC}C. + k_{AC}A. + k_{BC}B.) \tag{6-62}$$

The rate of change of monomer in the reactor will be the difference between the feed rate and the reaction rate, that is,

$$\frac{d}{dt}(VA) = FA - RA \tag{6-63}$$

$$\frac{d}{dt}(VB) = FB - RB \tag{6-64}$$

$$\frac{d}{dt}(VC) = FC - RC \tag{6-65}$$

where $V$ is the volume of the fluid phase, that is, excluding the volume occupied by the solid polymer particles.

$$V = VS + VA * MAR + VB * MBR + VC * MCR \tag{6-66}$$

The parameters $MIR$ in the above equation are density factors having units of liters/mole. $FJ$'s are the feed rates of each monomer $J$ in moles/min.

The rate of formation of polymer $P_w$ in weight units will be the sum of the monomer reaction rates times their respective monomer molecular weights:

$$\frac{dP_w}{dt} = R_A M_A + R_B M_B + R_C M_C \tag{6-67}$$

These equations are shown in the main program, Figure 6-22, which also shows the equations for the instantaneous polymer composition IJP (lines 21–23) and are the compositions of the polymer forming as a function of time. They are to be distinguished from the average polymer composition (AJP), which is the individual monomer in the polymer ratioed to the total

```
1*      C *** INITIATION AND DATA SECTION***
2*            REAL IAP,IBP,ICP,MMP
3*            DIMENSION FAT(4),FBT(4),FCT(4),FAF(4),FBF(4),FCF(4)
4*            DATA FAT/0.,60.,100.,140./FBT/0.,40.,70.,100./
5*            DATA FCT/0.,30.,70.,100./
6*            DATA FAF/7.,9.,9.,0./FBF/2.,4.,40.,0./FCF/1.,2.,2.,0./
7*            DATA VS/1000./VA/700./VB/200./VC/100./VI/.01/
8*      C ***DERIVATIVE SECTION***
9*          7 CALL REACT(VA,VB,VC,VI,VS,RA,RB,RC,RI)
10*     C      FEED RATES
11*            FA = FUN1(T,4,FAT,FAF)
12*            FB = FUN1(T,4,FBT,FBF)
13*            FC = FUN1(T,4,FCT,FCF)
14*            FI = 0.
15*            DAV = FA - RA
16*            DBV = FB - RB
17*            DCV = FC - RC
18*            DVI = FI - RI
19*            DPW = RA*50.+RB*70.+RC*110.
20*            SR  = RA + RB + RC
21*            IAP = RA/SR
22*            IBP = RB/SR
23*            ICP = RC/SR
24*            DP  = (RA+RB*RC)/RI
25*            CALL PRNTF(10.,400.,NF,T,IAP,IBP,ICP,AAP,ABP,ACP,RA,RB,RC)
26*            IF(NF .EQ. 2) STOP
27*     C ***INTEGRATION SECTION***
28*            CALL INTI(T,.5,2)
29*            CALL INT(VA,DAV)
30*            CALL INT(VB,DVB)
31*            CALL INT(VC,DCV)
32*            CALL INT(VI,DVI)
33*            CALL INT(TMAP,RA)
34*            CALL INT(TMBP,RB)
35*            CALL INT(TMCP,RC)
36*            CALL INT(PW,DPW)
37*     C   AVERAGE POLYMER COMPOSITIONS
38*            MMP = TMAP + TMBP + TMCP
39*            AAP = TMAP/MMP
40*            ABP = TMBP/MMP
41*            ACP = TMCP/MMP
42*            GO TO 7
43*            END
```

FIG. 6-22.  Main Program for Polymer Kinetics

moles of all the monomer in the polymer. Lines 38 → 41, Figure 6-22 show the program for the average polymer composition, where TMJP is the total moles of monomer J in the polymer, which is obtained by simply integrating the reaction rate RJ. The moles of monomer VJ in solution are obtained by integrating the derivative DVJ programmed in the derivative section. The monomer composition in solution J will be the total monomer left (VJ) divided by the volume of liquid phase V, that is, J = VJ/V. (lines 12–14 in subroutine REACT, Figure 6-21).

The degree of polymerization DP is a measure of the average number of moles in the polymer molecules, and it can be calculated as the ratio of the sum of the reaction rates to the initiation rate (line 24, Figure 6-22).

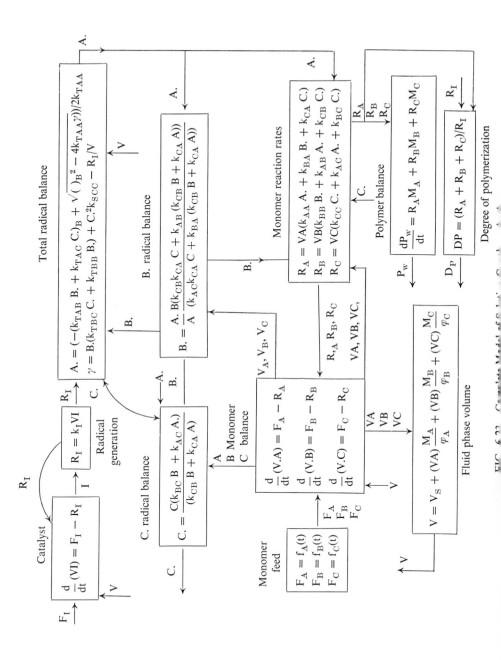

FIG. 6.22 *Complete Model of Solution Copolymerization*

All the equations comprising this kinetic model are shown in Figure 6-23 in signal flow form. The corresponding computer program consists of the main program (Figure 6-22) and the subroutine REACT (Figure 6-21). This subroutine is called on the first line of the derivative section and contains the following parts of the model; and the lines on which they appear:

1. Data for propagation and termination constants, 4 to 7
2. Catalyst balance, 9
3. Monomer balances and fluid phase volume, 11 to 14
4. Radical Balances (Figure 6-20), 16 to 22
5. Component reaction rates, 25 to 27

The argument list to the REACT subroutine contains the total moles in solution of each component and catalyst VA → VI and the solvent volume VS. The calculated reaction rates are the outputs from this subroutine and are the last four arguments in the list RA → RI.

The main program consists of the usual three sections, with an additional section at the end of the integration section. The data section contains the initial quantities of monomer and initiator (moles) and the total volume (VS, liters) on line 7. The continuous additions of monomer are specified as functions of time and are entered into arrays FAT, FBT, and so forth on lines 4, 5, and 6. These arrays are used in the arbitrary function generators to define the monomer feed rates on lines 11 → 13. The monomer and catalyst derivative are calculated on lines 15 to 18, and the instantaneous polymer compositions on lines 20 to 23. The degree of polymerization DP is calculated on line 24.

All the necessary integrations are performed on lines 28 → 36. Before returning to the first line of the derivative section (line 9, REACT) the average polymer compositions are calculated (line 28 → 41). These values cannot be calculated before the integration because on the first pass the total moles of polymer MMP is zero, and since it is in the denominator, it would have caused a computer-error termination.

## 6-5-4 Results

The numerical results for the stated conditions are shown in Figure 6-24. The first column shows the time variable in 10-min increments up to 400 min. The next three columns show the instantaneous polymer composition, the next three, the average polymer composition. The last three columns show the monomer concentration in the solvent phase. It will be seen that there is a significant variation in the instantaneous polymer composition. Such a result is generally undesirable since it produces a polymer of poor uniformity. Clearly, then, the monomer feed functions (synthesized arbitrarily) need to be modified in order to maintain the instantaneous polymer composition

| TIME | IAP | IBP | ICP | AAP | ABP | ACP | RA | RB | RC |
|---|---|---|---|---|---|---|---|---|---|
| .00000 | .65569+00 | .25099+00 | .93316-01 | .00000 | .00000 | .00000 | .94184+01 | .36052+01 | .13404+01 |
| .10000+02 | .65090+00 | .25556+00 | .93540-01 | .65331+00 | .25349+00 | .93208-01 | .89408+01 | .35104+01 | .12849+01 |
| .20000+02 | .64624+00 | .25752+00 | .96245-01 | .65098+00 | .25550+00 | .93934-01 | .85723+01 | .34160+01 | .12767+01 |
| .30000+02 | .64180+00 | .25708+00 | .10112+00 | .64874+00 | .25586+00 | .95406-01 | .82932+01 | .33219+01 | .13066+01 |
| .40000+02 | .63844+00 | .25503+00 | .10655+00 | .64664+00 | .25593+00 | .97437-01 | .80796+01 | .32275+01 | .13482+01 |
| .50000+02 | .63682+00 | .25205+00 | .11113+00 | .64490+00 | .25544+00 | .99615-01 | .79141+01 | .31324+01 | .13810+01 |
| .60000+02 | .63674+00 | .24827+00 | .11499+00 | .64360+00 | .25466+00 | .10174+00 | .77896+01 | .30372+01 | .14068+01 |
| .70000+02 | .63755+00 | .24411+00 | .11834+00 | .64273+00 | .25352+00 | .10375+00 | .76869+01 | .29433+01 | .14268+01 |
| .80000+02 | .64017+00 | .24082+00 | .11900+00 | .64224+00 | .25222+00 | .10554+00 | .75772+01 | .28504+01 | .14085+01 |
| .90000+02 | .64589+00 | .23906+00 | .11505+00 | .64230+00 | .25095+00 | .10676+00 | .74504+01 | .27576+01 | .13271+01 |
| .10000+03 | .65451+00 | .23860+00 | .10690+00 | .64299+00 | .24984+00 | .10717+00 | .73100+01 | .26648+01 | .11959+01 |
| .11000+03 | .66159+00 | .24046+00 | .97954-01 | .64424+00 | .24899+00 | .10678+00 | .70941+01 | .25784+01 | .10503+01 |
| .12000+03 | .66312+00 | .24575+00 | .91132-01 | .64556+00 | .24855+00 | .10589+00 | .67547+01 | .25033+01 | .92829+00 |
| .13000+03 | .65918+00 | .25474+00 | .86081-01 | .64657+00 | .24863+00 | .10480+00 | .63103+01 | .24387+01 | .82406+00 |
| .14000+03 | .64929+00 | .26811+00 | .82605-01 | .64703+00 | .24931+00 | .10365+00 | .57753+01 | .23848+01 | .73474+00 |
| .15000+03 | .63593+00 | .28417+00 | .79908-01 | .64682+00 | .25062+00 | .10256+00 | .52306+01 | .23373+01 | .65726+00 |
| .16000+03 | .62221+00 | .30051+00 | .77274-01 | .64606+00 | .25243+00 | .10152+00 | .47439+01 | .22912+01 | .58916+00 |
| .17000+03 | .60818+00 | .31712+00 | .74702-01 | .64487+00 | .25461+00 | .10053+00 | .43083+01 | .22464+01 | .52918+00 |
| .18000+03 | .59386+00 | .33394+00 | .72195-01 | .64344+00 | .25707+00 | .99586-01 | .39175+01 | .22029+01 | .47624+00 |
| .19000+03 | .57929+00 | .35095+00 | .69754-01 | .64157+00 | .25975+00 | .98688-01 | .35662+01 | .21605+01 | .42941+00 |
| .20000+03 | .56451+00 | .36811+00 | .67380-01 | .63958+00 | .26259+00 | .97830-01 | .32497+01 | .21191+01 | .38789+00 |
| .21000+03 | .54955+00 | .38539+00 | .65075-01 | .63744+00 | .26555+00 | .97009-01 | .29640+01 | .20787+01 | .35100+00 |
| .22000+03 | .53441+00 | .40275+00 | .62840-01 | .63517+00 | .26860+00 | .96223-01 | .27057+01 | .20391+01 | .31816+00 |
| .23000+03 | .51918+00 | .42015+00 | .60674-01 | .63281+00 | .27172+00 | .95470-01 | .24717+01 | .20003+01 | .28886+00 |
| .24000+03 | .50386+00 | .43756+00 | .58579-01 | .63036+00 | .27489+00 | .94748-01 | .22594+01 | .19620+01 | .26267+00 |
| .25000+03 | .48851+00 | .45494+00 | .56554-01 | .62786+00 | .27808+00 | .94055-01 | .20663+01 | .19243+01 | .23921+00 |
| .26000+03 | .47315+00 | .47225+00 | .54599-01 | .62532+00 | .28129+00 | .93389-01 | .18906+01 | .18870+01 | .21816+00 |
| .27000+03 | .45782+00 | .48947+00 | .52715-01 | .62275+00 | .28450+00 | .92750-01 | .17303+01 | .18499+01 | .19924+00 |
| .28000+03 | .44256+00 | .50654+00 | .50900-01 | .62017+00 | .28770+00 | .92136-01 | .15841+01 | .18131+01 | .18219+00 |
| .29000+03 | .42741+00 | .52344+00 | .49153-01 | .61757+00 | .29088+00 | .91546-01 | .14506+01 | .17765+01 | .16682+00 |
| .30000+03 | .41240+00 | .54012+00 | .47475-01 | .61498+00 | .29404+00 | .90979-01 | .13285+01 | .17399+01 | .15293+00 |
| .31000+03 | .39758+00 | .55656+00 | .45863-01 | .61240+00 | .29717+00 | .90434-01 | .12168+01 | .17033+01 | .14036+00 |
| .32000+03 | .38297+00 | .57271+00 | .44317-01 | .60983+00 | .30026+00 | .89910-01 | .11145+01 | .16668+01 | .12897+00 |
| .33000+03 | .36861+00 | .58855+00 | .42835-01 | .60728+00 | .30331+00 | .89906-01 | .10210+01 | .16302+01 | .11864+00 |
| .34000+03 | .35454+00 | .60404+00 | .41415-01 | .60477+00 | .30631+00 | .88922-01 | .93533+00 | .15936+01 | .10926+00 |
| .35000+03 | .34078+00 | .61916+00 | .40057-01 | .60228+00 | .30926+00 | .88456-01 | .85694+00 | .15569+01 | .10073+00 |
| .36000+03 | .32737+00 | .63387+00 | .38758-01 | .59983+00 | .31216+00 | .88007-01 | .78521+00 | .15204+01 | .92962-01 |
| .37000+03 | .31432+00 | .64816+00 | .37516-01 | .59742+00 | .31501+00 | .87576-01 | .71959+00 | .14839+01 | .85887-01 |
| .38000+03 | .30167+00 | .66200+00 | .36330-01 | .59505+00 | .31779+00 | .87162-01 | .65958+00 | .14475+01 | .79436-01 |
| .39000+03 | .28941+00 | .67539+00 | .35198-01 | .59273+00 | .32051+00 | .86763-01 | .60474+00 | .14112+01 | .73547-01 |
| .40000+03 | .27758+00 | .68830+00 | .34118-01 | .59045+00 | .32317+00 | .86379-01 | .55462+00 | .13753+01 | .68169-01 |

FIG. 6-24. Results of Trial Run on Solution Copolymerization Reaction

within tolerance for the entire conversion. This can be achieved by embarking on a series of runs, readjusting the monomer feed rates for each run. Since only a few seconds of computer time per run are required, this would provide a reasonable approach to the problem.

## 6-6  PARTICLE AGE DISTRIBUTION IN CSTR (Ref. 13)

One of the important reactions of industrial interest involves a solid and liquid phase continuously flowing through a vessel. The solid phase, generally in the form of particles, resides in the vessel for a period of time, reacts with the liquid phase and decreases in size, and finally leaves in the outflow. Such a situation cannot be characterized in a straightforward manner because the reacting particles do not all have the same residence times. Some particles exit the reactor almost immediately, while others remain for a much longer time. Since they are reacting and consequently decreasing in size, the reactor will contain a variety of different-sized particles. If the reaction rate of each individual particle is proportional to its surface area, the total reaction for the entire volume will be proportional to the sum of the areas for all the particles. To determine this, it is necessary to evaluate the age distribution of all the particles in the vessel. The next section provides the elementary concepts required to develop the model for this reaction vessel.

### 6-6-1  Age Distribution—Ideal Flow (Figure 6-25)

Suppose a quantity of particles $(N_i * V)$ is dumped into a well-agitated vessel at time $\theta = 0$. The volume of the contents of the vessel is $V$ and a continuous flow $S$ (ft³/min) is flowing in and out of the vessel. The concentration of particles ($N$ particles/unit volume) in the vessel and in the exit stream is established from the solution to an accumulation balance equation

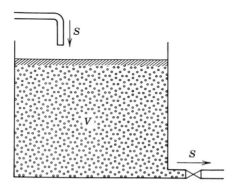

FIG. 6-25

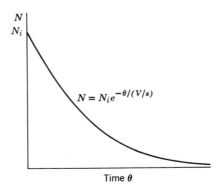

FIG. 6-26

on the particles, that is,

rate of change of particles in vessel = inflow − outflow

$$\frac{d}{d\theta}(N * V) = 0. - S * N \tag{6-68}$$

or, since $V$ = constant,

$$\frac{dN}{d\theta} = -\frac{S}{V} * N \tag{6-69}$$

The solution is (Figure 6-26)

$$N = N_i * e^{-\theta/(V/S)} \tag{6-70}$$

All the particles in the exit stream at any time $\theta$ will have the same age; that is, they will all have resided in the vessel for the same time $\theta$.

The total number of particles introduced initially is $N_iV$. The total number of particles that will have left up to any time $\theta$ will be

$$\int_0^\theta S.N. \, d\theta$$

substituting for $N$ (Equation 6-70) we have

$$\text{total particles that have left up to time } \theta = \int_0^\theta SN_i e^{-\theta/(V/S)} \, d\theta$$

Since the particles in the vessel at $\theta = V * N$, we have by integration:

initial particles − particles remaining = particles that have left
$$N_i * V - V * N = N_i * V[1. - e^{-\theta/(V/S)}] \tag{6-71}$$

The term in the bracket represents the fraction of the total initial particles that have left up to $\theta$. It is common to normalize these characteristics by defining the particles in the vessel as a fraction of the initial quantity, that is, $\bar{N} = N/N_i$, and by normalizing the time as $\bar{\theta} = \theta/H$ where $H = V/S$ the vessel holdup time. Applying these normalizing factors to equations 6-70 and 6-71 we obtain the normalized age distribution commonly known as the $E$ curve and the exit fraction curve, known as the $I$ curve (Figure 6-27). Now the $\int_0^\infty \bar{N}\, d\bar{\theta} = 1$; thus the area under the curve up to any time $\bar{\theta}$, represents the fraction of particles that have left up to that (normalized) time. This value is shown plotted as the $I$ curve.

It will be observed that both characteristic curves are independent of the initial quantity of particles $N_i$ introduced. Consequently, they are readily extended to the limiting case of an infinite number of infinitely small additions made over a period of time, that is, a continuous addition of particles. Since each infinitely small addition has the same age distribution, the characteristic curves also apply to a continuous flow of particles. They would be interpreted as follows. Suppose that the exit stream of a vessel, having a continuous flow of $N_i$ uniform particles/unit time, is sampled and the particles analyzed. It will be found that a few particles will be relatively "fresh," while a few will be quite "old," and all the others will lie between these two extremes. If the vessel is ideally mixed, then the age distribution curve (Figure 6-27) will characterize the exit particles. For example, the area under the curve up to $\bar{\theta}$, represents the fraction of particles that have resided in the vessel for less than $\bar{\theta}$ time. For practical purposes in such an ideal case, virtually all the particles will have normalized residence times less than 5. The ordinate of the $I$ curve (Figure 6-27), since it is $\int E\, d\bar{\theta}$, represents the fraction of particles with a residence time smaller than the corresponding value of $\bar{\theta}$ of the abscissa.

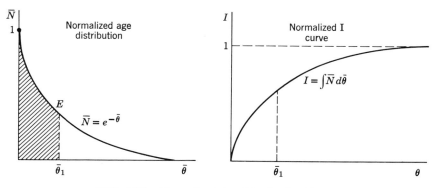

FIG. 6-27.  Normalized Age Distribution Curves

### 6-6-2  Nonideal Mixing

Unfortunately, most industrial processes involving dual-phase reactions are never ideally mixed. The ideal characteristic shown in Figure 6-26 is because the particles in the feed immediately achieve a uniform dispersion at the time of introduction. Since, in reality, a finite time period is required for adequate dispersion throughout the vessel, an ideal age distribution curve will not typify the age characteristics of the particles in the vessel and the exit stream. An age distribution curve for a nonideal vessel is usually determined experimentally by a variety of methods (Ref. 12), such as radioactive tracers, and could be as shown in Figure 6-28.

The most significant departure of this nonideal $E$ distribution from the ideal case is at low values of $\bar{\theta}$, where a rise to a maximum develops in the region of $\bar{\theta} = 1$. Figure 6-29 shows the corresponding residence fraction plot of $1 - I$ or $1 - \int E \, d\bar{\theta}$ for the age distribution shown in Figure 6-28.

This curve represents the fraction of the initial particles still residing in the vessel at time $\bar{\theta}$ and characterizes the nonideal mixing in a form that is

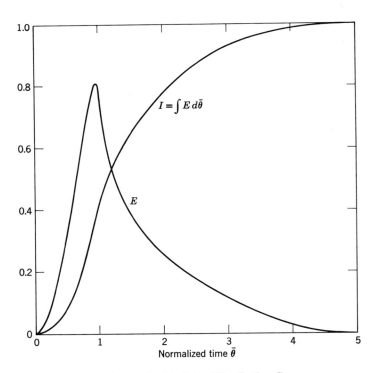

FIG. 6-28.  Nonideal Age Distribution Curves

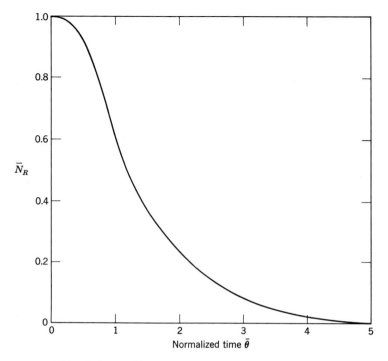

FIG. 6-29.   Residence Fraction for Nonideal Distribution

most convenient for use in a model of a reactor which is developed in the next section.

### 6-6-3   Heterogeneous Reaction in a CSTR

This example will demonstrate how a reaction in CSTR is characterized with the use of the age distribution relationships developed in the previous section.

Solid particles $P$ flow continuously at a rate of $M_{Pi}$ moles/min into a vessel (Figure 6-30) with a flow of solvent containing reagent $A$. The inlet concentration of $A$ in the solvent is $C_{Ai}$ (moles/ft³). The total fluid flow is $S$ ft³/min, and the holdup in the vessel will be a constant value of $V$ ft³. The particles react with reagent $A$ to form a soluble compound $B$, that is,

$$A + P \rightarrow B$$

The reaction rate per particle is proportional to the surface area and the

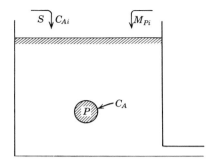

FIG. 6-30.   Heterogeneous Reactor

concentration of $A$ in the fluid phase $C_A$, that is,

$$R_p = kC_A(A_p) \qquad (6\text{-}72)$$

where  $R_p$ = reaction rate/particle (moles/min particle)

$A_p$ = particle area (ft²)

If the particle is assumed to be spherical with a radius $r$ (ft), then $A_p = 4\pi r^2$. The particle volume will be $\frac{4}{3}\pi r^3$. If the density is $\varphi$ moles/ft³, then the moles of solid/particle $P_p$ will be

$$P_p = \tfrac{4}{3}\pi r^3 \varphi \qquad (6\text{-}73)$$

rearranging, the radius $r$ will be

$$r = \sqrt[3]{3P_p/(4\pi\varphi)} \qquad (6\text{-}74)$$

Substituting for $r$ in the expression for the area we obtain the relationship between the area and the mass (moles) of the particle, that is,

$$A_p = (36\pi P_p{}^2/\varphi^2)^{\frac{1}{3}} \qquad (6\text{-}75)$$

It is convenient for the moment to consider the continuous feed of particles $M_{pi}$ (moles/min) as a discrete batch of $M_{pi}$ moles of particles dumped into the reactor at time $\bar{\theta} = 0$, where $\bar{\theta} = \theta/(V/S)$. The number of particles this represents will be

$$N_i = \frac{M_{pi}}{\tfrac{4}{3}\pi r_i{}^3 \varphi} \qquad (6\text{-}76)$$

where $r_i$ is the radius (ft) of the feed particles (specified). The fraction of these initial particles remaining in the reactor at any time $\bar{\theta}$ after introduction will be the arbitrary relationship $\bar{N}_R(\bar{\theta})$ obtained from experimental data and shown in Figure 6-29. The reaction rate due to all the $N_i$ particles as a function of time $\bar{\theta}$ will be

$$R = R_p(\bar{N}_R(\theta) * N_i) \qquad (6\text{-}77)$$

The total reaction due to this rate will be

$$RT = \int_0^\infty R \, d\theta \approx \int_0^5 N_i R_p \bar{N}_R(\bar{\theta}) \, d\bar{\theta} \cdot \frac{V}{S} \qquad (6\text{-}78)$$

This will be the total reaction created by the $N_i$ particles, and since $N_i$ is the feed rate/min, then $RT$ will be the reaction rate (moles/min) in the vessel.

The mole balance on a particle will be

$$\frac{d(p_p)}{d\theta} = -R_p \qquad (6\text{-}79)$$

Since $\bar{\theta} = \theta/(V/S)$ the normalized equation will be

$$\frac{d(P_p)}{d\bar{\theta}} = -R_p\left(\frac{V}{S}\right) \qquad (6\text{-}80)$$

The steady-state mole balance on component $A$ is based on the total reaction rate obtained above, that is,

$$\text{flow in} = \text{flow out} + \text{reaction}$$
$$SC_{Ai} = SC_A + RT \qquad (6\text{-}81)$$

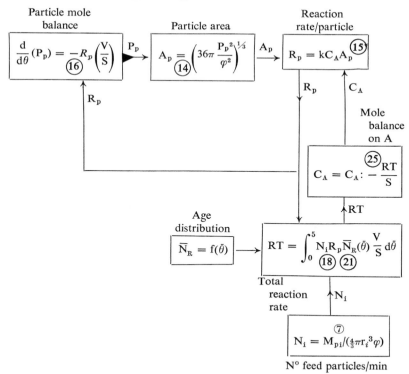

FIG. 6-31.  Information Flow Model of CSTR

Rearranging

$$C_A = C_{Ai} - \frac{RT}{S} \tag{6-82}$$

The equations developed from the analysis are shown in model form in Figure 6-31. It should be realized that the concentration of the $A$ component is a constant value. However, it is calculated from the total reaction $RT$. To calculate this total reaction, however, the constant value $C_A$ is required. Obviously, it will be necessary to conduct an iterative type of calculation in order to converge to the final values of $RT$ and $C_A$.

### 6-6-4   Computer Program

The numerical values for the parameters in the model are shown in the data statements in the program in Figure 6-32. The organization of the program is:

1. Lines 1 to 10    data section and calculations of constants.
2. Lines 11 to 13    initiation section. Integrated variables are set to their initial values at $\bar{\theta} = 0$.
3. Lines 14 to 18    derivative section.
4. Lines 20 to 22    integration section.

```
 1*     100 FORMAT(10E12.5)
 2*         DIMENSION AT(16),AN(16)
 3*         REAL MPI,K,NR
 4*         DATA AT/0.,.2,.4,.55,.75,1.,1.2,1.4,1.7,2.,2.5,3.,3.5,4.,4.5,5./
 5*         DATA AN/1.,.99,.96,.9,.8,.6,.5,.4,.3,.24,.14,.08,.04,.02,.01,0./
 6*         DATA MPI/2./RI/.005/RO/5./V/50./S/5./CAI/.5/K/.002/
 7*         NI=MPI/(4./3.*3.14*RI**3*RO)
 8*         PI=MPI/NI
 9*         VOS=V/S
10*         CA=CAI/2.
11*       5 TH=0.
12*         PP=PI
13*         RT=0.
14*       7 AP=(36.*3.14*PP**2/RO**2)**(1./3.)
15*         RP=K*CA*AP
16*         DPP=-RP*VOS
17*         NR=FUN1(TH,16,AT,AN)
18*         DRT=NI*RP*NR*VOS
19*         IF(TH.GE.5.) GO TO 6
20*         CALL INTI(TH,.01,1)
21*         CALL INT(RT,DRT)
22*         CALL INT(PP,DPP)
23*         IF(PP.LT.0.) PP=0.
24*         GO TO 7
25*       6 CAC=CAI-RT/S
26*         CALL CONV(CA,CAC,1,NC)
27*         GO TO (8,5),NC
28*       8 CNV=RT/MPI*100.
29*         PRINT 100 RT,CA,CNV
30*         STOP
31*         END
```

FIG. 6-32.   Main Program-Heterogeneous CSTR Reactor

The integration is performed for the interval $\bar{\theta}$ (TH) going from 0 to 5, by which time the final value for the total reaction RT has been achieved. At this time ($\bar{\theta} = 5$) the A concentration $C_A$ is calculated on line 25, and the CONV subroutine estimates a new value of $C_A$ if CAC is not close to the value of $C_A$ used for the integration. With this new value of $C_A$, the calculation recycles through the initiation section (lines 11–13), and the integration is repeated. This process continues until $C_A$ has achieved a converged limit, at which time it proceeds to address 8, where the conversion (CNV) is calculated and the total rate, $C_A$ and conversion are printed as the principal result of interest. For this run the results were:

$$RT = 1.1981 \text{ moles/min}$$
$$C_A = 0.26037 \text{ moles/ft}^3$$
$$CNV = 59.907\%$$

## Problems

1. The following reaction takes place in an isothermal reactor:

$$A + B \underset{k_2}{\overset{k_1}{\rightleftharpoons}} C + D$$
$$C + B \xrightarrow{k_3} E + F$$

Starting with 0.5 moles/ft³ of $A$ and 0.7 moles/ft³ of $B$ in a solvent, determine the time required to achieve 80% conversion given

$$k_1 = 15,600 \quad (\text{ft}^3/\text{mole min})$$
$$k_2 = 5.2 * 10^6 \quad (\text{ft}^3/\text{mole min})$$
$$k_3 = 1.2 \quad (\text{ft}^3/\text{mole min})$$

Assume the rate expressions correspond to the stoichiometric relationships.

2. The autoclave for the heterogeneous reactions of Example 6-4 has a maximum pressure rating of 10 atm. Determine the maximum charge of component $A$ that can be accommodated with this restriction.

3. In the polymer kinetics, Example 6-5, determine the optimum initial charge and continuous feeds of each monomer required to achieve a uniform polymer product having the following average compositions.

$$\text{Monomer } A = 0.6 \text{ mole fraction}$$
$$B = 0.3 \text{ mole fraction}$$
$$C = 0.1 \text{ mole fraction}$$

4. In the heterogeneous reactor Example 6-6-3 determine if it is possible to achieve a higher conversion for this reaction by using two vessels in series instead of the single vessel as in the example. Assume that the sum of the

volumes of the vessels equals the volume of the vessel in the example. (For an extension of the age distribution theory to multiple vessels in series see Ref. 13.)

## REFERENCES

The following references are suitable for more extensive reading on computer studies on chemical kinetics.

1. "Analog Simulation of Chemical Reaction Kinetics," T. Mathews, *Chem. Eng.*, 1964.
2. "Analog Computer Design of an Ethylene Glycol System," W. A. Parker and J. W. Prados, *Chem. Eng. Progress*, **60**, No. 6, 1964.
3. "Analog Computer Simulation of a Chemical Reactor," T. L. Batke, R. G. Franks, and E. W. James, *ISA J.*, **4**, 14–18, 1957.
4. "Process Development," T. J. Williams, *Chem. Eng.*, April 1960.
5. "Analog Methods Aid Simulation of Reaction Kinetics," W. F. Wagner, *Chem. Eng.*, April 1963.
6. "Programming Chemical Kinetics Problems for Analog Computers," R. C. H. Wheeler and G. F. Kinney, *IRE Trans. PGIE* IE3-70.
7. "Design of a Chemical Reactor by Dynamic Simulation," W. J. Dassau and G. H. Wolfgang, *Chem. Eng. Progr.*, **59**, No. 4, 1964.
8. "Simulating the Dynamics of Reaction Systems," T. Mathews, *Chem. Eng.*, August 1964.
9. "Analog Simulation of a Chemical Reactor Temperature Control System," F. X. Mayer and E. H. Spencer, *Proc. Inst. Aut. Conf.*, **15**, Part 2, 1960.
10. "Electronic Analogs in Reactor Design," J. Beutler and J. B. Roberts, *Chem. Eng. Progr.*, **52**, No. 2, 1957.
11. "Computer Simulation of Chemical Reactions," T. J. Williams, *Chem. Eng. News*, **40**, 1962.

### Textbooks

12. *Reaction Kinetics for Chemical Engineers*, S. M. Walas, McGraw-Hill, New York, 1959.
13. *Chemical Reaction Engineering*, O. Levenspiel, Wiley, New York, 1962.
14. *Chemical Process Principles III*, J. O. Hougen, K. M. Watson, Wiley, New York, 1947.

# CHAPTER VII

# FLUID FLOW

## GAS FLOW SYSTEMS (1)

The flow of gas through an orifice or restriction can be represented by the equation

$$Q = C_V \sqrt{\bar{P}(P_1 - P_2)}$$

where $P_1$ = upstream pressure

$P_2$ = downstream pressure

$\bar{P}$ = average absolute pressure across restriction

The capacity factor for the restriction, $C_V$, has units consistent with $Q$, the flow rate. The average pressure $\bar{P}$ can be assumed to be the arithmetic mean of $P_1$ and $P_2$ and reflects the density variations due to changes in pressure level. If the pressure drop $(P_1 - P_2)$ is small compared with the average pressure level $(P_1 + P_2)\frac{1}{2}$, it is convenient to use either $P_1$ or $P_2$ as an approximation for $\bar{P}$.

If $P_2 \leq 0.53\, P_1$, then critical flow is obtained. This means that if the downstream pressure is less than about half the upstream pressure, the gas will flow through the orifice (usually a valve) with sonic velocity, and the downstream pressure will have no influence on the flow rate. The relationship for critical flow at standard temperature is

$$Q = C_V P_1 0.85$$

If the dynamic flow variations cross the boundary between critical and subcritical flow, the following expression will define the flow at all times:

$$Q = \frac{C_V}{\sqrt{2}} \cdot P_1 \cdot F\left(\frac{P_2}{P_1}\right)$$

$$F\left(\frac{P_2}{P_1}\right) = \left[1 - \left(\frac{P_2}{P_1}\right)^2\right]^{\frac{1}{2}} \quad \text{for} \quad P_2 > 0.53 P_1$$

$$= \sqrt{0.72} = 0.85 \quad \text{for} \quad P_2 < 0.53 P_1$$

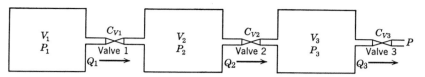

FIG. 7-1. Three-Volume Gas Flow System

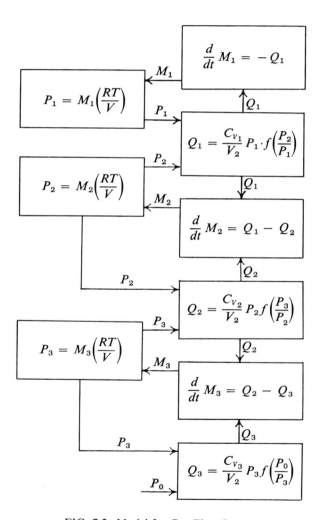

FIG. 7-2. Model for Gas Flow System

EXAMPLE 7-1

Suppose a process system contains three chambers with volumes $V_1$, $V_2$, and $V_3$ connected by a pipe and valves as in Figure 7-1. The first chamber is pumped to a high pressure $P_1$ with valve 1 closed. The pressure downstream of valve 3 is maintained at $P_0$, and both valves 2 and 3 are open and have capacity factors of $C_{V2}$ and $C_{V3}$. At time zero valve 1 is opened, and the gas in the first chamber is allowed to expand into chambers 2 and 3 and out through valve 3; to show this a mathematical model is required to relate the pressures and flows through the system.

If an ideal gas and isothermal operation is assumed, the pressure in each chamber is established by the ideal gas law

$$P_i = M_i\left(\frac{RT}{V}\right)$$

The moles of gas $M_i$ in each chamber $i$ is obtained by a mass balance equation

$$\frac{d}{dt} M_i = Q_{in} - Q_{out}$$

With the flow equation developed, the flows $Q$ are obtained from the pressure in each chamber; for example,

$$Q = \frac{C_V}{\sqrt{2}} \cdot P_U \cdot F\left(\frac{P_D}{P_U}\right)$$

Figure 7-2 shows the model, reflecting cause and effect, for the three-chamber system, the assembly of which is a straightforward process.

## HYDRAULIC TRANSIENTS

Complex problems dealing with the transport of fluids through pipes can be classified into two groups. In the first the pipes are of relatively short lengths and small diameter, and the interest is in determining the pressure at various points in the piping network. Problems of this type are based on the steady-state friction drop formulas; there are programs on digital computers available to compute the flow and pressure drops for extensive networks of piping (2, 3). The second class of problem contains those cases in which the momentum of the flowing body of liquid must be considered in the force balance as a significant term. The procedure for defining a set of equations in a cause-and-effect sequence for a complex, interconnected piping network may appear to be a formidable task; however, a simple, methodical approach combined with a grasp of the fundamentals invariably results in a logical mathematical model. In the following development the compressibility of fluids is neglected, and the so-called "rigid water column" analysis will be employed. To understand the meaning of this, consider the situation in the following example.

### EXAMPLE 7-2 (5–7)

In this case water is flowing in a pipeline connecting two reservoirs (Figure 7-3). The velocity of the water in the pipe is the same at all points, hence the term "rigid water column." The force exerted *on* the water in the pipe is the difference in pressure at either end of the column; that is, Force $= (H_1 - H_2) \cdot a\phi$, where $a =$ area of pipe. A friction force also acts against the direction of flow; using the Hazen-Williams formula, we can express this friction as

$$h_f = KL \frac{Q^{1.85}}{D^{4.87}} \cdot a \qquad \begin{aligned} D &= \text{pipe diameter} \\ K &= \text{friction factor} \\ Q &= \text{flow rate} \end{aligned}$$

If the positive direction is from left to right (Figure 7-3), this friction force can be redefined as

$$h_f = -KL \frac{Q \cdot |Q|^{0.85} a}{D^{4.87}}$$

That is, if $Q$, the flow rate, changes direction, then $h_f$ changes sign. The net force will be the sum of these two forces; that is,

$$\text{net force} = a\phi(H_1 - H_2) - KL \frac{Q|Q|^{0.85}}{D^{4.87}} \cdot a$$

This force will change the momentum of the column of water, that is,

$$\text{net force} = \frac{d}{dt}(\text{mass} \times \text{velocity}) = \frac{d}{dt}\left(\frac{La\phi}{g} \cdot v\right) = \frac{d}{dt}\left(\frac{L\phi Q}{g}\right)$$

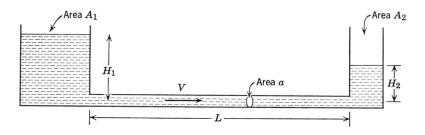

FIG. 7-3. Hydraulic Transient Between Two Reservoirs

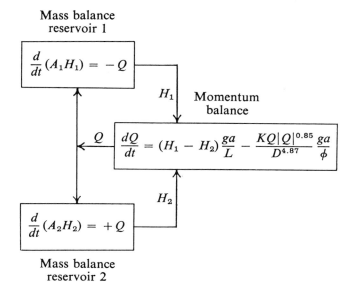

FIG. 7-4. Model for Reservoir

When these forces are equated, the dynamic equation describing the motion of the water in the pipe is

$$(H_1 - H_2)\phi - KL\frac{Q|Q|^{0.85}}{D^{4.87}} = \frac{d}{dt}\left(\frac{LQ\phi}{ga}\right)$$

which must be used in the following manner:

$$H_1 \longrightarrow \boxed{\frac{d}{dt}Q = (H_1 - H_2)\frac{ga}{L} - K\frac{Q\cdot|Q|^{0.85}}{D^{4.87}}\cdot\frac{ga}{\phi}} \longrightarrow Q$$
$$H_2 \longrightarrow$$

The difference in heads $H_1$ and $H_2$ causes the flow $Q$. The rest of the system is defined by the conservation equations around the reservoirs.

For Reservoir 1,

$$\frac{d}{dt}(A_1H_1) = -Q$$

For Reservoir 2,

$$\frac{d}{dt}(A_2H_2) = +Q$$

These three equations form a simple model (Figure 7-4) from which can be deduced the following generality, useful in many similar cases: "The

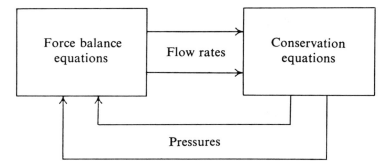

FIG. 7-5. General Model for Hydraulic Systems

force-balance equations establish the flow rates, and the conservation equations establish the accumulation, hence the hydraulic pressures that cause the flow." This concept is illustrated in Figure 7-5 and can be applied to the following situation:

EXAMPLE 7-3

A pump located several feet above a reservoir pumps water at a high velocity through a pipe of large diameter, up a hill, and down the other side (Figure 7-6). The pump fails, and a "break" in the water column occurs at the pump exit. The flow in leg $L_1$ continues for a while but gradually slows down, reverses direction, and accelerates back to the

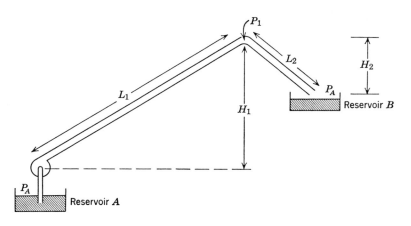

FIG. 7-6. Pumping System of Example 7-3

pump. To design a suitable protection system, the following information is required:

(a) Volume of water in leg $L_1$ returning to pump.
(b) Velocity immediately before impact with pump.

It can be assumed that when the pump fails the pressure at the pump will be $P_A$ (atmospheric pressure) and that at the other end of the pipe the water discharges into an open tank; hence the pressure at this point is also $P_A$. The pressure at the top of the hill is $P_1$. This pressure cannot fall below the vapor pressure of the water in the pipe (established by its temperature) $P_0$.

A mass balance on the $L_1$ leg is

$$\frac{d}{dt} M_1 = -Q_1\phi \qquad Q_1 > 0 \qquad M_1 = \text{mass of water in } L_1$$

$$Q_1 = \text{flow rate (volumetric)} \qquad (1)$$

$$\frac{dM_1}{dt} = 0 \qquad Q_1 < 0 \qquad \phi = \text{density}$$

When the flow rate reverses direction (sign convention: positive when flowing away from pump), a break occurs at the top of the hill (this assumes that $H_2 > 30'$), hence the need for the second equation

$$\frac{dM_1}{dt} = 0 \qquad Q_1 < 0$$

The force balance on the $L_1$ leg of water is

$$\text{pressure force} = (P_a - P_1)a \qquad a = \text{pipe area}$$

$$\text{friction force} = -Kl_1 \frac{Q_1|Q_1|^{0.85}}{D^{4.87}} \cdot a \qquad l_1 = \begin{array}{l}\text{length of water column} \\ \text{in } L_1 \text{ at any time}\end{array}$$

$$\text{gravity level} = -H_1\left(\frac{l_1}{L_1}\right)\phi a$$

$$\text{momentum} = \frac{d}{dt}\left(\frac{M_1 Q_1}{ga}\right)$$

Equating these forces results in the following equation:

$$\frac{d}{dt}\left(\frac{M_1 Q_1}{ga}\right) = (Pa - P_1)a - Kl_1 \frac{Q_1|Q_1|^{0.85}}{D^{4.87}} a - H_1\left(\frac{l_1}{L_1}\right)\cdot\phi a \qquad (2)$$

Because $P_1$ is a variable defining the period in which both legs $L_1$ and $L_2$

are joined together, another set of equations is required to define the second leg $L_2$.

$$\text{pressure force} = (P_1 - P_a)a$$

$$\text{friction force} = -Kl_2 \frac{Q_2|Q_2|^{0.85}}{D^{4.87}} a$$

$$\text{gravity level} = -H_2\left(\frac{l_2}{L_2}\right)\phi a$$

$$\text{momentum} = \frac{d}{dt}\left(\frac{M_2 Q_2}{ga}\right)$$

After assembling the terms

$$\frac{d}{dt}\left(\frac{M_2 Q_2}{ga}\right) = (P_1 - P_a)a - Kl_2 \frac{Q_2|Q_2|^{0.85}}{D^{4.87}} a - H_2\left(\frac{l_2}{L_2}\right)\phi a \qquad (3)$$

the mass balance on leg $L_2$ is

$$\frac{d}{dt} M_2 = \phi(Q_1 - Q_2) \quad \text{for} \quad \begin{matrix} Q_1 > 0 \\ Q_2 > 0 \end{matrix}$$

$$\frac{d}{dt} M_2 = -Q_2\phi \quad \text{for} \quad \begin{matrix} Q_2 > 0 \\ Q_1 < 0 \end{matrix} \qquad (4)$$

Assembling equations (1)–(4) into a model presents a major difficulty, for although there are only four equations there are five variables ($Q_1$, $Q_2$, $M_1$, $M_2$, $P_1$). A new relationship must be defined to establish the value of $P_1$, and this is obtained by considering the phenomena involved. If the pressure at $P_1$ is greater than the minimum value (vapor pressure of water at operating temperature), the water column is continuous and the velocities of both legs are the same; that is, $Q_1 = Q_2$. In other words, because water is virtually incompressible, $P_1$ is automatically forced to a value that will maintain continuity in the water column. This can be stated symbolically as

$$Q_1 \longrightarrow \boxed{Q_1 = Q_2} \longrightarrow P_1$$
$$Q_2 \longrightarrow$$

In this illustration $P_1$ "acts" in both force-balance equations, so that computationally a value of $P_1$ can be found iteratively that will achieve the condition $Q_1 = Q_2$. Essentially, this reproduces the force-balance mechanism actually occurring in the pipe and is similar to the multicomponent boiling temperature loop of Figure 5-2.

Unfortunately, such a scheme cannot be implemented on a computer because an algebraic loop iteration in this case does not exist. Iterating on $P_1$ requires recomputing the equations that lead back to $Q_1$ and $Q_2$, thus eventually satisfying the iteration criterion $Q_1 = Q_2$. These equations, however, are differential equations and are computed for each time interval so that $Q_1$ and $Q_2$ are fixed values at each increment. It must be realized, then, that iterations at each time increment to satisfy a particular criterion, must have at least one feedback loop back to the input of the criterion that is purely algebraic.

To accomplish this function it is necessary to convert the equation to a differential equation, that is,

$$\frac{dP_1}{dt} = G(Q_1 - Q_2)$$

The value of $G$ can be chosen arbitrarily to maintain $Q_1 \approx Q_2$; however, it should not be so large that excessive computer time is incurred (large $G$ forces the program to smaller step sizes).

The technique is similar to that used on an analog computer in which a high-gain amplifier is used with a small condenser in the feedback (Figure 7-7) to convert the amplifier into a high-gain integrator. Water has some compressibility and therefore the differential equation is a closer parallel to the natural mechanism than to the algebraic criterion. The equation can now be assembled into the consistent model shown in Figure 7-8.

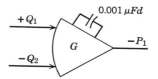

FIG. 7-7. Analog Computer Mechanization of $dP_1/dt = G(Q_1 - Q_2)$

Solution of this model requires an initial value of $Q_1$ ($= Q_2$), and a computer mechanization of this model automatically iterates to achieve the equilibrium value of $P_1$. This would be the value during steady-state operation of the pump and therefore at the instant the pump fails. From this point on $P_1$ drops gradually and finally reaches $P_0$, at which point $Q_1$ ceases to equal $Q_2$, and a cavity is formed at $P_1$.

This arrangement (Figure 7-8), although comprehensible and mathematically stable, requires unnecessarily long digital computer runs because of the large value of $G$ in the pressure equilibrium equation and small integrating step sizes. If this requirement becomes troublesome, it can be avoided by a modified treatment. Although $P_1 = P_0$, both legs $l_1$ and $l_2$ can be considered as a single mass described by one mass balance equation and one force balance equation; $P_1$ is derived from the force balance equation for leg 2, for $M_2$ is constant and $Q_2 = Q_1 = Q$ (obtained from total force balance equation). As soon as $P_1 = P_0$, the model switches back

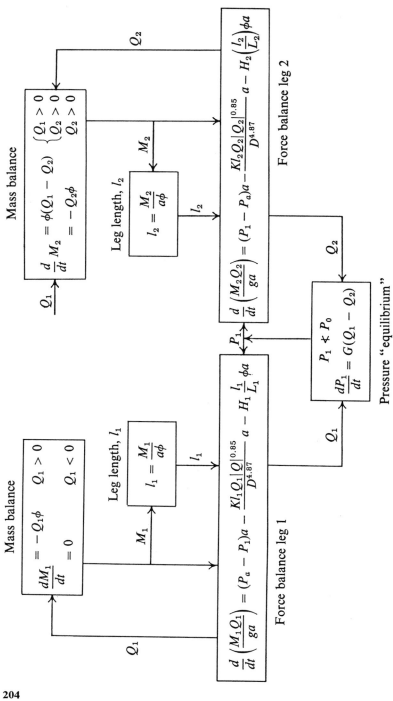

FIG. 7-8. Model for Pumping System Example 7-3

204

to that shown in Figure 7-8. In this way, the implicit method for obtaining $P_1$, as in Figure 7-8, is avoided. (For further refinements, see the exercise problem at the end of this chapter.)

EXAMPLE 7-4 (4)

Figure 7-9 shows a bank of pumps divided into two groups $R$ (running) and $T$ (tripped), pumping water through a pipe that branches into two legs $L_3$ and $L_2$. At a given time ($t = 0$) the $T$ group of pumps trip; that is, the driving power fails, and the pump rotor slows down and eventually reverses direction. The equation describing the tripped-pump rotation is

$$I \frac{d}{dt} N_{TP} = M_{TP}$$

where $N$ = rotation speed,
   $M$ = torque,
   $I$ = momentum of inertia of rotor.

The torque is a function of speed and flow through the pump.

$$M_{TP} = N_{TP}^2 \cdot f_2\left(\frac{Q_{TP}}{N_{TP}}\right) \quad \text{or} \quad Q_{TP}^2 \cdot f_4\left(\frac{N_{TP}}{Q_{TP}}\right)$$

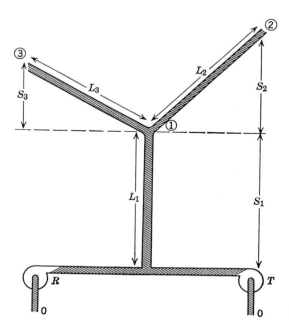

FIG. 7-9. Pumping System for Example 7-4

Because $N$ passes through zero, two equations are required; these functions are the hydraulic characteristics of the pumps. Each pump in the group of tripped pumps acts identically with the others.

The head across a tripped pump is a function of rotation speed and flow

$$H_P = N_{TP}{}^2 \cdot f_1\!\left(\frac{Q_{TP}}{N_{TP}}\right) \quad \text{or} \quad Q_{TP}{}^2 \cdot f_3\!\left(\frac{N_{TP}}{Q_{TP}}\right)$$

This head also exists across the running pumps. The flow rate through a running pump is a function of the head $H_P$:

$$Q_{RP} = f(H_P)$$

If we assume that there are $A$ running pumps and $B$ tripped pumps, the flow in $L_1$ will be

$$Q_1 = A \cdot Q_{RP} + B \cdot Q_{TP}$$

The continuity equation at the junction of $L_1$, $L_2$, and $L_3$ is

$$Q_1 = Q_2 + Q_3$$

If the discharge pressure of $L_2$ and $L_3$ and the suction pressure of all the pumps are at atmospheric pressure $H_0$, the equations for the momentum balance in each leg are as follows:

$$\frac{L_1}{ag}\frac{dQ_1}{dt} = -H_1 - H_0 + H_P + S_1 - K_1 Q_1|Q_1|^{0.85}$$

$$\frac{L_2}{ag}\frac{dQ_2}{dt} = (H_1 - H_0) - S_2 - K_2 Q_2|Q_2|^{0.85}$$

$$\frac{L_3}{ag}\frac{dQ_3}{dt} = (H_1 - H_0) - S_3 - K_3 Q_3|Q_3|^{0.85}$$

There are now eight equations and eight unknowns ($Q_1$, $Q_2$, $Q_3$, $N_{TP}$, $Q_{TP}$, $M_{TP}$, $H_P$, $H_1$), and there is a need to assemble the equations into a "reasonable" model. If we adhere to the general scheme shown in Figure 7-5, each of the momentum balance equations will establish $Q_i$, the flow rate involved. The inertia equation for the tripped pumps establishes the speed $N$ because the imposed torque controls the rate of speed change.

$$M_{TP} \longrightarrow \boxed{M_{TP} = I\,\frac{dN_{TP}}{dt}} \longrightarrow N_{TP}$$

The remaining four equations can be assigned as follows:
(a) Pump head

$$H_P = Q_{TP}{}^2 f_3\left(\frac{N_{TP}}{Q_{TP}}\right)$$

$$H_P = N_{TP}{}^2 f_1\left(\frac{Q_{TP}}{N_{TP}}\right)$$

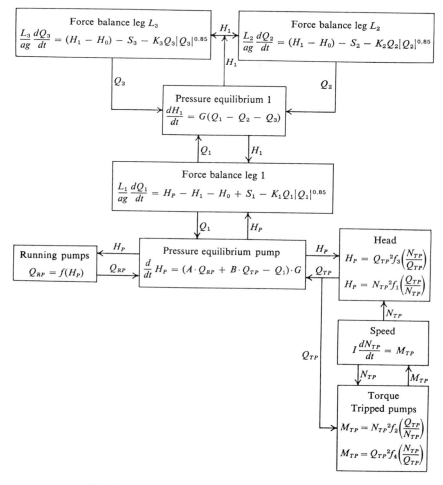

FIG. 7-10. Model for Pumping System Example 7-4

(b) Flow through running pumps

$$H_P \longrightarrow \boxed{Q_{RP} = f(H_P)} \longrightarrow Q_{RP}$$

The remaining two "continuity" equations

$$Q_1 = Q_2 + Q_3$$
$$Q_1 = A \cdot Q_{RP} + B \cdot Q_{TP}$$

are used to establish $H_1$ and $H_P$, respectively, for in the actual system the pressures automatically equilibrate to maintain continuity. The assembled model is shown in Figure 7-10.

Difficulties can occur in the solution of such a model, for although the pressures $H_1$ and $H_P$ can be obtained from the continuity equations by programming the high-gain differential equations the method suffers the same inconvenience of requiring excessive computer time. A rearrangement similar to the one carried out in Example 1-2 can be made by using the force equation for leg 1 to establish $H$, for $Q_1 = Q_2 + Q_3$ (and $dQ_1/dt = dQ_2/dt + dQ_3/dt$), and the running pump equation to establish $H_P$, $Q_{RP}$ being obtained by $Q_{RP} = (Q_1 - B \cdot Q_{TP})1/A$. Computer runs would involve solving the equations and plotting the variables of interest, such as flow rates, pump speed $N_{TP}$, and head $H_P$, as functions of time.

**Exercise Problems**

1. An open reservoir feeds water through a long pipeline to an enclosed vessel, compressing the gas space. Because the feed line has a large diameter, the momentum of the water is significant and will cause an overshoot beyond the equilibrium level, to compress the gas further. Construct a mathematical model that defines the transient pressure surge in the enclosed vessel (Figure 7-11).

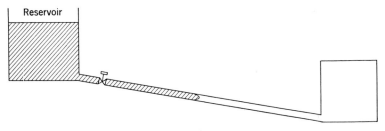

FIG. 7-11

Assume that a valve can be installed at the outlet of the reservoir to throttle the flow and prevent an overshoot with excessive pressure surges. How would this valve be included in the model?

2. A reactor is supplied with a fluid from a centrifugal pump located in a storage area (Figure 7-12a). Under certain operating conditions surging in the line connecting the pump to the reactor is suspected. A flow control maintains a constant flow from the reactor which also has an enclosed gas space above the liquid surface. The relationship between output pressure and flow for the pump is known (Figure 7-12b).

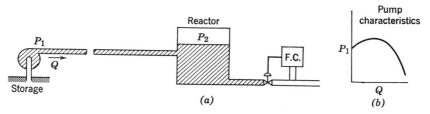

FIG. 7-12 a and b

Construct a model of the system, which, when simulated on a computer, will reveal the possibility that the surging is caused by the interaction of the pump and the inertance of the liquid in the feed line.

3. At the end of the reaction cycle in the autoclave (Figure 7-13) the vapor valve I closes and the liquid valve II opens to allow the hot fluid contents to flow from the autoclave to the separator. The autoclave vapor space contains only the vapor of the liquid contents at a high pressure. As the autoclave empties, the pressure falls, accompanied by a fall in temperature of the liquid. The separator volume is initially filled with an inert gas at a pressure below the autoclave pressure. The hot liquid passing through

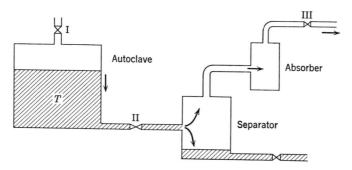

FIG. 7-13

valve II then flashes; a fraction turns into vapor and sweeps out the inert gas from the separator into the absorber, where all the vapor is absorbed into a solvent. The inert gas continues on and passes out through valve III. Assuming that the separator gas space is well mixed at all times and neglecting the liquid heel at the bottom of the separator, construct a mathematical model that will define the pressure surge in the absorber and the flow of gas through valve III. Assume negligible pressure drop through the separator and absorber and subcritical flows through the valves.

4. Molten metal is pumped continuously from a storage up to a reactor (Figure 7-14). The pressure in the reactor is $P_R$ and in the storage, $P_0$. At time $t = 0$ the pump fails, but because of the momentum of the flowing mass the flow rate decreases over a period of time. It is possible for a vacuum to occur at $D$ when the mass in leg $(b + L_2)$ reverses direction. When the mass in $L_1$ reverses direction because of $P_R$, it will eventually strike $E$ if the $(b + L_2)$ leg has emptied from the $b$ portion of the pipe. This could possibly cause a rupture. Assume that the column breaks at the reactor end and that pump resistance with stopped rotor is proportional to the square of the velocity $V$. Establish a mathematical model which, when solved on a computer, provides quantitative answers to the following problems:

1. What is the mass in $L_1$ that returns?
2. Where will the $(b + L_2)$ mass be when the $L_1$ mass strikes $E$?

*Data.*    $a$ = cross section of pipe
$\phi$ = density of liquid metal
$b, c, d$ = elevation as shown in Figure 7-14
$b, L_1, L_2$ = pipe lengths shown in Figure 7-14
$F_P$ = friction factor for pipe
$F_R$ = friction factor for pump rotor when stopped

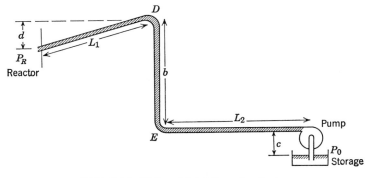

FIG. 7-14. Molten Metal Pumping System

## NOMENCLATURE

$A$    Number of running pumps (Example 7-4), reservoir area (Example 7-2)

$a$    pipe area (ft²)

$B$    number of tripped pumps

$C_V$    valve constant

$D$    pipe diameter (ft)

$G$    gain factor

$g$    acceleration of gravity (32 ft/sec²)

$H$    head (ft)

$h_f$    friction force

$I$    moment of inertia

$K$    friction factor

$L$    length of pipe (ft)

$l$    length of water column (ft)

$M$    gas vol. moles (Example 7-1), mass water lb (Example 7-3), torque (Example 7-4)

$N$    rotation speed

$P$    pressure (subscripts, $U$, upstream, $D$, downstream)

$Q$    flow rate mass/time

$S$    level (ft)

$t$    time

$v$    fluid velocity (ft/time)

$\phi$    density (lb/ft³)

## REFERENCES

1. "Mathematical Description of Gas Flow Processes for Control Purposes," G. V. Schwent and W. K. McGregor, *ISA J.*, August 1956.
2. "Evaluation of Pipeline Networks," R. J. Hunn, R. L. McIntire, and K. L. Austin, *Chem. Eng. Progr. Symp. Ser.*, **56**, No. 1, 1960.
3. "Digital Computer Solution of Gas Distribution System Network Flow Problems," D. V. Kniebes and G. G. Wilson, *Chem. Eng. Progr. Symp. Ser.*, **56**, No. 1, 1960.
4. *Analog Computer Solution of a Complex Transient Hydraulic Problem in the Power Industry*, E. H. Taylor, A. Reisman, E. C. Deland, and H. H. Baudistel, ASME Paper 60-WA-5, 1961.
5. *Pressure Surges Following Water Column Separation*, J. T. Kephart and K. Davis, ASME Paper 60-WA-120, 1961.
6. *Computer Representations of Engineering Systems Involving Fluid Transients*, F. D. Ezekiel and H. M. Paynter, M.I.T. H-3 P-10, 1956.
7. "Simulation of Plug Flow Systems," S. J. Ball, *Inst. Control Systems*, **36**,
8. "Model Simulation of Uni-Directional Fluid Dynamics," D. Childs, *Simulation*, **3**, No. 3, 1964.

# CHAPTER VIII

# STAGED OPERATIONS

Preceding chapters deal with systems of single, well-agitated vessels, completely defined by one set of algebraic/differential equations. Frequently in chemical processing, reactions or separations are carried out in a series of interconnected stages. The most common example of this is a distillation column in which vapor and liquid flowing countercurrently come into contact on a series of plates. This chapter will demonstrate that once a model has been formulated for a single stage, a series of identical stages is merely a repetition of the same number of models. Several additional subroutines will be added to the DYFLO library developed in Chapter 5. These routines are particularly suitable for simulating staged situations.

## 8-1  SUBROUTINE SPLIT  (J, K, M, RKJ)

This routine will be required whenever a stream J (Figure 8-1) splits into two streams (K and M) without changing total enthalpy and with the phase of the exit streams remaining the same as the phase of the feed stream. The split will be defined as the ratio of the flow in the K stream to the feed stream J, that is, RKJ. The routine is shown in Figure 8-2a, the items in the argument list being J, the feed stream number; K and M, the exit stream numbers; and RKJ, the flow ratio of K to J. Using this ratio, the routine

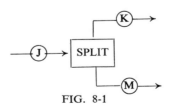

FIG.  8-1

212

```
 1 *        SUBROUTINE SPLIT(J,K,M,RKJ)
 2 *        COMMON/CD/STRM(300,24),DATA(20,10),RCT(22),NCF,NCL,LSTR
 3 *        STRM(K,21) = STRM(J,21) * RKJ
 4 *        STRM(M,21) = STRM(J,21) - STRM(K,21)
 5 *        STRM(M,22) = STRM(J,22)
 6 *        STRM(K,22) = STRM(J,22)
 7 *        STRM(K,23) = STRM(J,23)
 8 *        STRM(M,23) = STRM(J,23)
 9 *        DO 5 N = NCF,NCL
10 *        STRM(K,N) = STRM(J,N)
11 *      5 STRM(M,N) = STRM(J,N)
12 *        RETURN
13 *        END
```

(a)

FIG. 8-2a.  Listing for Subroutine SPLIT
Argument list:
J = feed stream number
K = exit stream number
M = exit stream number
RKJ = ratio K/J

establishes the flow rates of streams K and M (lines 3 and 4), then transfers to the K and M arrays all the temperature, enthalpy, and composition values from the feed stream J (lines 5–11).

## 8-2  SUBROUTINE SUM  (I, J, K, L)

Another useful package is the routine SUM that adds the contents of two streams I and J to create a third stream K. This procedure merely adds the flow rates (line 3, Figure 8-2b), then makes a component balance on each component (line 5), that is,

$$X_k = \frac{F_i X_i + F_J X_J}{F_k}$$

```
 1 *        SUBROUTINE SUM(I,J,K,L)
 2 *        COMMON/CD/STRM(300,24),DATA(20,10),RCT(22),NCF,NCL,LSTR
 3 *        STRM(K,21) = STRM(I,21) + STRM(J,21)
 4 *        DO 5 N = NCF,NCL
 5 *        STRM(K,N)=(STRM(I,21)*STRM(I,N)+STRM(J,21)*STRM(J,N))/STRM(K,21)
 6 *      5 CONTINUE
 7 *        ENIN=STRM(I,23)*STRM(I,21)+STRM(J,23)*STRM(J,21)
 8 *        STRM(K,23)=ENIN/STRM(K,21)
 9 *        CALL TEMP(K,L)
10 *        RETURN
11 *        END
```

(b)

FIG. 8-2b.  Listing for Subroutine SUM
Argument list:
I = input stream number
J = input stream number
K = output stream number
L = phase: 3 = liquid, 0 = vapor

where

$$F_i = \text{flow rate } i,$$

$$X_i = \text{composition in stream } i$$

The enthalpy of the exit stream K is calculated by adding the inlet enthalpy flows (line 6) and dividing by the total flow (line 7). Now that the stream molal enthalpy and composition have been established, the temperature is obtained by calling the subroutine TEMP for stream K. The fourth item in the argument list of SUM is L, specifying whether the exit stream K phase is liquid (L = 3) or vapor (L = 0).

### 8-3 EXAMPLE: COUNTERCURRENT EXTRACTION

The first example to be studied is a series of agitated vessels, each containing a settling section that allows two solvents to flow countercurrent to each other while mixing intimately at each stage (Figure 8-3). The purpose of this liquid-liquid extraction is to concentrate a solute from a concentration $X_{\alpha F}$ in solvent $\alpha$ to a concentration $X_{\beta 1}$ in solvent $\beta$ and to reduce the solute in solvent $\alpha$ to as low a concentration as possible. Each stage is numbered 1 to $N$; the general stage is $n$. This notation is shown as a subscript for the compositions (Figure 8-4).

In the general dynamic case being considered the flows and compositions at any stage may vary with time. Also, it is assumed that each stage achieves equilibrium between the two solvents according to the distribution coefficient $D$

$$D = \frac{X_{\beta n}}{X_{\alpha n}}$$

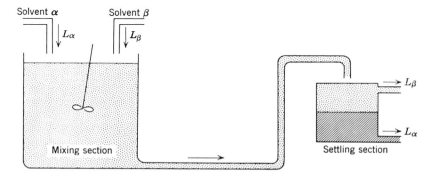

FIG. 8-3. Single Stage for Countercurrent Extraction

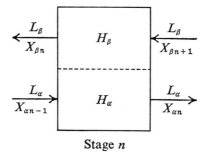

Stage $n$

FIG. 8-4. Extraction Stage $n$

Only two equations are required for this case; the first is a component mass balance for both phases:

rate of change of solute holdup $= $ [inflow] $-$ [outflow]

$$\frac{d}{dt}[(H_\alpha X_{\alpha n}) + (H_\beta X_{\beta n})] = (L_\alpha X_{\alpha n-1} + L_\beta X_{\beta n+1}) - (L_\alpha X_{\alpha n} + L_\beta X_{\beta n})$$

The second is the equilibrium equation

$$X_{\beta n} = DX_{\alpha n}$$

The component mass balance equation should be used to solve for either $X_{\alpha n}$ or $X_{\beta n}$, whichever is the larger. If they are the same order of magnitude, it does not matter which one is selected. In this case it is assumed to be $X_{\beta n}$. The equilibrium equation is then used to solve for the other variable $X_{\alpha n}$. The model for this stage $n$ is Figure 8-5. The overall signal flow diagram for the whole series of stages is shown in Figure 8-6.

For the simple, steady-state operation the differentials in the differential equation can be equated to zero and the equation reduces to

$$L_\alpha X_{\alpha n-1} + L_\beta X_{\beta n+1} = L_\alpha X_{\alpha n} + L_\beta X_{\beta n}$$

For $N$ stages of steady-state operation, the overall relationship of $X_{\alpha F}$ to $X_{\beta 1}$ can be established by manipulating the equations that result in a single expression involving the ratio of the solvent flows $L_\alpha/L_\beta$, the total number of stages $N$, and the distribution coefficient $D$. Such a regression to a simplified case is typical of the approach of conventional analysis. The technique of computerized model building, however, permits adding greater complexity to the system. For example, suppose that a component $C$ is produced by the reaction

$$A + B \leftrightarrows C + D$$

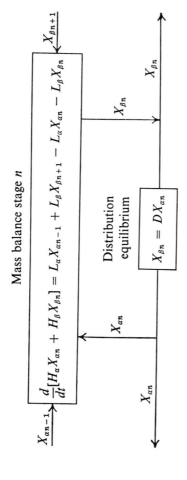

Mass balance stage $n$

$$X_{\alpha n-1} \xrightarrow{\hspace{1cm}} \frac{d}{dt}[H_\alpha X_{\alpha n} + H_\beta X_{\beta n}] = L_\alpha X_{\alpha n-1} + L_\beta X_{\beta n+1} - L_\alpha X_{\alpha n} - L_\beta X_{\beta n} \xleftarrow{X_{\beta n+1}}$$

Distribution
equilibrium

$$X_{\beta n} = D X_{\alpha n}$$

FIG. 8-5.   Model for Extraction Stage $n$

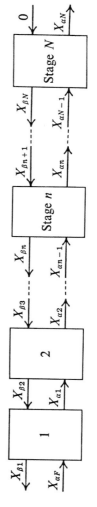

FIG. 8-6.   Signal Flow for Extraction Stage $1 \to n$

216

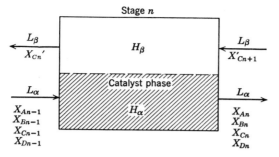

FIG. 8-7. Signal Flow for Reaction Stage $n$

and that this reaction takes place in solvent $\alpha$ (containing the catalyst). Of the four components in the system, only component $C$ is soluble in the $\beta$ solvent with a distribution coefficient

$$D = \frac{X'_{Cn}}{X_{Cn}}$$

where $X'_{Cn}$ is the mole fraction of $C$ in the $\beta$ phase and $X_{Cn}$ is the mole fraction in the $\alpha$ phase.

As in the preceding example, a model for a countercurrent staged system can be synthesized by considering a general section $n$ (Figure 8-7). The equations for this stage are as follows:

1. Reaction rate in phase $\alpha$

$$R = H_\alpha(k_F X_A X_B - k_R X_C X_D) \tag{8-1}$$

2. Mass balance on component $A$

$$\text{accumulation} = \text{inflow} - \text{outflow} - \text{reaction}$$

$$\frac{d}{dt}(H_\alpha X_A) = L_\alpha X_{An-1} - L_\alpha X_{An} - R \tag{8-2}$$

3. Mass balance on component $B$

$$\frac{d}{dt}(H_\alpha X_{Bn}) = L_\alpha X_{Bn-1} - L_\alpha X_{Bn} - R \tag{8-3}$$

4. Mass balance on component $C$ (both phases $\alpha$ and $\beta$)

$$\frac{d}{dt}[(H_\alpha X_{Cn}) + (H_\beta X'_{Cn})]$$

$$= L_\alpha X_{Cn-1} - L_\alpha X_{Cn} + L_\beta X'_{Cn+1} - L_\beta X'_{Cn} + R \tag{8-4}$$

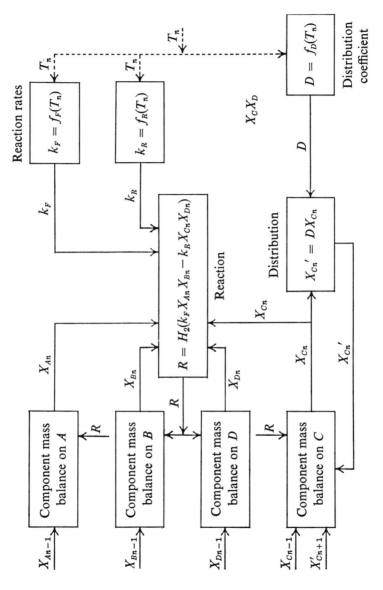

FIG. 8-8. Model for Reaction Stage $n$

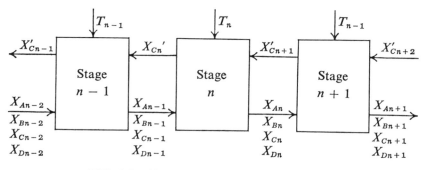

FIG. 8-9. Signal Flow for Stages $n - 1 \rightarrow n + 1$

5. Mass balance on component $D$

$$\frac{d}{dt}(H_\alpha X_{Dn}) = L_\alpha X_{Dn-1} - L_\alpha X_{Dn} + R \tag{8-5}$$

6. Equilibrium on component $C$

$$X'_{Cn} = DX_{Cn} \tag{8-6}$$

The arrangement for these equations is shown in model form in Figure 8-8. The arrangement is quite simple, for the component material balances are used to establish the concentrations. If the reaction rates $k_F$ and $k_R$ and the distribution coefficient $D$ are functions of temperature, the model shown in Figure 8-8, duplicated for $n$ stages and programmed on a computer, can be used to optimize by rapid trial and error the temperature conditions in each stage to give maximum yield (Figure 8-9).

## 8-4 COMPUTER PROGRAM

Using the DYFLO structure introduced in Chapter 5, the extraction process modeled in Section 8-3 can be programmed with a few modifications to the model. First, since the holdups $H_{\alpha,\beta}$ are constant, the component mass balances will be calculated as

$$\frac{d}{dt}(X_{in}) = \frac{L_\alpha(X_{in-1} - X_{in}) - R}{H_\alpha}$$

where $i$ applies to components $A$, $B$, and $D$. For component $C$, the following

```
 1*            SUBROUTINE REACT(JO,HA)
 2*            COMMON/CD/STRM(300,24),DATA(20,10),RCT(22),NCF,NCL,LSTR
 3*     C      DATA AR,BR,AF,BF/...................../
 4*            RT=1.98*STRM(JO,22)
 5*            RK=AR*EXP(BR/RT)
 6*            RF=AF*EXP(BF/RT)
 7*            R=HA*(FK*STRM(JO,1)*STRM(JO,2)-RK*STRM(JO,3)*STRM(JO,4))
 8*            RCT(1)=-R
 9*            RCT(2)=-R
10*            RCT(3)=R
11*            RCT(4)=R
12*            RETURN
13*            END
```

FIG. 8-10. Listing for Subroutine REACT

transformation is made by substituting equation 8-6 in equation 8-4.

$$\frac{d}{dt}(H_\alpha X_{Cn} + H_\beta X'_{Cn}) = \frac{d}{dt}(H_\alpha X_{Cn} + H_\beta DX_{Cn}) = (H_\alpha + DH_\beta)\frac{d}{dt}(X_{Cn})$$

$$\therefore \quad \frac{d}{dt}(X_{Cn}) = \frac{L_\alpha(X_{Cn-1} - X_{Cn}) + L_\beta(X'_{Cn+1} - X_{Cn}) + R}{H_\alpha + DH_\beta}$$

The program will consist of a main program calling on two subroutines, REACT and STGR. The REACT subroutine is shown in Figure 8-10. The input and output streams to stage $N$ are symbolized as shown in the Figure 8-11. The arguments to REACT are the number of the $\alpha$ phase streams

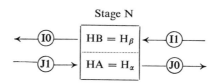

```
 1*            SUBROUTINE STGR(J1,I1,JO,IO,HA,HB)
 2*            COMMON/CD/STRM(300,24),DATA(20,10),RCT(22),NCF,NCL,LSTR
 3*            DIMENSION XA(10),YA(10)
 4*     C      DATA XA/............./YA/............./
 5*            D=FUN1(STRM(JO,22),10,XA,YA)
 6*            STRM(IO,4)=D*STRM(JO,4)
 7*            DO 5 J=1,3
 8*            DX=(STRM(J1,21)*(STRM(J1,J)-STRM(JO,J))+RCT(J))/HA
 9*          5 CALL INT(STRM(JO,J),DX)
10*            DX4=(STRM(J1,21)*(STRM(J1,4)-STRM(JO,4))+RCT(4)+
11*          1     STRM(I1,21)*(STRM(I1,4)-STRM(IO,4)))/(HA+D*HB)
12*            CALL INT(STRM(JO,4),DX4)
13*            RETURN
14*            END
```

FIG. 8-11. Listing for Subroutine STGR

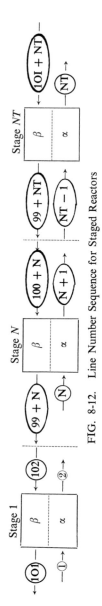

FIG. 8-12. Line Number Sequence for Staged Reactors

leaving section $N$, and the molar holdup $H_\alpha$ of the $\alpha$ phase. The temperature of the reaction phase is known and is placed in STRM(J0,22) in the main program. Starting with this temperature, the subprogram calculates the reaction coefficients $RK$ and $RF$, and the reaction rate $R$. It then places this rate with the appropriate sign in the RCT array for each of the components $1 \rightarrow 4$ corresponding to components $A$, $B$, $D$, $C$, respectively. This rate is used as a flow in the STGR subroutine, which is shown in Figure 8-11.

In the argument list for STGR, all four streams associated with stage $N$ are symbolized, as well as the molar holdup of both phases $HA$ and $HB$. The distribution coefficient $D$ is expressed as an arbitrary function of temperature (line 5). The composition $C$ in the $\beta$ phase (STRM(I0,4)) is calculated as the equilibrium value to the $\alpha$ phase composition (line 6). The component mass balance derivatives are formed on line 8 and integrated on line 9. The derivative for component $C$, since it is also present in the $\beta$ phase, has a different formulation and is calculated on line 10 and integrated on line 12.

Figure 8-12 shows the number sequence for the streams in the first, last ($NT$), and general stage $N$. It will be observed that the integration for each component is included in the subroutine for each stage. Now, in order to coordinate the integration sequence properly, it is necessary to list the program sequence in the direction opposite to the physical flow of information. Since most of the information is carried from stage to stage with the $\alpha$ phase flow as compositions, the calling sequence in the main program will be from stage $NT$ to stage 1. The $C$ composition in the $\beta$ phase travels with the program sequence. However, since this is calculated for each stage in STGR *before* the component mass balance is performed, it will also be correctly sequenced.

The main program for this process is shown in Figure 8-13. After completing the initial data section, the calculation proceeds to the print and

```
 1*    C** INITIATION SECTION**
 2*        COMMON/CD/STRM(300,24),DATA(20,10),RCT(22),NCF,NCL,LSTR
 3*        LOGICAL NF
 4*    C   DATA HA,HB,DT,NT,NCF,NCL/................./
 5*        DATA(STRM(J,21),J=1,11)/11*10./
 6*        DATA(STRM(J,22),J=1,11)/11*363./
 7*        GO TO 7
 8*    C** DERIVATIVE SECTION**
 9*      6 DO 5 N=NT,2,-1
10*        CALL REACT(N,HA)
11*      5 CALL STGR(N-1,100+N,N,99+N,HA,HB)
12*      7 CALL PRL(1.,10.,NF,NT,101,2,3,4,5,6,7,8,9,105)
13*        IF(NF) GO TO 8
14*    C** INEGRATION SECTION**
15*        CALL INTI(TIM,DT,4)
16*        GO TO 6
17*      8 STOP
18*        END
```

FIG. 8-13. Listing for Main Program of Countercurrent Staged Extractor

integration of the independent variable INTI that sets the pass and call counters for all the succeeding integrations buried in the STGR subroutines. It then proceeds to the derivative section, where the REACT and STGR subroutines are called sequentially from the last section NT to the first. The print routine PRL will print the compositions in the output flows NT and 1O1, as well as some of the intermediate flows. Actual numerical results and data are not provided for this example, since it has been used to demonstrate the programming techniques in the digital program.

## 8-5 DISTILLATION COLUMNS

Perhaps the most common unit process in the chemical and petroleum industry is the distillation column. It is also a typical example of a staged countercurrent separation system and, because of the nature of its operation, can sometimes be a great source of difficulty. The complexity of distillation columns can range from a simple binary separation with constant molal overflow to a multicomponent, nonideal separation with reaction on each stage and multiple feeds and sidestreams (Figure 8-14).

The approach adopted for defining the dynamic (as opposed to the steady-state) mathematical model for the column is to establish the model for a single stage and to duplicate it in the computer for all stages in the column, also incorporating any exceptional parts such as the feed plate, reboiler, and condenser. The following examples develop the dynamic equations for a general stage, starting with a simple binary and proceeding to the more complex case of multicomponent, nonideal distillation.

### Binary Distillation

The first approximation that is made in most simulations of distillation columns is to describe the column in terms of its theoretical stages. A 20-plate column with a plate efficiency of 70% can be adequately represented by a 14-theoretical equilibrium stage column, each stage having a holdup of $(1/0.7) \times$ holdup per tray in the actual column. Figure 8-15 is a diagrammatic representation of an equivalent theoretical tray, with the effluent vapor in equilibrium with the reflux leaving the stage, which for a binary system (for stage $n$) can be expressed as

$$Y_n = f(X_n)$$

The component material balance is

$$\frac{d}{dt}(HX_n) = V_{n-1}Y_{n-1} + L_{n+1}X_{n+1} - V_nY_n - L_nX_n$$

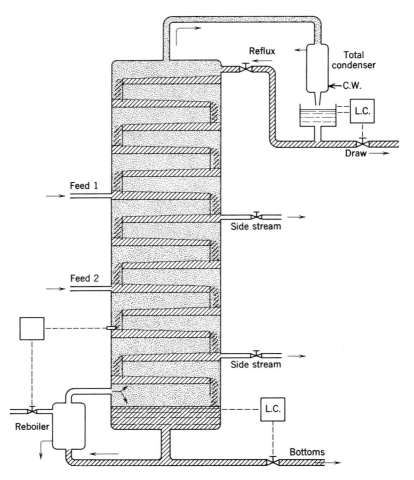

FIG. 8-14.   Distillation Column with Two Feeds and Two Sidestreams

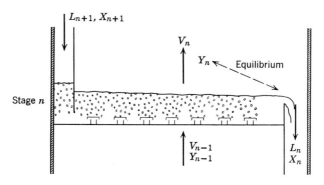

FIG. 8-15.   Equilibrium Stage for Distillation Column

With the assumption of constant molal overflow and negligible delay in the vapor flow, the following relationships hold:

$$V_{n-1} = V_n$$

$$\tau \frac{dL_n}{dt} = L_{n+1} - L_n.$$

The second differential equation represents the time delay that occurs between the reflux flowing onto the plate $(L_{n+1})$ and flowing off the plate $(L_n)$. The value of $\tau$, the hydraulic time constant, is obtained from the dimensions of the plate and weir. The equations defined above apply to each stage; the only exceptions necessary are the feed plate where the feed term $FX_F$ appears in the mass balance equation and the reboiler (see Figure 8-16) where the component mass balance becomes

rate of change of holdup = flow from stage 1 — bottoms flow — boilup

$$\frac{d}{dt}(HX_0) = L_1 X_1 - L_0 X_0 - V Y_0$$

$$\frac{dH}{dt} = L_1 - L_0 - V$$

boilup = heat flux latent heat

$$V = \frac{Q}{\lambda}$$

The last equation assumes a heat input of $Q$ to the reboiler. The model for this simple binary column is shown in Figure 8-16, with just one stage given in detail. Also, the value of the bottoms flow $L_0$ is determined as a boundary condition and is shown as an input in Figure 8-16.

## 8-6 MULTICOMPONENT SEPARATIONS

The more general case of distillation is the separation of multicomponent mixtures. For each component mass balance the equation has the same form as the one described in the preceding section; namely, for component $i$ on plate $n$

$$\frac{d}{dt}(_iX_n L_n) = L_{n+1} \cdot {}_iX_{n+1} + V_{n-1} \cdot {}_iY_{n-1} - L_n \cdot {}_iX_n - V_n \cdot {}_iY_n$$

The vapor-liquid composition relationship for each component is

$$_iY_n = \left[\frac{K_i(T_n)}{\pi_n}\right](_i\gamma_n \, _iX_n)$$

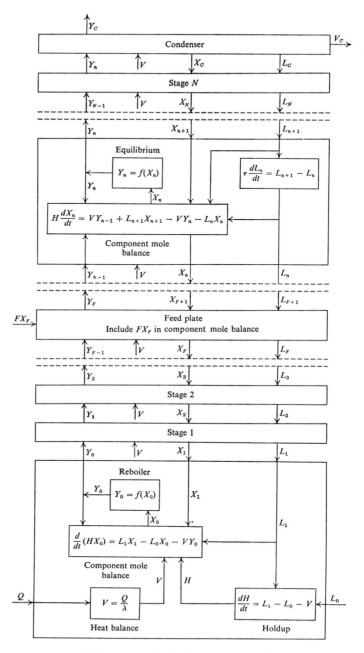

FIG. 8-16. Signal Flow for $N$ Column Stages

where $K_i(T_n)$ is the vapor pressure of the pure component, $\gamma$, the activity, and $\pi_n$, the total pressure on the plate.

This multicomponent system is treated just as described in Chapter 5; namely, the component mass balance equations yield the liquid compositions $_iX_n$, whereas the equilibrium equations are used to define the vapor compositions $_iY_n$ and the temperature on the plate, $T_n$. The model arrangement for this equilibrium system, as shown in Chapter 4, is adequate as long as the key component used to establish the temperature can be changed in different parts of the column. The more general approach, particularly suitable to distillation columns, is to establish the temperature from the balance $\sum Y = 1$, shown in model form in Figure 8-17 and described in Chapter 5.

The assumption of constant molal overflow, in which, under steady-state conditions, the mole flow of vapor onto the plate equals the vapor flow out and, similarly, the mole flow in the reflux onto the plate equals the reflux flow out (steady-state conditions), does not hold for some cases of distillation. For such cases a heat balance equation is required and is used to establish the vapor flow leaving the plate $V_n$. The overall mass balance should be used to establish the reflux flow $L_n$. The heat balance equation relates enthalpy in the ingoing and outgoing streams as follows:

$$\frac{d}{dt}(H_h T_n) = Q_{n-1} + q_{n+1} - Q_n - q_n$$

where $Q$ is the vapor enthalpy, $q$ is the liquid enthalpy, and $H_h$ is the heat capacity on the plate.

The accumulation term

$$\frac{d}{dt}(H_h T_n)$$

is negligible compared with the heat fluxes $Q$ through the plate, so that this term can usually be eliminated from the equation. The enthalpies of the vapor and liquid streams are defined in terms of the composition and temperature as follows:

$$Q = V \sum Q_i Y_i$$
$$Q_i = \alpha_i + \beta_i T$$

where $Q_i$ is the component enthalpy.

The complete model for a single stage for a four-component system is shown in Figure 8-17. The arrangement is typical of equilibrium systems; namely, the component mass balances establish the liquid composition and the total mass balance establishes the liquid flow (which is passed through a delay representing the approximation to the "hydraulic" dynamics of the stage). The vapor liquid equilibrium equations establish the vapor

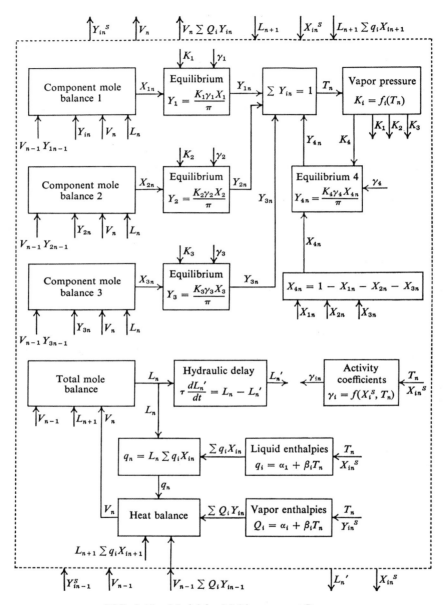

FIG. 8-17.  Model for Multicomponent Stage $n$

compositions, whereas the overall heat balance defines the vapor flow leaving the plate $V_n$.

## 8-7 GENERALIZED COLUMN PROGRAM

The foregoing discussion described the relations involved in multicomponent separations. We already have accumulated a series of subroutines that perform simple operations, such as flashing, heat exchange, and so on; now we will embark on creating several more routines that can be assembled to simulate a distillation column. The objective will be to develop a program capable of simulating the dynamic behavior of a column that, when at steady state, will naturally provide a profile of compositions, temperature, and flow rates through the column. Unfortunately, a simulation program that essentially mechanizes the model shown in Figure 8-17, although quite satisfactory for some situations, can be literally prohibitive for other cases that need to be studied. Consequently, a special feature is developed to overcome this latent difficulty in the next section.

## 8-8 STIFFNESS ASPECTS OF COLUMN SIMULATION

Previous work in the field of digital simulation of distillation column dynamics has revealed a frustrating tendency for the calculation to become unstable (Ref. 25). We know from the discussion in Chapter 3 that this is due to the presence in the system of differential equations having very small time constants compared with the dominant time constants of the system. Moreover, these time constants vary with the conditions within the column, which means that the program will be conditionally stable, a very unsatisfactory state of affairs. Naturally, a small integration step size can be specified that would accommodate any high stiffness situation, but this leads to very long running times. A more satisfactory solution follows.

The principal dynamic characteristics of a distillation column are due to the accumulation and balance of all the components on all the stages in the column. A general program is constructed so that each of the components is solved on each stage by integrating a differential equation, that is,

$$\frac{dX_n}{d\theta} = \frac{(V_{n-1}Y_{n-1} + L_{n+1}X_{n+1}) - V_nY_n - L_nX_n}{H}$$

where
H = holdup
X = liquid composition
Y = vapor composition

Since equilibrium is assumed between $Y_n$ and $X_n$, we can define an equilibrium ratio

$$\frac{Y_n}{X_n} = K_n$$

Substituting for $Y_n$ in the differential equation, we obtain

$$\frac{dX_n}{d\theta} = \frac{(V_{n-1}Y_{n-1} + L_{n+1}X_{n+1}) - (V_nK_n + L_n)X_n}{H}$$

This equation can be regrouped to have the form

$$\frac{dX_n}{d\theta} = \frac{1}{\tau}(F_i - X_n)$$

where the time constant $\tau$ will be

$$\tau = \frac{H}{V_nK_n + L_n}$$

This time constant establishes the critical step size for the differential equation. It contains the stage holdup $H$, the boilup $V_n$ and liquid flow $L_n$, items which are common to all the components in the column. It also contains the vapor/liquid composition ratio $K_n$, which is specific for each component. This $K$ can vary greatly, from values close to zero (high boilers, tars, etc.) to very large values for inerts and low boilers. A small $K$ makes $\tau \to H/L_n$ and is no problem, but a very high $K$ makes $\tau \to 0$, creating a very small time constant, which forces the programmer to specify a very small step size.

It is stressed in Chapter 3 that the proper approach to such stiff situations, is not to run the simulation with a tiny step size, but simply to eliminate these differential equations having small time constants and solve them as algebraic equations instead. In this situation, it will be accomplished as follows:

1. Calculate $K_i$ for component $i$ ($K_i = Y_i/X_i$ on stage $n$).
2. If $K_i < 5$ calculate the derivative and integrate.
3. If $K_i > 5$ solve for $Y_i$ algebraically and obtain $X_i$ by equilibrium. (The threshold ratio $K_i = 5$ is chosen arbitrarily.)

The algebraic solution for $Y_i$ is simply a steady-state component balance on the plate $n$, that is, (subscript $i$ omitted).

$$0 = \qquad \text{IN} \qquad - \qquad \text{OUT}$$
$$0 = (L_{n+1}X_{n+1} + V_{n+1}Y_{n+1}) - (L_nX_n + V_nY_n)$$

Substitute $K_n = Y_n/X_n$

$$Y_n\left(\frac{L_n}{K_n} + V_n\right) = L_{n+1}X_{n+1} + V_{n-1}Y_{n-1}$$

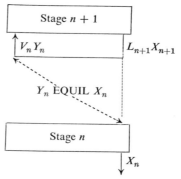

FIG. 8-18.  Flashing of
Volatile Component

and finally $Y_n = (L_{n+1}X_{n+1} + V_{n-1}Y_{n-1})/(L_n/K_n + V_n)$ followed by $X_n = Y_n/K_n$.

Physically, these equations say that in a case of a relatively volatile component, most of the flow from the plate above $(L_{n+1}X_{n+1})$ flashes and returns to stage $n+1$ in the vapor $(V_nY_n)$ as shown in Figure 8-18. The preceding explanation of this acceleration feature is necessary for an understanding of the structure of the following subroutines.

## 8-9  SUBROUTINE STAGE

This subroutine is essentially a program for the model of a multicomponent stage shown in Figure 8-17. For the sake of convenience the following modifications have been made to the basic equations.

1. Reaction flow RCT(n) is included in the component (RCT(n)), total (RCT(21)) and heat balance (RCT(22)) equations as was done for subroutine HLDP and VVBOIL in Chapter 5.

2. The total mole holdup HL is considered constant, but the imbalance in the input and output flows is accounted for in the component and heat balance equations

$$\frac{d(HL)}{d\theta} = DHL = FLIN - V_n - L_n \tag{8-9a}$$

$$\text{exit liquid flow } \frac{d}{d\theta}(L_n) = \frac{DHL}{HTC} \tag{8-9b}$$

where $FLIN = L_{n+1} + V_{n+1} + RCT(21)$, total inflow,
$HTC$ = hydraulic time constant.

### Heat Balance

Defining the input heat HIN as

$$HIN = (VH_v)_{n-1} + (Lh_L)_{n+1} + RCT(22) + H \qquad (8\text{-}9c)$$

where $H$ is an additional heat flux (if any) applied to the stage, a correct heat balance on stage $n$ will be:

$$\text{Accumulation} = \text{IN} - \text{OUT}$$

$$\frac{d}{d\theta}(HL \cdot h_L) = HIN - (VH_v + Lh_L)_N \qquad (8\text{-}9d)$$

where $HL$ = liquid holdup on stage (mole). Differentiating we obtain

$$HL\frac{dh_L}{d\theta} + h_L\frac{d(HL)}{d\theta} = HIN - (VH_v + Lh_L)_N \qquad (8\text{-}9e)$$

Now the change in holdup $d(HL)/d\theta = DHL$ and is defined in equation 8-9a. Substituting for $d(HL)/d\theta$ in 8-9e, neglecting the first term $HL\,dh_L/d\theta$, and rearranging terms, we obtain

$$V_n = \frac{HIN - FLIN * h_L}{H_v - h_L} \qquad (8\text{-}9f)$$

which is the equation programmed on line 9, in the subroutine in Figure 8-20. Neglecting the first term (i.e., sensible heat change) is justified, since the change in liquid enthalpy $dh_L/d\theta$ is usually small and inconvenient to obtain (requires differentiation, storage, etc.), whereas the second term $h_L \cdot d(HL)\,d\theta$ is not neglected, because even though small, it is not inconvenient to include in the balance. The case where the sensible heat change is not neglected is programmed as STGH in Section 8-13.

### Component Balance

Starting with the standard definition

$$\text{ACCUMULATION} = \text{IN} - \text{OUT}$$

that is,

$$\frac{d}{d\theta}(HL \cdot X_n) = (LX)_{n+1} + (VY)_{n-1} + RCT(N) - (LX)_n - (VY)_n \qquad (8\text{-}9g)$$

by differentiating, substituting for $d(HL)/d\theta$ (equation 8-9a), and regrouping, we obtain

$$\frac{dX_n}{d\theta} = \frac{(LX)_{n+1} + (VY)_{n-1} + RCT(N) - FLIN \cdot X_n - V_n(Y_n - X_n)}{HL}$$

$$(8\text{-}9h)$$

In the subroutine STAGE (Figure 8-20) the first three terms of equation 8-9h are summed as CNIN (line 14).

The above derivative is calculated for those components with an HK (= $Y/X$) ratio less than 5. For those components with HK > 5, the algebraic balance is calculated on lines 20 to 22, establishing the instantaneous value of the vapor and liquid composition. In order not to disturb the sequence for the integration counters, the calculation is returned to the integration routine INT, with the derivative (DERN) set to zero.

Figure 8-19 shows a brief information flow diagram for the subroutine that is listed in Figure 8-20. The items in the argument list of subroutine STAGE are:

I1, number of liquid stream from stage $n + 1$
I2, number of vapor stream from stage $n - 1$
IL, number of liquid stream leaving stage $n$
IV, number of vapor stream leaving stage $n$
H, external heat or cooling load
HL, holdup (mole)
HTC, hydraulic time constant (same time units as the flow rates)

The extra heat flux $H$ represents possible heating or cooling coils on the stage, other than reaction heat, which is accounted for by RCT(22).

The first operation is to establish the equilibrium composition and temperature in the vapor stream $IV$ for the compositions in the liquid stream $IL$. Then follow the liquid and vapor enthalpies (lines 5, 6) required for the heat balance (line 9). The total mass balance is then calculated (line 11), followed by the DO loop for the component mass balances. The integrations are the last operations before returning to the main program.

On the initial pass (LSTR = .TRUE.) only the equilibrium and vapor flow are calculated (lines 4 to 9).

## 8-10 FEED STAGE SUBROUTINE STGF

It will be realized that a feed stage can be readily simulated by adding a vapor feed to the vapor flow entering a stage from below (see Figure 8-21) or by adding a liquid feed to the liquid flow from the stage above. This addition can be made in either case by using the subroutine SUM developed earlier in this chapter. Unfortunately, in using this approach a rather cumbersome line-numbering system results. It is more elegant to use a modified version of the routine STAGE developed in the previous section, called STGF. Figure 8-22 shows a listing for STGF, together with a definition of the items in the argument list, which is the same as for STAGE but now contains the feed flow I3. This feed is included in the mass balance (line 7),

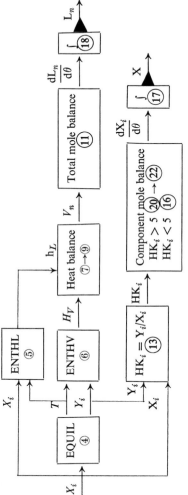

FIG. 8-19. Information Flow for Stage Subroutine

```
 1 *        SUBROUTINE STAGE(I1,I2,IL,IV,H,HL,HTC)
 2 *        COMMON/CD/STRM(300,24),DATA(20,10),RCT(22),NCF,NCL,LSTR
 3 *        LOGICAL LSTR
 4 *        CALL EQUIL(IL,IV)
 5 *        CALL ENTHL(IL)
 6 *        CALL ENTHV(IV)
 7 *        FLIN = STRM(I1,21) + STRM(I2,21)+RCT(21)
 8 *        HIN = STRM(I1,23)*STRM(I1,21)+STRM(I2,23)*STRM(I2,21)+H+RCT(22)
 9 *        STRM(IV,21)=(HIN-FLIN*STRM(IL,23))/(STRM(IV,23)-STRM(IL,23))
10 *        IF (LSTR) RETURN
11 *        DL = (FLIN-STRM(IV,21)-STRM(IL,21))/HTC
12 *        DO 7 N = NCF,NCL
13 *        HK=STRM(IV,N)/STRM(IL,N)
14 *        CNIN=STRM(I1,21)*STRM(I1,N)+STRM(I2,21)*STRM(I2,N)+RCT(N)
15 *        IF(HK.GT.5.) GO TO 5
16 *        DERN=(CNIN-FLIN*STRM(IL,N)-STRM(IV,21)*(STRM(IV,N)-STRM(IL,N)))/H
17 *      7 CALL INT(STRM(IL,N),DERN)
18 *        CALL INT(STRM(IL,21),DL)
19 *        RETURN
20 *      5 STRM(IV,N) = CNIN/(STRM(IV,21)+STRM(IL,21)/HK)
21 *        STRM(IL,N)=STRM(IV,N)/HK
22 *        DERN=0.
23 *        GO TO 7
24 *        END
```

FIG. 8-20. Listing for Subroutine Stage

Argument list:

I1 = input stream number
I2 = input stream number
IL = output liquid stream number
IV = output vapor stream number
H = external heat (or cooling) flux
HL = holdup (Moles)
HTC = hydraulic time constant

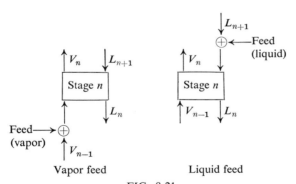

Vapor feed          Liquid feed

FIG. 8-21

```
 1  *        SUBROUTINE STGF(I1,I2,I3,IL,IV,H,HL,HTC)
 2  *        COMMON/CD/STRM(300,24),DATA(20,10),RCT(22),NCF,NCL,LSTR
 3  *        LOGICAL LSTR
 4  *        CALL EQUIL(IL,IV)
 5  *        CALL ENTHL(IL)
 6  *        CALL ENTHV(IV)
 7  *        FLIN = STRM(I1,21) + STRM(I2,21)+STRM(I3,21) +RCT(21)
 8  *        HIN = STRM(I1,23)*STRM(I1,21)+STRM(I2,23)*STRM(I2,21)+H+RCT(22)+
 9  *       1STRM(I3,21)*STRM(I3,23)
10  *        STRM(IV,21)=(HIN-FLIN*STRM(IL,23))/(STRM(IV,23)-STRM(IL,23))
11  *        IF (LSTR) RETURN
12  *        DL = (FLIN-STRM(IV,21)-STRM(IL,21))/HTC
13  *        DO 7 N = NCF,NCL
14  *        HK=STRM(IV,N)/STRM(IL,N)
15  *        CNF=STRM(I3,21)*STRM(I3,N)
16  *        CNIN=STRM(I1,21)*STRM(I1,N)+STRM(I2,21)*STRM(I2,N)+RCT(N)+CNF
17  *        IF(HK.GT.5.) GO TO 5
18  *        DERN=(CNIN-FLIN*STRM(IL,N)-STRM(IV,21)*(STRM(IV,N)-STRM(IL,N)))/HL
19  *      7 CALL INT(STRM(IL,N),DERN)
20  *        CALL INT(STRM(IL,21),DL)
21  *        RETURN
22  *      5 STRM(IV,N) = CNIN/(STRM(IV,21)+STRM(IL,21)/HK)
23  *        STRM(IL,N)=STRM(IV,N)/HK
24  *        DERN=0.
25  *        GO TO 7
26  *        END
```

FIG. 8-22. Subroutine STGF for A Feed Stage
Argument list:

$I1$ = input liquid stream number

$I2$ = input vapor stream number

$I3$ = feed stream number

$IL$ = output liquid stream number

$IV$ = output vapor stream number

$H$ = external heat flux

$HL$ = holdup moles

$HTC$ = hydraulic time constant

the heat balance (line 8), and the component balances (lines 15 and 16). Otherwise, the arrangement is the same as for STAGE. This feed stream I3 can be liquid or vapor, or even a mixture of both. It will not require any special qualifications, since its phase quality is implicitly determined by the enthalpy, composition, and pressure, information already contained in the stream array for I3. For this reason, the STRM array I3 must be completely specified before calling STGF.

## 8-11 SIDESTREAM STAGE SUBROUTINE STGS

As for the feed stage, a sidestream can be readily simulated by taking a split from either an exit vapor or liquid flow, using the routine SPLIT developed earlier in the chapter. But again, this results in an awkward numbering scheme, so another modified version of STAGE will be used called STGS. The listing for STGS is shown in Figure 8-23, where the

```
 1 *      SUBROUTINE STGS(I1,I2,IL,IV,IS,H,HL,HTC)
 2 *      COMMON/CD/STRM(300,24),DATA(20,10),RCT(22),NCF,NCL,LSTR
 3 *      LOGICAL LSTR
 4 *      CALL EQUIL(IL,IV)
 5 *      CALL ENTHL(IL)
 6 *      CALL ENTHV(IV)
 7 *      FLIN = STRM(I1,21) + STRM(I2,21)+RCT(21)
 8 *      HIN = STRM(I1,23)*STRM(I1,21)+STRM(I2,23)*STRM(I2,21)+H+RCT(22)
 9 *      STRM(IV,21)=(HIN-FLIN*STRM(IL,23))/(STRM(IV,23)-STRM(IL,23))
10 *      IF (LSTR) RETURN
11 *      DL = (FLIN-STRM(IV,21)-STRM(IS,21)-STRM(IL,21))/HTC
12 *      DO 7 N = NCF,NCL
13 *      HK=STRM(IV,N)/STRM(IL,N)
14 *      CNIN=STRM(I1,21)*STRM(I1,N)+STRM(I2,21)*STRM(I2,N)+RCT(N)
15 *      IF(HK.GT.5.) GO TO 5
16 *      DERN=(CNIN-FLIN*STRM(IL,N)-STRM(IV,21)*(STRM(IV,N)-STRM(IL,N)))/HL
17 *    7 CALL INT(STRM(IL,N),DERN)
18 *      CALL INT(STRM(IL,21),DL)
19 *      DO 8 N=NCF,NCL
20 *    8 STRM(IS,N)=STRM(IL,N)
21 *      DO 9 N=22,24
22 *    9 STRM(IS,N)=STRM(IL,N)
23 *      RETURN
24 *    5 STRM(IV,N) = CNIN/(STRM(IV,21)+(STRM(IL,21)+STRM(IS,21))/HK)
25 *      STRM(IL,N)=STRM(IV,N)/HK
26 *      DERN=0.
27 *      GO TO 7
28 *      END
```

FIG. 8-23.  Subroutine STGS for Sidestream Stage
Argument list:
I1 = input liquid stream number
I2 = input vapor stream number
IL = output liquid stream number
IV = output vapor stream number
IS = sidestream number
H = external heat flux
HL = holdup moles
HTC = hydraulic time constant

additional item in the argument list IS is the number of the sidestream. The subroutine transfers the liquid composition on the stage to the sidestream array (lines 20), as well as the enthalpy, temperature, and pressure (line 22). The sidestream flow appears only in the mass balance (line 11) since the exit liquid flows have been eliminated from the heat and component balance equations by algebraic manipulations.

## 8-12  COLUMN BOTTOMS SUBROUTINE BOT

The base of the column requires special treatment for the following reasons:

1. There is an input heat flux Q establishing the boilup vapor flow V.
2. The holdup is variable and considerable, and changes in sensible heat cannot be neglected (Figure 8-24).

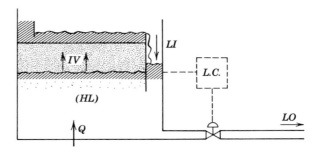

FIG. 8-24. Column Bottoms

3. The outflow of liquid from the bottoms is determined externally, eventually to be controlled by a bottoms level controller.

The equations for the bottoms follow (a listing for the subroutine BOT is shown in Figure 8-25).

### Holdup HL

The rate of change of moles in the holdup is

$$\frac{d(HL)}{d\theta} = LI - LO - IV + RCT(21) = DHL$$

where the flow lines are shown in Figure 8-24.

### Heat Balance

The basic relationship is

$$\frac{d}{d\theta}(HL \cdot h_L) = Q - LO \cdot h_L - IV \cdot H_v$$

Differentiating and rearranging, we obtain

$$IV = \frac{Q - LO \cdot h_L - HL \cdot \dfrac{dh_L}{d\theta} - h_L \dfrac{d(HL)}{d\theta}}{H_v}$$

Substituting $d(HL)/d\theta = DHL$ from the mass balance

$$IV = \frac{Q - h_L(LO + DHL) - DENL}{H_v}$$

where $DENL = HL \cdot (dh_L)/(d\theta)$. This term was neglected for the various stage simulations, but will be accounted for here by forming a derivative of the enthalpy $h_L$. The subroutine performs this function on lines 5 and 9,

```
 1 *      SUBROUTINE BOT(LI,LO,IV,Q,HL)
 2 *      COMMON/CINT/T,DT,JS,JN,DXA(500),XA(500),IO,JS4
 3 *      COMMON/CD/STRM(300,24),DATA(20,10),RCT(22),NCF,NCL,LSTR
 4 *      LOGICAL LSTR
 5 *      IF((JS4.EQ.4).OR.(JS.EQ.2).OR.LSTR) STRM(LO,20)=STRM(LO,23)
 6 *      CALL EQUIL(LO,IV)
 7 *      CALL ENTHL(LO)
 8 *      CALL ENTHV(IV)
 9 *      DENL=(STRM(LO,23)-STRM(LO,20))*HL/DT
10 *      QP=Q+STRM(LI,21)*STRM(LI,23)+RCT(22)
11 *      DHL = STRM(LI,21)-STRM(LO,21)-STRM(IV,21)+RCT(21)
12 *      DEN=STRM(IV,23)-STRM(LO,23)
13 *      STRM(IV,21)=(QP-STRM(LO,23)*(STRM(LI,21)+RCT(21))-DENL )/DEN
14 *      IF (STRM(IV,21).LT.0.) STRM(IV,21) = 0.
15 *      IF (LSTR) RETURN
16 *      DO 9 N = NCF,NCL
17 *      HK=STRM(IV,N)/STRM(LO,N)
18 *      IF(HK.GT.5.) GO TO 5
19 *      FNI=STRM(LI,21)*STRM(LI,N)+RCT(N)
20 *      FNO=STRM(LO,21)*STRM(LO,N)+STRM(IV,21)*STRM(IV,N)
21 *      DERN=(FNI-FNO-STRM(LO,N)*DHL)/HL
22 *    9 CALL INT(STRM(LO,N),DERN)
23 *      CALL INT(HL,DHL)
24 *      RETURN
25 *    5 DN=STRM(IV,21)+STRM(LO,21)/HK
26 *      STRM(IV,N)=(STRM(LI,21)*STRM(LI,N)+RCT(N))/DN
27 *      STRM(LO,N)=STRM(IV,N)/HK
28 *      DERN=0.
29 *      GO TO 9
30 *      END
```

FIG. 8-25. Subroutine BOT

LI = input liquid stream number
LO = output bottoms stream number
IV = vapor boilup stream number
Q = heat flux
HL = liquid holdup

where a backward derivative is obtained by storing the enthalpy of the liquid holdup on each legitimate pass in location 20 of STRM LO. This stored enthalpy is subtracted from the current value after equilibration (lines 7, 8, and 9) on line 9, providing the incremental change. Multiplying this by the holdup HL and dividing by the step size DT yields the derivative DENL. This procedure is similar to the one developed for the subroutine VVBOIL described in Section 5-12.

### Component Balance

The component balance is performed on lines 19 to 21, where FNI is the input flow of component N, including the reaction flow RCT(N) if any. FNO is the outlet flow in the vapor and bottoms flow stream. The derivative for the composition $N$, DERN is calculated on line 21 and integrated on line 22, followed by the integration for the holdup on line 23. The acceleration feature for volatile components ($HK > 5$) is also used in the routine and the procedure (lines 25 to 28) is the same as described for STAGE.

## 8-13 SUBROUTINE STGH

One of the unfortunate facts of generalized macrocomputer programs like those being developed in this text is that they never accommodate all possible situations. One of the assumptions in the formulation of the model and subroutine for the generalized equilibrium stages STAGE, STGF, and STGS is that the sensible heat change on each stage is small compared with the heat content of the upflowing vapor and could, therefore, be considered as a minor term. This results in treating the derivative of the sensible heat in the heat balance equation as an algebraic term and using the heat balance to define algebraically the vapor flows leaving the stage. This approach is valid in most typical situations where there is always a substantial vapor flow up the entire length of the column and where the temperature gradient from top to bottom of the column is not large. Neglecting the sensible heat change is invalid, however, for cases where the temperature gradient is large and sharp; that is, where most of the change occurs across just a few stages, coupled with liquid enthalpies that are large when compared with the vapor enthalpies. It is possible in these situations for the vapor flow virtually to disappear temporarily during a column upset. Figure 8-26 shows a section of column where the temperature profile changes rapidly in the region of stage n. During a column disturbance the equilibrium conditions will force a movement of this profile up or down the column. The amount of heat required to change the temperature of the stage and its liquid contents across a large range is now comparable to the heat content of the vapor and will result in significant changes in vapor and liquid flows. If such a situation is simulated by treating the derivative of the sensible heat as an algebraic term in the heat balance, the computation will become unstable. It is pointed out in Chapter 4 that a differential equation can only be solved algebraically when the derivative is small, compared with the other terms in the equation. Since, now, the derivative is no longer small, we are forced to treat the heat balance equation in the normal manner by integrating the input and exit heat fluxes and solving for the liquid enthalpy on the stage.

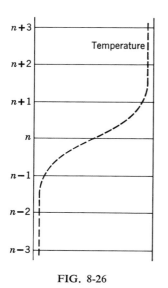

FIG. 8-26

This sounds straightforward, but it raises the problem of how to determine the vapor flow leaving the stage. There appears no direct way of establishing the vapor flow, so it has to be done by an indirect method. This method consists of calculating the equilibrium temperature for the stage, given the liquid composition and pressure, then establishing a pseudo enthalpy based on this temperature and composition. The vapor flow can now be calculated from the concept that the vapor flow should be of such a magnitude as to match the equilibrium enthalpy with the actual enthalpy from the heat balance. Since the vapor flow influences these two enthalpies only through differential equations, that is, the heat balance and the component material balances, it is not possible to provide an exact match, as can be done with algebraic loops. Thus a compromise is used, whereby the vapor flow is calculated from the difference between these enthalpy values times a suitable gain factor.

This sequence may be more clearly understood by referring to the model shown in Figure 8-27a, where the subprograms with the solid triangle output indicate an integration of a differential equation and hence a potential starting point in the program. The number contained in the boxes refers to the line numbers in the subprogram STGH shown in Figure 8-27b. This model, when translated to the sequential structure of the STGH subroutine (line numbers in parentheses), follows:

1. Argument list—This routine is a generalized stage that will accommodate a feed or sidestream flow designated as stream I3. For a sidestream, the flow rate (STRM (I3,21)) is specified as a negative number, and the other extensive properties, composition, temperature, and so on are the same as the reflux flow leaving the stage IL. This information transfer is made in the main program prior to calling the sidestream stage. For stages without the third feed, a dummy line number is used. All the other items in the argument list are the same as for the other stage subroutines discussed previously except the parameter HC, which is the holdup capacity of the stage in terms of

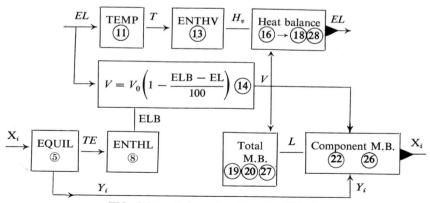

FIG. 8-27a. Information Flow for STGH

```
 1 *        SUBROUTINE STGH(I1,I2,I3,IL,IV,H,HL,HC,HTC)
 2 *        COMMON/CD/STRM(300,24),DATA(20,10),RCT(22),NCF,NCL,LSTR
 3 *        LOGICAL LSTR
 4 *        FLIN = STRM(I1,21) + STRM(I2,21)+STRM(I3,21) +RCT(21)
 5 *        CALL EQUIL(IL,IV)
 6 *        IF(LSTR) GO TO 6
 7 *      8 EL=STRM(IL,23)
 8 *        CALL ENTHL(IL)
 9 *        ELB=STRM(IL,23)
10 *        STRM(IL,23)=EL
11 *        CALL TEMP(IL,3)
12 *        STRM(IV,22)=STRM(IL,22)
13 *        CALL ENTHV(IV)
14 *        STRM(IV,21) = STRM(IV,20)*(1-(ELB-EL)/100.)
15 *        IF (STRM(IV,21).LT.0.) STRM(IV,21) = 0.
16 *        HIN = STRM(I1,23)*STRM(I1,21)+STRM(I2,23)*STRM(I2,21)+H+RCT(22)+
17 *       1STRM(I3,21)*STRM(I3,23)
18 *        DENL=(HIN-EL*(FLIN-STRM(IV,21))-STRM(IV,21)*STRM(IV,23))/HC
19 *        DHL = FLIN - STRM(IV,21)-STRM(IL,21)
20 *        DL=DHL/HTC
21 *        IF (LSTR) RETURN
22 *        DO 7 N = NCF,NCL
23 *        CNF=STRM(I3,21)*STRM(I3,N)
24 *        CNIN=STRM(I1,21)*STRM(I1,N)+STRM(I2,21)*STRM(I2,N)+RCT(N)+CNF
25 *        DERN=(CNIN-FLIN*STRM(IL,N)-STRM(IV,21)*(STRM(IV,N)-STRM(IL,N)))/HL
26 *      7 CALL INT(STRM(IL,N),DERN)
27 *        CALL INT(STRM(IL,21),DL)
28 *        CALL INT(STRM(IL,23),DENL)
29 *        IF(STRM(IL,21).LT.0.)STRM(IL,21) = 0.
30 *        RETURN
31 *      6 STRM(IV,20)=FLIN-STRM(IL,21)
32 *        CALL ENTHL(IL)
33 *        GO TO 8
34 *        END
```

(b)

FIG. 8-27b.  Subroutine STGH for an Equilibrium Stage with Heat Capacity

ARGUMENT LIST

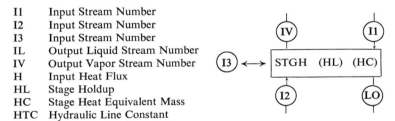

| I1 | Input Stream Number |
| I2 | Input Stream Number |
| I3 | Input Stream Number |
| IL | Output Liquid Stream Number |
| IV | Output Vapor Stream Number |
| H | Input Heat Flux |
| HL | Stage Holdup |
| HC | Stage Heat Equivalent Mass |
| HTC | Hydraulic Line Constant |

equivalent liquid moles. This includes not only the actual liquid moles but also the extra moles equivalent to the heat capacity of the metal of the stage and, possibly, part of the shell, that is,

$$HC = \text{moles holdup} + \frac{\text{stage mass} * \text{specific heat}}{\text{liquid specific heat}}$$

A definition of the argument list is also provided in Figure 8-27b.

2. After calculating the total input flow (line 4), the equilibrium temperature (TE) for liquid exit stream IL is determined (and stored in STRM

(IL,22)), and the vapor composition in equilibrium with IL is calculated and stored in IV by calling EQUIL (IL,IV) on line 5.

3. The actual enthalpy established from the heat balance is temporarily retained as EL from its storage STRM (IL,23) (line 7), followed by determining the psuedo enthalpy for the equilibrium temperature (line 8). This value is transferred to the variable ELB (line 9), and the actual enthalpy is transferred back to the IL array on line 10.

4. The temperature for the IL stream is now determined, based on its composition and enthalpy (line 11) and is assigned to the vapor stream IV (line 12).

5. The actual vapor enthalpy is now calculated (line 13), based on its composition and temperature.

6. The vapor flow is calculated on line 14, proportional to the difference between the actual (EL) and equilibrium (ELB) enthalpies. The nominal level will be the initial vapor flow that is determined on the first pass (LSTR = .TRUE.) from a material balance (line 31) and stored in location 20 of STREAM IV for convenience.

7. The derivative of the liquid enthalpy is established on line 18 and is derived as follows:

$$\uparrow V_o \qquad \downarrow L_i$$

| STGH EL, HC | EL = liquid enthalpy (PCU/mole °C) |
| | HC = holdup capacity (moles) |

$$\uparrow V_i \qquad \downarrow L_o$$

Heat balance equation:

$$\frac{d}{dt}(EL \cdot HC) = (\text{heat-in}) - (\text{heat-out})$$

$$= HIN - L_o(EL) - V_o H_v$$

Differentiating and rearranging

$$\frac{d}{dt}(EL) = \frac{HIN - L_o(EL) - V_o H_v - EL \cdot \dfrac{d}{dt}(HC)}{HC}$$

substituting $d(HC)/dt = FLIN - L_0 - V_0$

$$\frac{d}{dt}(EL) = \frac{HIN - EL(FLIN - V_0) - V_o H_v}{HC}$$

This equation is programmed on line 18 and subsequently integrated on line 28.

8. The total material balance derivative is specified on lines 19 and 20 and is integrated on line 27. The component material balances are performed

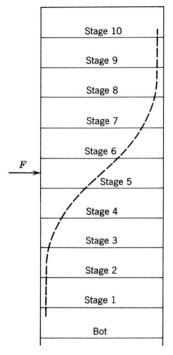

FIG. 8-28. Recommended Use of STGH and STAGE

from lines 23 to 25 and should be self-evident. The feature used in previous stage routines for circumventing troublesome high-volatility components is not used in STGH, since it causes trouble-some instability.

### Characteristics of STGH

The numerical stability of STGH is largely controlled by the gain factor used in the vapor equation. In Figure 8-27, line 14, this gain factor is 1/100, which is suitable for the average case. If this is increased, the stiffness (see Chapter 3) of the simulation is increased, requiring a smaller integration step size. If it is decreased, the stiffness characteristic is improved, but the accuracy deteriorates; thus, in general, a compromise has to be made. If necessary, this gain factor can be optimized for each problem.

It should also be realized that STAGE and STGH can be freely intermixed. For example, Figure 8-28 shows a 10-stage column, where it is expected that the sharp gradient profile could traverse a region from stages 4 to 7. The stages between these limits are then simulated with the STGH subroutine, while in the other region, where the gradient will always be moderate, the STAGE subroutine will be a more accurate representation.

### 8-14 REBOILER

The reboiler for a distillation column is the unit that generates the continuous vapor stream to be sent up the column as boilup. A common design for this unit is a single-pass shell-and-tube heat exchanger (Figure 8-29). The liquid from the column bottoms enters the base of the exchanger on the tube side and is partially vaporized, then the two-phase mixture flows back to the column, where it separates. Since the two-phase mixture in the heat-exchanger tubes has a lower density than the single-phase level on the column side, a hydraulic head is automatically created that forces the bottom fluid to flow through the reboiler (Thermosyphon). This flow is usually quite high, especially in the two-phase region, providing effective heat transfer on the tube side. The heating medium, usually steam or Dowtherm vapor, is introduced on the shell side via a control valve. The vapor condenses on the outside of the tubes at a higher temperature ($T_c$) than the temperature ($T_B$)

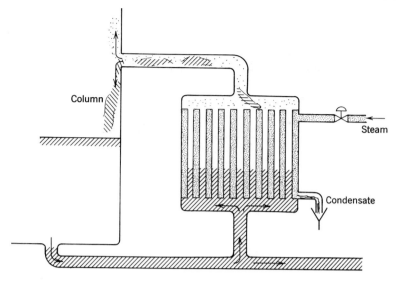

FIG. 8-29.   Column Reboiler

at the bottoms, thus developing the temperature gradient that drives the heat flux $H$ into the bottom fluid.

A column simulation should include the operating and dynamic characteristics of the reboiler. The two most important features are as follows:

1. Capacity—the flow of heating vapor through the control valve is

$$W = AC_v\sqrt{P_s(P_s - P_c)}$$

where  $A$ = fractional valve opening
   $C_v$ = valve capacity (moles/(min atm))
   $P_s$ = vapor supply pressure (atm)
   $P_c$ = condensate vapor pressure (atm)

This expression for the vapor flow is valid for subcritical flow (see Chapter 7), that is, $P_c > P_s/2$, which is generally the case. The condensate pressure $P_c$ will be the usual Antoine function of temperature, that is,

$$P_c = \text{EXP}\left(C_1 + \frac{C_2}{T_c + C_3}\right)$$

so that as the heat load $H$ increases, $T_c$ rises, increasing $P_c$ and, hence, decreasing the pressure drop across the valve ($P_s - P_c$). This change in valve gain influences the control characteristics at various operating levels. In fact,

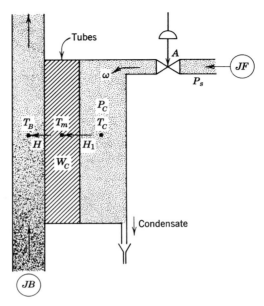

FIG. 8-30.  Reboiler "Lumped" Equivalent

should the temperature in the column bottoms rise to a high level, it could significantly decrease the heat flux and boilup, even with the vapor valve wide open.

2. Reboiler dynamics—the vapor volume on the shell side of the reboiler is usually quite small compared with the vapor throughput, so that the time response of the pressure due to the volume capacity is insignificant. What should be considered, though, is the heat capacity of all the tubes and part of the shell. If this is "lumped" as shown in Figure 8-30, then the tube temperature $T_m$ will be established by the differential equation

$$\frac{d}{d\theta} (T_m \cdot W_c) = (H_1 - H)$$

where  $W_c$ = heat capacity of tubes and $\frac{1}{2}$ the shell (PCU/°C)
$H_1$ = heat flux shell side to tube wall (PCU/min)
$H$ = heat flux tube wall to bottom liquid (PCU/min)

As a first approximation, the total resistance to heat transfer will be divided equally on both sides of the tube, that is,

$$H_1 = U(T_c - T_m)$$

$$H = U(T_m - T_b)$$

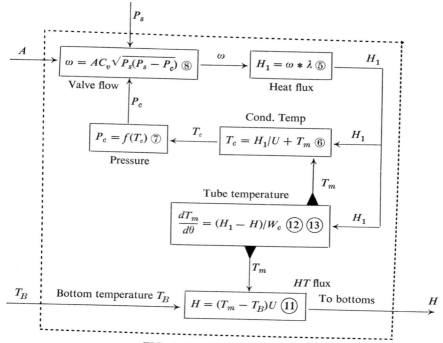

FIG. 8-31.  Model for Reboiler

A mathematical model of these reboiler equations is shown in Figure 8-31. The input variables to this model will be the valve area fraction $A$, the bottom temperature $T_B$, and the vapor supply pressure $P_s$ (generally constant). The output will be the heat flux $H$ that is used by the bottoms to generate the vapor boilup $V$.

Since the storage capacity of the vapor volume has been eliminated, the shell-side model becomes a series of simultaneous algebraic equations. The heat flux to the tube wall $H_1$ is assumed to be equal to the latent heat content of the vapor flow, that is, $H_1 = \omega * \lambda$ where $\lambda = $ latent heat/mole. The condensate temperature is then derived from this heat flux, pivoting from the tube temperature $T_m$, that is, $T_c = H_1/U + T_m$. The remainder of the model and the direction of the information flow should be self-evident.

## 8-15  SUBROUTINE REB  (A, H, CV, WC, JF, JB)

This subroutine is a program for the reboiler model shown in Figure 8-31 with the listing shown in Figure 8-32. The items in the argument list

```
 1 *          SUBROUTINE REB(A,H,CV,WC,JF,JB)
 2 *          COMMON/CD/STRM(300,24),DATA(20,10),RCT(22),NCF,NCL,LSTR
 3 *          LOGICAL LSTR
 4 *          IF(LSTR) GO TO 5
 5 *        8 H1=STRM(JF,21)*STRM(JF,23)
 6 *          TC=H1/STRM(JF,20)+TM
 7 *          PC=EXP(STRM(JF,1)+STRM(JF,2)/(TC+STRM(JF,3)))
 8 *          W=A*CV*SQRT(STRM(JF,24)*ABS(STRM(JF,24)-PC))
 9 *          CALL CONV(STRM(JF,21),W,1,NC)
10 *          GO TO (7,8),NC
11 *        7 H=STRM(JF,20)*(TM-STRM(JB,22))
12 *          DTM=(H1-H)/WC
13 *          CALL INT(TM,DTM)
14 *          RETURN
15 *        5 W=H/STRM(JF,23)
16 *          PC=STRM(JF,24)-W**2/(A**2*CV**2*STRM(JF,24))
17 *          TC=STRM(JF,2)/(LOG(PC)-STRM(JF,1))-STRM(JF,3)
18 *          TM=(STRM(JB,22)+TC)/2.
19 *          STRM(JF,20)=H/(TC-TM)
20 *          STRM(JF,21)=W
21 *          RETURN
22 *          END
```

FIG. 8-32.  Subroutine Reb for Column Reboiler
Argument list:
A = fractional valve opening
H = heat flux to bottoms (PCU/Min)
CV = valve capacity (moles/min)
WC = tube heat capacity (PCU/°C)
JF = vapor supply stream number
JB = bottoms stream number

are:

$A$ = fractional valve opening
$H$ = heat flux to bottoms (PCU/min)
$C_v$ = valve capacity (mole/atm min)
WC = tube heat capacity (PCU/°C)
JF = vapor supply stream number
JB = bottoms stream number

For the sake of convenience, the thermal constants for the heating vapor are carried in the stream array as follows:

STRM (JF,1)  = Antoine coefficient $C_1$
STRM (JF,2)  = Antoine coefficient $C_2$
STRM (JF,3)  = Antoine coefficient $C_3$
STRM (JF,23) = latent heat at operating conditions
STRM (JF,24) = supply pressure $P_s$ (atm)
STRM (JF,21) = Flow rate (mole/min)
STRM (JF,20) = U, heat transfer coefficient (PCU/min °C)

The pressure and flow values are in their normal positions in the STRM array. All other numbers merely use the array for nonstandard storage, since the reboiler vapor is invariably a single component stream.

When this subroutine is called, initial values for all the items in the argument list must be supplied. The programmer must supply consistent values of A, H, and $C_v$ for the starting condition. On the first pass (LSTR = TRUE) the subroutine will calculate (lines 15–21) the vapor flow W, the condensate pressure $P_c$, the condensate temperature TC (Antoine equation), the initial tube temperature $T_m$, and, finally, the heat transfer coefficient U, which it stores in STRM (JF,20).

On subsequent passes (LSTR = .FALSE.) it will solve for the variables, as shown in the signal flow diagram (Figure 8-31). Numbers in boxes refer to line numbers in the listing in (Figure 8-32), starting and converging on the value of the vapor flow W in STRM(JF,21). The last function to be performed before leaving the subroutine is to integrate the derivative of the tube temperature $T_m$, and this is done on line 13.

## 8-16 DYFLO SIMULATION OF A DISTILLATION COLUMN

The previous section has developed several subroutines for simulating sections of a distillation column. A typical example will now be developed that will show how these subroutines are assembled in a master calling program for simulating a particular column. Case 5-1 in Chapter 5 (page 149) was a simulation of a batch distillation. The initial vapor flow (25 moles/min) and composition will be used as the feed to an 18-plate column having a plate efficiency of 0.5. The column is equivalent to nine equilibrium stages (Figure 8-33) plus a bottoms. The vapor boilup is generated in a reboiler, while the overhead vapor passes to a partial condenser, where a fraction is condensed and stored in a hold tank. A flow from this reflux tank back to the column constitutes the reflux flow, which will be assumed to be constant. The vapor feed to the column will enter below the fourth stage from the bottom, while a liquid sidestream will leave from the fifth stage. The subroutine equivalent of this column is shown in Figure 8-34, complete with the vapor and liquid line numbers connecting the various sections. The hold tank will be simulated with the HLDP subroutine, and for the partial condenser, the subroutine PCON will be used (Chapter 5) where a condensate exit equilibrium temperature is specified.

The main program for this entire column is shown in Figure 8-35. It will be recalled that in order for the integration procedure contained within the subroutine to function correctly, it is necessary for the calling program information flow to be countercurrent to the liquid flow which contains the dynamic lags. In this way, all the derivatives will be calculated at the same

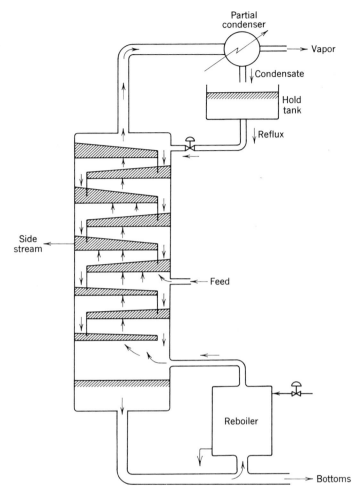

FIG. 8-33. Nine-Stage Column

time instant. Thus the calling program starts with the reboiler (line 10), then the bottoms, followed by all the stages up to the top of the column. The feed stage (line 14) and sidestream stage (line 15) are inserted in the middle between the two DO loops that calculate all the other stages. The last two calls are for the partial condenser and the holdup (lines 18 and 19). The end of the derivative section calls the output subroutine PRL and the repeat RPRL to print all the lines in the column.

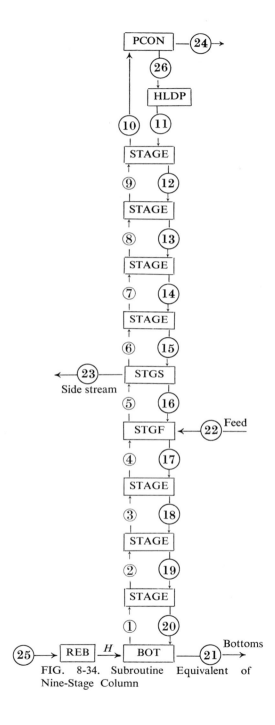

FIG. 8-34. Subroutine Equivalent of Nine-Stage Column

251

```
1*      C **DATA SECTION**
2*              COMMON/CD/STRM(300,24),DATA(20,10),RCT(22),NCF,NCL,LSTR
3*              COMMON/CINT/T,DT,JS,JN,DXA(500),XA(500),IO,JS4
4*              LOGICAL LSTR,NF
5*              CALL DAT
6*              DATA A,WC,CV,H,DT/.308,187.,2.5,37500.,,1/
7*              LSTR=.TRUE.
8*      C **DERIVATIVE SECTION**
9*            7 STRM(21,21)=STRM(20,21)-STRM(1,21)
10*             CALL REB(A,H,CV,WC,25,21)
11*             CALL BOT(20,21,1,H,40.)
12*             DO 6 N=1,3
13*           6 CALL STAGE(N,20-N,21-N,N+1,0.,5.,,1)
14*             CALL STGF(4,16,22,17,5,0.,5.,,1)
15*             CALL STGS(5,15,16,6,23,0.,5.,,1)
16*             DO 8 N=6,9
17*           8 CALL STAGE(N,20-N,21-N,N+1,0.,5.,,1)
18*             CALL PCON(10,24,26,76.)
19*             CALL HLDP(26,11,3,50.,0.)
20*             CALL PRL(10.,30.,NF,1,2,3,4,5,6,7,8,9,10,26,24)
21*             CALL RPRL(11,12,13,14,15,16,17,18,19,20,21,23)
22*             IF(NF) GO TO 10
23*      C **INTEGRATION SECTION**
24*             CALL INTI(T,DT,1)
25*             GO TO 7
26*          10 CONTINUE
27*             STOP
28*             END
```

FIG. 8-35.   Main Program for Distillation Column Simulation

```
1*              SUBROUTINE DAT
2*              COMMON/CD/STRM(300,24),DATA(20,10),RCT(22),NCF,NCL,LSTR
3*              DATA NCF,NCL/1,3/
4*              DATA(DATA(1,N),N=1,8)/13.46,-5210.,273.,8.,,.01,9020.,20.,,.02/
5*              DATA(DATA(2,N),N=1,8)/15.2,-6050.,273.,12.2,.02,11500.,32.,,.01/
6*              DATA(DATA(3,N),N=1,8)/15.4,-5312.,273.,6.5,.01,7500.,16.,,.03/
7*              DATA(STRM(25,J),J=1,24)/11.775,-3888.5,230.16,19*0.,8000.,7./
8*              DATA(STRM(22,J),J=1,24)/.2169,.0911,.692,17*0.,.25,.95.35,8994.,1./
9*              DATA STRM(23,21)/5.0/STRM(1,21)/5./
10*             DATA(STRM(J,21),J=11,21)/5*13.,5*8.,3./
11*             DATA(STRM(N,22),N=11,21)/11*80./
12*             DO 5 J=1,11
13*             JD=22-J
14*             STRM(JD,1)=.2169
15*             STRM(JD,2)=.0911
16*             STRM(JD,3)=.692
17*             STRM(J,24)=1.11-J*.01
18*             STRM(JD,24)=STRM(J,24)
19*             IF(J.EQ.11) GO TO 5
20*             CALL EQUIL(JD,J)
21*           5 CALL ENTHL(JD)
22*             STRM(26,24) = 1.
23*             CALL EQUIL(21,1)
24*             RETURN
25*             END
```

FIG. 8-36.   Subroutine DAT: Assigns Starting Values to STRM Arrays and Basic Data Arrays

252

The integration section consists of one call on INTI, using the first-order integration method with a step size OT = 0.1 min, which has been found to provide a reasonable balance between running time and accuracy. In order to make a computer run, it is necessary to specify the basic data for the components; the feed, sidestream, and bottoms flow rates; the reboiler heat (H); partial condenser temperature; reboiler control valve ($C_v$); reboiler heat capacity (WC); and valve opening A. This is done in the subroutine DAT (Figure 8-36) called in the data section of the main program (line 5), and the reboiler parameters are specified in the data statement on line 6.

### Subroutine DAT

Because of the structural sequence of the program, only the following initial conditions need to be specified. They are assembled here as a checklist that should be referred to when simulating a column.

1. Number of first (NCF) and last (NCL) components (line 3).
2. Basic data on equilibrium and enthalpy for all components, entered in DATA array (lines 4, 5, and 6).
3. Equilibrium data, latent heat, and supply pressure, entered in STRM array for reboiler vapor (line 7).
4. Feed composition, flow rate, enthalpy, and pressure, entered in feed STRM array (line 8).
5. Flow rate of sidestream specified (line 9).
6. Estimate of vapor boilup, entered into STRM array (line 9).
7. Enter estimated value of temperature and flow in all liquid STRM arrays (lines 10 and 11).
8. Enter initial compositions in all liquid STRM arrays (lines 14–16). It is not necessary to enter the vapor compositions, since these will be calculated on the first pass.
9. It is essential for the pressure to be entered into the vapor and liquid STRM arrays. This is done on lines 17 and 18 by assuming the bottoms pressure (1.11 atm) and successively subtracting the pressure drop per stage.
10. The pressure is also entered for the condensate flow out of the condenser (line 22). (Required for PCON.)
11. The temperature of the holdup tank is specified (line 11) and enthalpy calculated (line 21).
12. The first call in the main program is for the reboiler REB. This will calculate the heat transfer coefficients based on the specified heat flux and temperature gradient. Prior to calling REB, it is necessary to have the correct bottoms temperature, which is done by calling EQUIL on the last line of DAT (line 23) for the bottoms lines 21 and 1.
13. The reboiler data required for specifying the dynamic and capacity

TIME = .0000

| STRM NO | 1 | 2 | 3 | 4 | 5 | 6 | 7 | 8 | 9 | 10 | 26 | 24 |
|---|---|---|---|---|---|---|---|---|---|---|---|---|
| FLOW | .5688+01 | .5684+01 | .5679+01 | .5675+01 | .3355+02 | .3355+02 | .3355+02 | .3355+02 | .3355+02 | .3361+02 | .5937+01 | .2767+02 |
| TEMP | .8096+02 | .8074+02 | .8053+02 | .8031+02 | .8009+02 | .7986+02 | .7964+02 | .7941+02 | .7919+02 | .7896+02 | .7600+02 | .7600+02 |
| ENTHAL | .8240+04 | .8238+04 | .8236+04 | .8235+04 | .8233+04 | .8231+04 | .8229+04 | .8227+04 | .8225+04 | .8223+04 | .1474+04 | .8135+04 |
| PRESS | .1100+01 | .1090+01 | .1080+01 | .1070+01 | .1060+01 | .1050+01 | .1040+01 | .1030+01 | .1020+01 | .1010+01 | .1000+01 | .0000 |
| COMP 1 | .5597-01 | .5598-01 | .5599-01 | .5600-01 | .5601-01 | .5602-01 | .5604-01 | .5605-01 | .5606-01 | .5607-01 | .1519+00 | .3499-01 |
| COMP 2 | .1248-01 | .1247-01 | .1245-01 | .1243-01 | .1242-01 | .1240-01 | .1238-01 | .1237-01 | .1235-01 | .1233-01 | .4451-01 | .5262-02 |
| COMP 3 | .9315+00 | .9316+00 | .9316+00 | .9316+00 | .9316+00 | .9316+00 | .9316+00 | .9316+00 | .9316+00 | .9316+00 | .8024+00 | .9600+00 |

| STRM NO | 11 | 12 | 13 | 14 | 15 | 16 | 17 | 18 | 19 | 20 | 21 | 23 |
|---|---|---|---|---|---|---|---|---|---|---|---|---|
| FLOW | .1300+02 | .1300+02 | .1300+02 | .1300+02 | .1300+02 | .8000+01 | .8000+01 | .8000+01 | .8000+01 | .8000+01 | .3000+01 | .5000+01 |
| TEMP | .8000+02 | .7896+02 | .7919+02 | .7941+02 | .7964+02 | .7986+02 | .8009+02 | .8031+02 | .8053+02 | .8074+02 | .8096+02 | .0000 |
| ENTHAL | .1632+04 | .1609+04 | .1614+04 | .1619+04 | .1624+04 | .1629+04 | .1634+04 | .1639+04 | .1644+04 | .1649+04 | .1654+04 | .0000 |
| PRESS | .1000+01 | .1010+01 | .1020+01 | .1030+01 | .1040+01 | .1050+01 | .1060+01 | .1070+01 | .1080+01 | .1090+01 | .1100+01 | .0000 |
| COMP 1 | .2169+00 | .2169+00 | .2169+00 | .2169+00 | .2169+00 | .2169+00 | .2169+00 | .2169+00 | .2169+00 | .2169+00 | .2169+00 | .0000 |
| COMP 2 | .9110-01 | .9110-01 | .9110-01 | .9110-01 | .9110-01 | .9110-01 | .9110-01 | .9110-01 | .9110-01 | .9110-01 | .9110-01 | .0000 |
| COMP 3 | .6920+00 | .6920+00 | .6920+00 | .6920+00 | .6920+00 | .6920+00 | .6920+00 | .6920+00 | .6920+00 | .6920+00 | .6920+00 | .0000 |

TIME = .1000+02

| STRM NO | 1 | 2 | 3 | 4 | 5 | 6 | 7 | 8 | 9 | 10 | 26 | 24 |
|---|---|---|---|---|---|---|---|---|---|---|---|---|
| FLOW | .4298+01 | .4356+01 | .4383+01 | .4387+01 | .2945+02 | .2957+02 | .2980+02 | .2998+02 | .3046+02 | .3097+02 | .1266+02 | .1831+02 |
| TEMP | .9980+02 | .9725+02 | .9624+02 | .9607+02 | .9617+02 | .9503+02 | .9352+02 | .9110+02 | .8707+02 | .8142+02 | .7600+02 | .7600+02 |
| ENTHAL | .9136+04 | .8986+04 | .8939+04 | .8939+04 | .8952+04 | .8887+04 | .8791+04 | .8683+04 | .8495+04 | .8303+04 | .1453+04 | .8134+04 |
| PRESS | .1100+01 | .1090+01 | .1080+01 | .1070+01 | .1060+01 | .1050+01 | .1040+01 | .1030+01 | .1020+01 | .1010+01 | .1000+01 | .0000 |
| COMP 1 | .2640+00 | .2302+00 | .2192+00 | .2207+00 | .2267+00 | .2394+00 | .2410+00 | .2230+00 | .1737+00 | .1034+00 | .1884+00 | .4338-01 |
| COMP 2 | .9570-01 | .8083-01 | .7660-01 | .7642-01 | .7750-01 | .5944-01 | .4306-01 | .2848-01 | .1581-01 | .6946-02 | .1438-01 | .1700-02 |
| COMP 3 | .6403+00 | .6889+00 | .7042+00 | .7029+00 | .6955+00 | .7019+00 | .7154+00 | .7494+00 | .8098+00 | .8905+00 | .7985+00 | .9554+00 |

| STRM NO | 11 | 12 | 13 | 14 | 15 | 16 | 17 | 18 | 19 | 20 | 21 | 23 |
|---|---|---|---|---|---|---|---|---|---|---|---|---|
| FLOW | .1300+02 | .1249+02 | .1193+02 | .1190+02 | .1153+02 | .6510+01 | .6372+01 | .6392+01 | .6453+01 | .6171+01 | .2095+01 | .5000+01 |
| TEMP | .7615+02 | .8142+02 | .8707+02 | .9110+02 | .9352+02 | .9503+02 | .9617+02 | .9607+02 | .9624+02 | .9725+02 | .9980+02 | .9503+02 |
| ENTHAL | .1462+04 | .1650+04 | .1852+04 | .2014+04 | .2130+04 | .2225+04 | .2315+04 | .2311+04 | .2317+04 | .2353+04 | .2459+04 | .2225+04 |
| PRESS | .1000+01 | .1010+01 | .1020+01 | .1030+01 | .1040+01 | .1050+01 | .1060+01 | .1070+01 | .1080+01 | .1090+01 | .1100+01 | .1050+01 |
| COMP 1 | .1847+00 | .3602+00 | .4878+00 | .5365+00 | .5345+00 | .5040+00 | .4620+00 | .4561+00 | .4548+00 | .4647+00 | .4882+00 | .5040+00 |
| COMP 2 | .2113-01 | .4500-01 | .7973-01 | .1200+00 | .1653+00 | .2147+00 | .2696+00 | .2696+00 | .2709+00 | .2765+00 | .2963+00 | .2147+00 |
| COMP 3 | .7949+00 | .5948+00 | .4332+00 | .3428+00 | .3006+00 | .2801+00 | .2682+00 | .2740+00 | .2743+00 | .2591+00 | .2189+00 | .2801+00 |

TIME = .2000+02

| STRM NO | 1 | 2 | 3 | 4 | 5 | 6 | 7 | 8 | 9 | 10 | 26 | 24 |
|---|---|---|---|---|---|---|---|---|---|---|---|---|
| FLOW | .3304+01 | .3310+01 | .3319+01 | .3393+01 | .2850+02 | .2867+02 | .2887+02 | .2914+02 | .2956+02 | .3016+02 | .1325+02 | .1691+02 |
| TEMP | .1210+03 | .1177+03 | .1124+03 | .1046+03 | .9719+02 | .9589+02 | .9421+02 | .9162+02 | .8732+02 | .8144+02 | .7600+02 | .7600+02 |
| ENTHAL | .1102+05 | .1067+05 | .1013+05 | .9490+04 | .9003+04 | .8919+04 | .8820+04 | .8689+04 | .8505+04 | .8288+04 | .1444+04 | .8126+04 |
| PRESS | .1100+01 | .1090+01 | .1080+01 | .1070+01 | .1060+01 | .1050+01 | .1040+01 | .1030+01 | .1020+01 | .1010+01 | .1000+01 | .0000 |
| COMP 1 | .6659+00 | .6444+00 | .5356+00 | .3746+00 | .2478+00 | .2639+00 | .2670+00 | .2454+00 | .1888+00 | .1096+00 | .1935+00 | .4456-01 |
| COMP 2 | .3011+00 | .2433+00 | .1831+00 | .1217+00 | .8004-01 | .5708-01 | .3779-01 | .2227-01 | .1077-01 | .3911-01 | .7770-02 | .9186-03 |
| COMP 3 | .3331-01 | .1077+00 | .2813+00 | .5037+00 | .6712+00 | .6790+00 | .6952+00 | .7324+00 | .8005+00 | .8858+00 | .7975+00 | .9542+00 |

| STRM NO | 11 | 12 | 13 | 14 | 15 | 16 | 17 | 18 | 19 | 20 | 21 | 23 |
|---|---|---|---|---|---|---|---|---|---|---|---|---|
| FLOW | .1300+02 | .1240+02 | .1198+02 | .1172+02 | .1152+02 | .6359+01 | .6249+01 | .6142+01 | .6252+01 | .6087+01 | .2910+01 | .5000+01 |
| TEMP | .7600+02 | .8144+02 | .8732+02 | .9162+02 | .9421+02 | .9589+02 | .9719+02 | .1046+03 | .1124+03 | .1177+03 | .1210+03 | .9589+02 |
| ENTHAL | .1446+04 | .1632+04 | .1836+04 | .2001+04 | .2126+04 | .2235+04 | .2347+04 | .2607+04 | .2873+04 | .3052+04 | .3132+04 | .2235+04 |
| PRESS | .1000+01 | .1010+01 | .1020+01 | .1030+01 | .1040+01 | .1050+01 | .1060+01 | .1070+01 | .1080+01 | .1090+01 | .1100+01 | .1050+01 |
| COMP 1 | .1929+00 | .3833+00 | .5240+00 | .5798+00 | .5755+00 | .5383+00 | .4860+00 | .5633+00 | .6134+00 | .6199+00 | .5772+00 | .5383+00 |
| COMP 2 | .8955-02 | .2562-01 | .5396-01 | .9245-01 | .1409+00 | .1992+00 | .2665+00 | .2969+00 | .3253+00 | .3528+00 | .3863+00 | .1992+00 |
| COMP 3 | .7977+00 | .5918+00 | .4222+00 | .3275+00 | .2835+00 | .2615+00 | .2488+00 | .1402+00 | .5897-01 | .1931-01 | .5376-02 | .2615+00 |

FIG. 8-37. Numerical Results of Column Simulation

characteristics is entered in the data statement in the main program. These data are:

Valve opening A (0.5)
Heat capacity of reboiler tubes WC = 187 PCU/°C
The valve capacity $C_v$ = 1.5 moles/PSI
Initial heat flux H = 37500 PCU/min

For convenience, the bottoms holdup will be assumed to be constant. Thus, the bottoms flow will be the difference between the reflux from the first stage and the boilup (line 9).

The case that will be demonstrated here is a startup transient, assuming that initially the liquid composition throughout the column and in the reflux tank is the same as the feed composition. With a fixed reboiler valve position (0.5) and a constant condenser temperature of 76°C, the final steady state is reached in approximately $\frac{1}{2}$ hour. The initial transient output is shown in Figure 8-37 for two 10-min intervals, with the steady-state composition profile up and down the column at t = 20 min being very close to the final profile. Chapter 12 will deal with the simulation of automatic process controls, and this column will be used as a candidate for demonstrating how controls are connected to a typical process to automatically regulate it to the desired state.

### Problems

1. Solvent extraction tower. A vapor consisting of four components is passed up a tower containing 26 sieve plates having a plate efficiency of 69%.

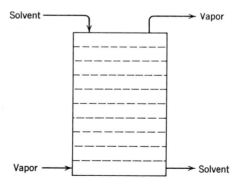

A solvent (component 5) flows in at the top and cascades down the column. Three of the components in the vapor are soluble in the solvent, while the fourth (No. 4) is insoluble. Assuming that the activity of components one to three are related by a three-suffix Margules equation, proceed to construct:

(a) The activity coefficients subprogram
(b) The main program, using DYFLO subroutines
(c) The DATA subroutine for the initial and boundary values and basic data, assuming that all numerical values are known

The program is to simulate the dynamic responses of temperature and composition of the exit streams to sudden feed changes.

2. Acquire data from an operating distillation column, either a laboratory or an industrial column, preferably in the form of a transient temperature measurement on a particular plate, that results from a specified disturbance, such as a feed-rate change, starting from a steady-state condition. Construct a computer program using DYFLO subroutines that will simulate the dynamic properties of this column, and determine how close the calculated transient temperature matches the measured temperature. If the correspondence is unsatisfactory, which assumption in the model can be justifiably modified to improve the match? It will be necessary initially to match the steady-state operation of the column by modifying the plate efficiency or activity correlations, if any.

3. Ten well-stirred jacketed reactors of volume $V$ are connected in series, sequentially, processing material as a liquid phase. The cooling water through

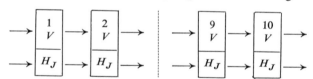

each jacket (holdup $H_J$) flows cocurrently with the reactor flows. The reaction taking place is

$$A \underset{k_2}{\overset{k_1}{\rightleftharpoons}} B \underset{k_4}{\overset{k_3}{\rightleftharpoons}} C$$

The reaction rates are first order, and the rate constants are Arrhenius temperature functions. The reactions are exothermic, hence the cooling jackets. Construct a computer program (omitting numerical data) using appropriate DYFLO subroutines that will simulate the dynamics of this reactor train. How would the program be modified if the cooling water were flowing in a countercurrent direction?

## REFERENCES

1. "Process Heat and Material Balances," L. M. Naphtali, *Chem. Eng. Progr.*, **60**, No. 9, 1964.

2. *SPADE, A Computer Cost Estimating System*, J. A. Ziebarth, American Association of Cost Engineers, 8th National Meeting, 1964.
3. *Using GIFS in the Analysis and Design of Process Systems*, W. H. Dodrill, Fall Joint Computer Conference, 1962.
4. *GIFS, Generalized Interrelated Flow Simulation*, Users Manual, Issued by Service Bureau Corporation, 1962.
5. "Heat and Mass Balancing on a Digital Computer," A. E. Ravilz and R. L. Norman, *Chem. Eng. Progr.*, **60**, No. 8, 1964.
6. "Improve Refining Operation with Process Simulation." H. A. Lindahl, *Chem. Eng. Progr.*, **61**, No. 4, April 1965.
7. *A Digital Computer Executive Program for Process Simulation and Design*, P. T. Shannon and H. Mosler, A.I.Ch.E. 53rd National Meeting, May 1964.
8. "Absorption and Stripping-Factor Functions for Distillation Calculation by Manual and Digital Computer Methods," W. C. Edmister, *A.I.Ch.E. J.*, **3**, No. 2, 1958.
9. "Generalized Material Balance," D. I. Rubin, *Chem. Eng. Progr. Symp. Ser.*, **58**, No. 37, pp. 54–61, 1962.
10. "A Machine Computation Method for Performing Material Balances," *Chem. Eng. Progr.*, **58**, No. 10, 1962.
11. *A Computer System for Process Simulation*, M. G. Kessler and P. R. Griffiths, American Petroleum Institute, May 13, 1963.
12. *A Study of a Contact Sulphuric Acid Plant Using PACER*, Department of Chemical Engineering, McMaster University, Hamilton, 1965.
13. "Analog Simulation of Countercurrent Crystallization Process," W. L. Godfrey and R. D. Benham, *Simulation*, **4**, No. 1, 1964.
14. "Hybrid Simulation of Multistage System Dynamics," R. G. Franks, *Chem. Eng. Progr.*, **60**, No. 3, 1964.
15. "Dynamic Simulation of a Distillation Tower," A. M. Peiser and S. S. Grover, *Chem Eng. Progr.*, **58**, No. 9, 1962.
16. *The Simulation of Multi-Component Distillation*, E. C. Deland and M. B. Wolf, AD 286794, 1962.
17. "High Speed Memory Analog Computers," S. H. Jury and J. M. Andrews, *Ind. Eng. Chem.*, **53**, No. 11, 1961.
18. "How to Analyze Control Program for Distillation Column," R. J. Ruszkay, *Chem. Eng.*, April 1963.
19. Analog Computer Simulation of a Solvent Recovery Column, R. J. Ruszkay, *ISA J.*, **6**, 2–264, 1964.
20. "Ester Exchange Equilibrium from Dimethyl-Terephthalate and Ethylene Glycol," *Recueil 1960 Communication No. 3*, Institute of Cellulose Research AKU, Utrecht, 1960.
21. *Approximation Models for the Dynamic Response of Large Distillation Columns*, J. S. Moczek and T. J. Williams, 1962.
22. "Dynamic Characteristics and Analog Simulation of Distillation Columns," D. E. Lamb, R. L. Pigford, and D. W. T. Rippin, *Chem. Eng. Progr. Symp. Ser.*, No. 36, 1961.
23. "Development and Application of a General Purpose Analog Computer Circuit to Steady State Multi-Component Distillation," N. G. O'Brien and R. G. Franks *Chem. Eng.*, **55**, No. 21, 1958.
24. "Steady State Heat and Material Balances," H. G. Garner, *Chem. Eng.*, April 1963.
25. "Dynamic Behavior of Multi-component, Multi-stage Systems. Numerical Methods of Solution," R. S. H. Mah, S. Michaelson, and R. W. H. Sargent, *Chem. Eng. Sci.*, **17**, 1962.

26. *Material and Energy Balance Computations*, E. J. Henley and E. M. Rosen, Wiley, New York, 1969.
27. "Cyclic Distillation Control," H. L. Wade, C. H. Jones, T. B. Rooney, and L. B. Evans, *Chem. Eng. Progr.*, **65,** No. 3, March 1969.
28. "Multi-component Batch Distillation Simulations on an IBM 1130 Computer," E. R. Robinson and M. R. Goldman, *Simulation*, **13,** No. 6, 1969.

# CHAPTER IX

# DISTRIBUTED SYSTEMS

The preceding chapters have demonstrated the formulation of mathematical models for systems consisting mostly of well-agitated, homogenous vessels. The equations involved in the mathematical models defined transient conditions in these vessels and were ordinary differential equations, usually nonlinear, with time as the independent variable. The extension of the model from the single vessel to multistaged systems has also been covered. The situation in which heat and mass transfer occurs distributed over a distance is a further logical extension from the multistaged system. Examples of such systems are heat exchangers, condensers, pipeline reactors, and so forth. For the most general transient case they are described by partial differential equations, with time and distance as independent variables. In this chapter only steady-state situations are considered, which means that the differentials with respect to time are eliminated and the variables in the system are defined in relation to the geometric dimensions in the equipment. The following examples are arranged in order of complexity, and the analysis of these examples follows the basic philosophy developed in the preceding chapters; namely, the mathematical model arranged in block diagram form is developed to describe the physical variables as a function of distance.

EXAMPLE 9-1. COUNTERCURRENT HEAT EXCHANGER

The simplest case in this class of problems is that of a countercurrent flow, double pipe, liquid-liquid heat exchanger (Figure 9-1). The known input conditions are the two inlet temperatures $T_{Si}$ and $T_{Ti}$, the two flow rates $F_S$ and $F_T$, the dimensions of the pipes, and the heat transfer coefficient. The two unknown values are the exit temperatures $T_{So}$ and $T_{To}$. A set of equations is required to relate these unknowns to known inlet

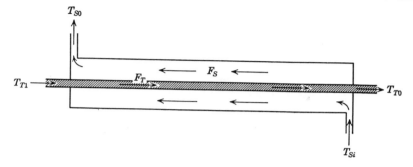

FIG. 9-1. Countercurrent Liquid-Liquid Heat Exchanger

temperatures and flows. The differential equations are developed by the classical approach of defining a steady-state heat balance across an infinitesimal section $\Delta X$ of the exchanger (Figure 9-2). The rate of change of temperature with distance $(dT/dX)$ in either stream varies along the whole length $X$ of the exchanger, but because the elemental section $\Delta X$ is very small it can be considered constant across this section. If the temperature on the inlet side of the section is $T$, the outlet temperature (see Figure 9-2) is

$$\left(T + \frac{dT}{dX} \cdot \Delta X\right)$$

NOTE. The sign convention makes $-\Delta X$ on the shell side.

The pipe surface area transferring heat between the two fluids is $\pi D \cdot \Delta X$, and if $U$ is the heat transfer coefficient the rate of heat transfer between the two fluids in the section $\Delta X$ will be

$$Q \cdot \Delta X = U(\pi D \cdot \Delta X)(T_S - T_T)$$

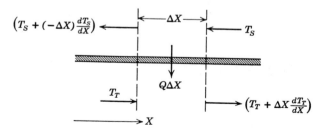

FIG. 9-2. $\Delta X$ Section of Liquid-Liquid Heat Exchanger

where $T_S$ and $T_T$ are the shell and tube side temperatures, respectively. A heat balance on the shell side section is

$$\text{heat in} = \text{heat out}$$

$$F_s C_s T = \left[ T_S + \frac{dT_S}{dX}(-\Delta X) \right] F_s C_s + Q \cdot \Delta X$$

NOTE. Because only the steady-state aspects are being considered, there is no accumulation rate term as in other chapters. Also, $X$ is considered positive with tube flow direction.

This equation can be reduced to the following by rearranging terms:

$$F_S C_s \frac{dT_S}{dX} = +Q$$

A heat balance on the tube side gives a similar equation

$$F_T C_T \frac{dT_T}{dX} = +Q$$

These equations, a model of which is shown in Figure 9-3, are solved on a computer by integrating the equations from $X = 0$ to $X = L$, where $L$ is the length of the exchanger. Because there are two differential equations, it is necessary to supply initial values for $T_S$ and $T_T$ at $X = 0$. At the inlet

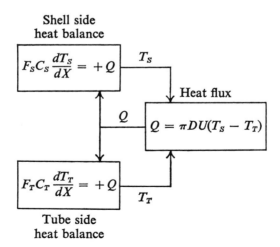

Shell side
heat balance

$$F_S C_S \frac{dT_S}{dX} = +Q \qquad T_S$$

Heat flux

$$Q$$

$$Q = \pi D U (T_S - T_T)$$

$$F_T C_T \frac{dT_T}{dX} = +Q \qquad T_T$$

Tube side
heat balance

FIG. 9-3. Model for Liquid-Liquid Heat Exchanger

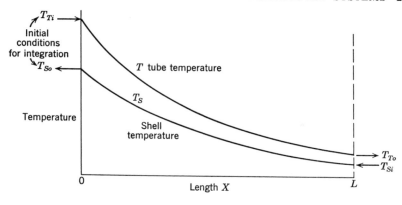

FIG. 9-4

of the tube side (where $X = 0$) the temperature $T_{Ti}$ is known, but the other initial value, the exit temperature on the shell side, $T_{So}$, is unknown. A similar problem is encountered if the computation is started at the other end of the heat exchanger. It becomes necessary then to "guess" the unknown initial value $T_{So}$, to integrate down to the other end, and to compare the end value $T_S$ computed with the known inlet temperature $T_{Si}$. If they are dissimilar, the computation is repeated, starting with a new initial value of $T_{So}$. This procedure is repeated until the computed value of $T_S$ matches the known value. This type of problem, in which the computation has to be repeated to satisfy boundary conditions, is termed a "split boundary value" problem. The results of this computation can be shown plotted as two curves of temperature versus length (Figure 9-4).

The iterations are performed by the computer to converge to a match on the boundary values, and a simple algorithm (method) can be used to converge automatically on the initial values. Because of the high speed of modern analog computers, the computation can be cycled repetitively, and convergence is obtained in a fraction of a second. A large digital machine would be almost as fast on a problem of this magnitude.

The results of this simple case indicate the overall heat capacity that can be obtained from the exchanger and the variations of the capacity with changes of flow, inlet temperature, length, and diameter of the pipe.

EXAMPLE 9-2. PIPELINE GAS FLOW

The flow of a compressible gas down a pipe exhibits a high degree of nonlinearity because the rate of pressure drop is proportional to the square of the velocity. As the pressure decreases in the direction of flow, the gas density will also drop and in turn will cause the velocity to increase. The equation for the isothermal pressure drop is

$$\frac{dP}{dl} = K_1 \frac{V^2 \phi}{D^5}$$

$$V = \frac{M}{\phi}$$

$$\phi = kP$$

where $D$ = diameter,
$\quad V$ = volume flow rate,
$\quad K_1$ = friction factor
$\quad M$ = mole flow rate,
$\quad \phi$ = density,
$\quad P$ = pressure.

The model for the arrangement is shown in Figure 9-5. The foregoing model assumes isothermal expansion. Suppose that an exchange of heat from the inside to the outside of the pipe causes a significant change in the temperature. The model would now have to include the following energy balance:

$$\frac{d}{dl}(MCT) = U\pi D(T_j - T)$$

where $C$ = heat capacity. $T_j$ = jacket temperature.

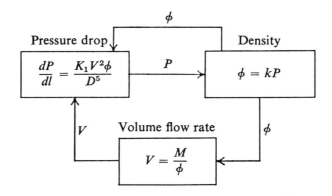

FIG. 9-5. Model for Tubular Gas Flow

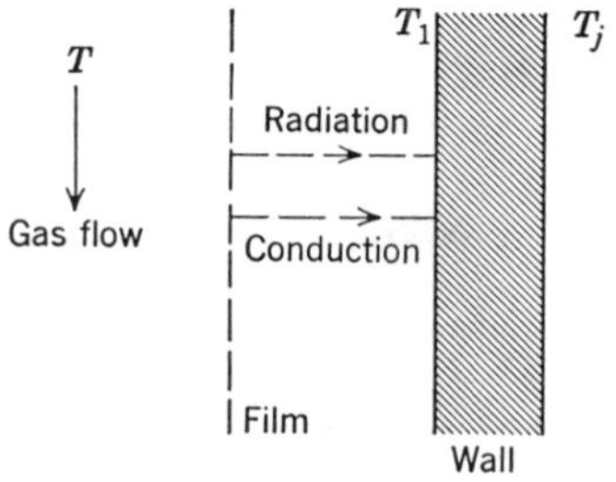

FIG. 9-6. Heat Flow from Gas to Wall

The temperature $T$ must also be considered in the density term, that is,

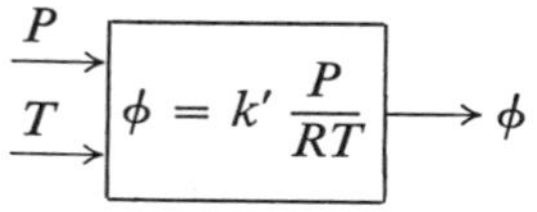

where $R$ = gas constant.

A further refinement can now be added, namely, the variation of the heat transfer coefficient $U$ which results from the variation of the resistance of heat flow in the film on the inside of the pipe because of changes in gas velocities; that is, $U = f(V)$. If high temperatures are involved, a more accurate representation would include the heat of radiation from the pipe wall across the gas film (Figure 9-6).

Heat flux across gas film:

$$Q = \pi D[\overset{\text{conduction}}{h(T - T_1)} + \overset{\text{radiation}}{\alpha(T^4 - T_1{}^4)}]$$

Where $T_1$ = temperature on the inside surface of the pipe. The heat balance equation now becomes

$$\frac{d}{dl}(MCT) = \pi D[h(T - T_1) + \alpha(T^4 - T_1{}^4)]$$

The best way to obtain the temperature $T_1$ at the interface of the wall and the gas film is to equate the flux through the gas film to that through the wall; that is,

$$Q = \pi D[h(T - T_1) + \alpha(T^4 - T_1{}^4)] = U_w \pi D(T_1 - T_j) \rightarrow T_1$$

The complete model for the system is shown in Figure 9-7, which includes the variation of pipe diameter with length; that is, $D = f(l)$. To predict the temperature as a function of length in a proposed design of a gas cooler, these equations have to be solved. The effect of variations in pipe diameter and jacket temperature can also be predicted.

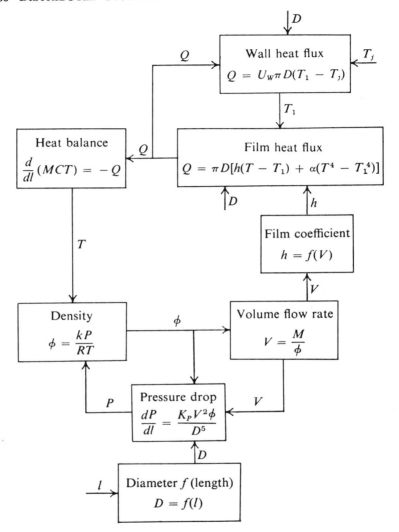

FIG. 9-7. Model for Adiabatic Tubular Gas Flow

EXAMPLE 9-3 (1)

An example mathematically similar to a countercurrent heat exchanger involves separating a mixture of gases by permeating through a semi-permeable material. The apparatus consists of a thin-walled glass tube enclosed in a larger tube, through which flows a mixture of gases $A$ and

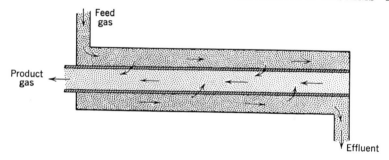

FIG. 9-8. Tubular Permeation Process

$B$ at high pressure (Figure 9-8). Gas permeates from the shell side, flows through the wall of the tube, and out while the excess gas on the shell side flows out at the other end. This arrangement allows the gases on the tube and shell side to flow countercurrently. Suppose that gas $A$ permeates through the wall of the glass tube much faster than gas $B$; it follows then that the gas flowing out of the inner tube will be greatly enriched in component $A$. The rate of permeation of each component depends on the diffusional properties of the glass, the dimensions of the tube, and the partial pressures of that component on either side of the tube wall.

The rate of permeation in an elemental section length $\Delta X$ for component $i$ (Figure 9-9) is

$$\text{flow rate} = (\text{permeation coefficient})(\text{area})\left(\frac{\text{pressure difference}}{\text{wall thickness}}\right)$$

$$Q_{Pi} = \beta_i(\pi D_m \, \Delta X)\left(\frac{P_{Si} - P_{Ti}}{t}\right)$$

where $Q$ = flow rate SCFM
$\quad D_m$ = mean diameter
$\quad t$ = wall thickness

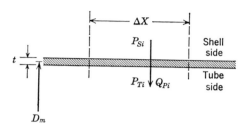

FIG. 9-9. $X$ Section of Tubular Permeation Process

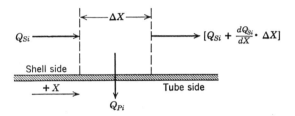

FIG. 9-10. Flowrate Change Across $X$ Section

The partial pressure of the gas is the mole fraction of the component times the total gas pressure; that is,

$$P_{Si} = Y_{Si}P_S \quad \text{and} \quad P_{Ti} = Y_{Ti}P_T$$

The mass balance for component $i$ on the shell side can be derived in a manner similar to that of the differential equation for the heat exchanger (Figure 9-10):

$$\text{in} = \text{out}$$

$$Q_{Si} = \left(Q_{Si} + \frac{dQ_{Si}}{dX} \cdot \Delta X\right) + Q_{Pi}$$

NOTE. $X$ is positive with shell side flow.

This equation reduces to

$$\frac{dQ_{Si}}{dX} = -\frac{Q_{Pi}}{\Delta X}$$

Substituting for $Q_{Pi}$ and $P_{Si}$ and $P_{Ti}$ in the above equations,

$$\frac{dQ_{Si}}{dX} = -\beta_i\left(\frac{\pi D_m}{t}\right)(P_S Y_{Si} - P_T Y_{Ti})$$

A similar equation is obtained for the tube side flow

$$-\frac{d}{dX} Q_{Ti} = \beta_i\left(\frac{\pi D_m}{t}\right)(P_S Y_{Si} - P_T Y_{Ti})$$

The compositions $Y$ (mole fraction) can be defined in terms of the volumetric flow because moles and standard volume are equivalent

$$Y_{TA} = \frac{Q_{TA}}{Q_{TA} + Q_{TB}} \qquad Y_{SA} = \frac{Q_{SA}}{Q_{SA} + Q_{SB}}$$

(similar expressions are obtained for component $B$). The pressure on the shell side can be considered constant, but, because of the restricted confines inside the tube, there is a significant pressure drop which has to be computed; that is, $P_T$ is a function of $X$. This pressure drop is defined by the laminar flow equation

$$\frac{dP_T}{dX} = k_f \frac{\mu W}{\phi_M D^4}$$

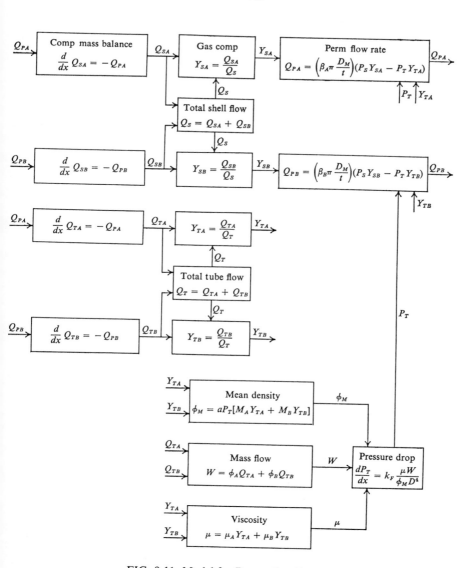

FIG. 9-11. Model for Permeation Process

where $k_f$ is the friction factor, $\mu$ is the viscosity, $\phi_M$ is the mean density, and $D$ is the internal diameter of the tube.

Auxiliary equations defining these variables in terms of components $A$ and $B$ are (assuming ideal mixtures)

$$\text{viscosity } \mu = \mu_A Y_{TA} + \mu_B Y_{TB}$$

$$\text{mass flow } W = \phi_A Q_{TA} + \phi_B Q_{TB}$$

$$\text{density } \phi_M = aP_T(M_A Y_{TA} + M_B Y_{TB})$$

where $M_i$ = molecular weight for component $i$.

A complete arrangement for these equations is shown in Figure 9-11. The arrangement is quite logical; the difference in partial pressures establishes the permeation flow rate, whereas the mass balance equations define the flows in the shell side and tube side. As in the preceding example, the equations are solved by integrating from $X = 0$ to $X = L$. When $X = 0$, the composition and flow on the shell side are known because this is the shell side inlet. The initial conditions for the two differential equations defining shell side flow $Q_{SA}{}^0$ and $Q_{SB}{}^0$ are therefore known, but the initial conditions for the tube flows $Q_{TA}{}^0$ and $Q_{TB}{}^0$ must be estimated. The boundary value criterion is that the tube flows are zero at $X = L$; the initial estimates are adjusted accordingly. The initial tube pressure at $X = 0$ is also known and will increase to reach a maximum at $X = L$. A set of typical results is shown in Figure 9-12.

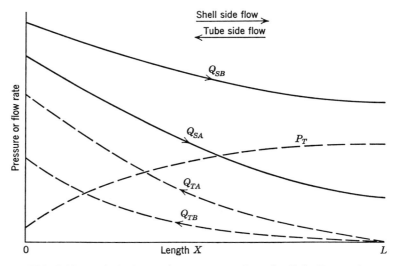

FIG. 9-12. Typical Flow and Pressure Gradients for Tube Permeation

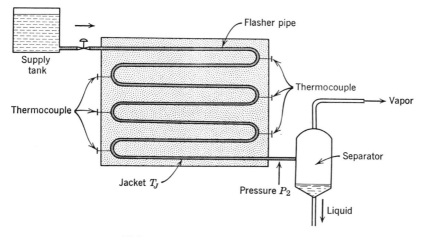

FIG. 9-13a. Pipeline Flasher Process

The amount and composition of permeate product gas varies with the tube diameter, length, flow rate of supply gas, and pressure levels on either side of the tube. By varying these conditions in the mathematical model and computing various cases the optimum conditions and tube dimensions can be found.

EXAMPLE 9-4. FLASHER DESIGN

The pipeline "flasher" in this process is basically a heat exchanger whose function is to boil partially a multicomponent mixture. This is accomplished by flowing the mixture down a heated pipe, the vapor formed being removed at the end of the pipe. The remaining fluid must achieve a desired composition at the end of the pipe, and it must be realized that this is a complex function of the temperature-length history developed by the fluid while it is flowing down the pipe. Figure 9-13a shows the general scheme of such a pipe (simplified in 9-13b) through which flows the mixture to be flashed, heated on the outside by a suitable jacket maintained at a

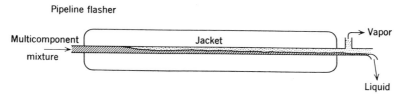

FIG. 9-13b. Pipeline Flasher

constant temperature. Experience obtained on a small scale experimental apparatus allows the following analysis to be made:

The first observable change that occurs as the fluid enters the tube is a rise in fluid temperature from the inlet temperature to the boiling point. The equation that covers this first phase is

$$\left.\begin{array}{c}\text{rate of change of heat} \\ \text{with respect to length} \\ \text{PCU/ft time}\end{array}\right\} = \left\{\begin{array}{c}\text{heat transferred through} \\ \text{the pipewall per unit length} \\ \text{PCU/ft time}\end{array}\right.$$

$$\frac{d}{dl}(WCT) = UA(T_j - T)$$

During this phase the pressure drop that occurs is described by the laminar flow equation.

$$\frac{dP}{dl} = \frac{K_1 W}{D^4}$$

where $W$ = total mass flow rate,
$\quad D$ = pipe diameter,
$\quad K_1$ = friction coefficient $\left[= \dfrac{32\mu}{\phi g(\pi/4)}\right]$.

When the boiling temperature is reached, flashing starts, and from this point on a mixture of vapor and liquid flows down the pipe. The fraction of vapor continually increases as boiling proceeds. The rate of pressure drop with pipe length for a mixture of vapor and liquid is expressed by the following equation (5, 2, 3):

$$\frac{dP}{dl} = \frac{K_2 W^a V^b}{D^5}$$

where $V$ is the mass flow of vapor. As $V$ increases, then, the rate of pressure drop $dp/dl$ also increases.

The temperature after boiling starts is determined by the pressure and composition of the liquid, and the conventional vapor-liquid relationship developed in preceding chapters has to be used; that is,

$$Y_i = X_i \frac{f_i(T)\gamma_i}{P}$$

where $Y_i$ = vapor composition of component $i$,
$\quad X_i$ = liquid composition of component $i$,
$\quad f_i(T)$ = vapor pressure of pure component $i$,
$\quad \gamma_i$ = activity coefficient.

The temperature is obtained by the balance

$$\Sigma \, Y_i = 1 \to T$$

As the liquid travels down the pipe, its composition and pressure $P$ change; both contribute to the temperature change. The heat balance equation after vaporization begins, becomes

rate of change of heat content = jacket heat flux − heat of vaporization

$$\frac{d}{dl}(W^\circ CT) = UA(T_j - T) - v\lambda$$

where $v =$ vaporization rate,
$\quad \lambda =$ latent heat,

and is now used to establish the vaporization rate $v$. The total vapor flow at any point down the length of the pipe is obtained from the simple relationship

$$\frac{dV}{dl} = v$$

Also, the mass balance on the liquid gives

$$L = L_0 - v.$$

The vapor composition $Y_i$ used in the vapor-liquid equilibrium equations is the composition of the vaporizing flow $v$ in equilibrium with the liquid composition $X$. It is not necessary here to make the assumption that the composition of the main body of vapor $V$ is in equilibrium with the liquid. Hence the vapor composition $Y_{vi}$ is obtained by the following component mass balance:

$$\frac{d}{dl}(VY_{vi}) = vY_i$$

Similarly, the liquid composition is obtained from the component balance

$$\frac{d}{dl}(LX_i) = -vY_i$$

Assembling these equations into a model suitable for computer programming raises an interesting point that has not yet been covered. Most process systems can be classified into either boiling or nonboiling systems; in this case, however, both regimes occur in the same process. This can be handled in two ways; in the first we use a "logic" statement or "relay" that switches from one equation to another as soon as the actual temperature has reached the equilibrium temperature, as in Figure 9-14. In this scheme the actual temperature is obtained from the heat balance equation when $T < T_E$, but as soon as $T = T_E$ the logic statement "switches" so that $T_{ACT}$ is obtained from the equilibrium equation. At the same time the criterion activates a new heat balance equation that includes the vaporization term to compute for $v$, the vaporization flow.

A more elegant approach, which, incidentally, is closer to the "natural" microscopic mechanism (see Chapter 5), is to use the complete heat balance equation for both regimes to solve for temperature, "compare" this temperature with the equilibrium temperature, and derive a vapor flow $v$ by defining

$$v = 0 \qquad T < T_E$$

$$v = G(T - T_E) \quad T > T_E$$

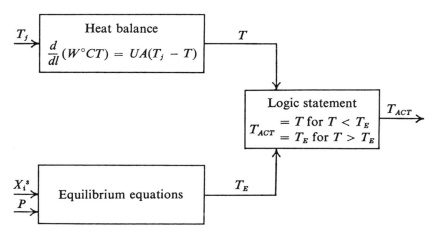

FIG. 9-14. Model for Establishing Temperature

If $G$ is sufficiently large, $T$ will not differ significantly from $T_E$. The model now becomes that shown in Figure 9-15, an arrangement that is valid for both regimes. The equation $v = G(T - T_E)$ must be translated into a suitable algorithm when programmed for a digital computer, as described in preceding chapters.

This arrangement, although straightforward, is often costly in computer time, for the large value of $G$ necessary to force $T \approx T_E$ reduces the effective "time constant" of the heat balance equation to a small value. This automatically forces the integration step size in a digital computer to a very small value and causes excessive run times for a desired integration interval. For such situations it is advisable to program the macroscopic model of Figure 9-14.

The most significant and critical parameter in this system is the heat transfer coefficient $U$. This coefficient is composed of the inside and outside film resistance plus the resistance through the pipe wall, which is usually small. The outside film resistance can also be considered small if the heating medium is condensing steam. The inside film resistance is a function of the nature of the two-phase flow that occurs inside the tube. In a flashing operation such as this sufficient heat is being transmitted through the walls of the pipe to vaporize the liquid. Hence the liquid is, to a large extent, flowing through the center of the pipe inside a vapor envelope coating the walls. As part of the liquid stream touches the walls, flashing occurs and the vapor envelope is replenished. Using this concept, we can

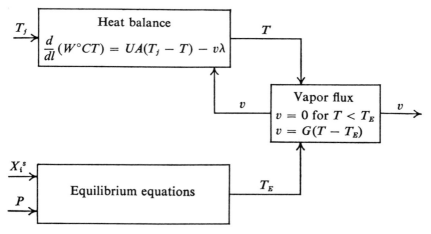

FIG. 9-15. Alternative Model for Establishing Temperature

form an empirical equation to reflect an increased resistance to heat transfer as the ratio of vapor to liquid increases; that is,

$$U = \frac{U_L}{1 + k_4(v/W)^n}$$

$$U_L = K_3 \frac{W^{a1}}{D^{b1}}$$

By adjusting the parameter $K_4$ and exponent $n$ we can achieve values that provide a match with experimental data.

A complete assembly of the equation for this system appears in Figure 9-16$a$ as an information flow diagram. Attention is drawn to the logical use

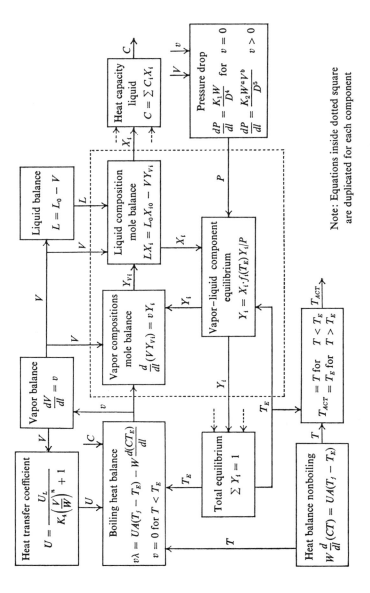

FIG. 9-16a. Model for Pipeline Flasher

Note: Equations inside dotted square are duplicated for each component

276

of each equation; that is, the component mass balances are used to establish the liquid and vapor compositions, the equilibrium sets the temperature after flashing by rebalancing through the heat-balance equation, and so on.

Mechanizing such a mathematical model for a computer presents little difficulty, because it can be treated as a straightforward integration down the length of the flasher. If a downstream pressure has to be achieved at the end of the computation, then repeated runs must be made in which the starting pressure is varied until the end condition is achieved.

EXAMPLE 9-5. TUBULAR REACTOR

Many commercial gaseous reactions are carried out by feeding the gases through a long tube packed with catalyst. As the gases flow through the tube, they will react on the surface of the catalyst. The determination of the actual kinetics for any of these processes could answer many questions of economic interest, such as the best possible yield and the determination of the optimum thermal history. The case considered in this example is the determination of rate constants for the decomposition of acetylene and ethylene. Ethylene decomposes to form acetylene and hydrogen, and acetylene decomposes to form carbon and hydrogen.

$$\text{ethylene} \quad C_2H_4 \rightarrow C_2H_2 + H_2$$

$$\text{acetylene} \quad C_2H_2 \rightarrow C_2 + H_2$$

Experimental data were obtained from an apparatus (Figure 9-16$b$) that consisted of a single tube heated in a resistance furnace. Hydrocarbon and diluent feeds were metered into the tube. The off-gas was quenched with a water spray downstream of the reaction zone, and gas samples for chemical and mass-spectrometer analysis were taken from a gas-liquid separator. Thermocouples measured the gas temperature within the tube, and a complete spectrum of materials found in the off-gas was measured by wet chemical methods. The second set of thermocouples measured the wall temperature on the inside of the tube. The mathematical model was based on the assumption of a first-order reaction for the decomposition of ethylene and a first-order reaction for the decomposition of acetylene. It was

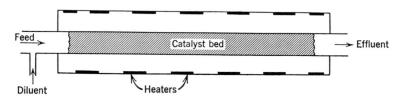

FIG. 9-16$b$. Pipeline Reactor

assumed that the reaction rate coefficients followed the Arrhenius relation $k = Ae^{-B/T_G}$, where $T_G$ is the absolute temperature of the gas. The reaction rate coefficients for each reaction are defined as

$$\text{ethylene} \quad k_1 = A_1 e^{-B_1/T_G}$$

$$\text{acetylene} \quad k_2 = A_2 e^{-B_2/T_G}$$

and the reaction rates are

$$R_1 = k_1 P Y_1$$

where $P$ = gas pressure,
$\quad Y_1$ = mole fraction of ethylene.

NOTE. The units of $R$ in this case would be lb moles/ft³ hr. Similarly,

$$R_2 = k_2 P Y_2$$

where $Y_2$ = mole fraction of acetylene.

The mass balance equation for each component in the gas is developed in the classical manner as follows (Figure 9-17):

flow rate of component into section $\Delta X = MY$

flow rate of component out of section $\Delta X = MY + \Delta X \dfrac{d}{dX}(MY)$

where $M$ = total mole flow.

reaction rate in section $\Delta X$ = reaction rate × volume

$$= R \cdot \left( \frac{\pi D^2}{4} \cdot \Delta X \right)$$

where $D$ = tube diameter.

Mass balance: in = out

$$MY = \left[ MY + \Delta X \cdot \frac{d}{dX}(MY) \right] + R\left( \Delta X \cdot \frac{\pi D^2}{4} \right)$$

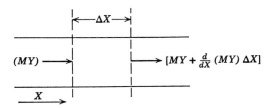

FIG. 9-17. $\Delta X$ Section of Pipeline Reactor-Material Balance

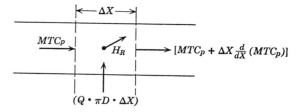

FIG. 9-18. $\Delta X$ Section of Pipeline Reactor-Heat Balance

This equation reduces to

$$\frac{d}{dX}(MY) = -R\left(\frac{\pi D^2}{4}\right)$$

So, for ethylene, we have

$$\frac{d}{dX}(MY_1) = -R_1\left(\frac{\pi D^2}{4}\right)$$

and for acetylene

$$\frac{d}{dX}(MY_2) = (R_1 - R_2)\frac{\pi D^2}{4}$$

For total moles

$$\frac{d}{dX}(M) = R_1\left(\frac{\pi D^2}{4}\right)$$

because two moles of gas ($C_2H_2$ and $H_2$) are formed from one mole of ethylene in the first reaction.

The temperature of the gas is obtained by solving a heat balance equation obtained as follows (Figure 9-18):

heat in = heat out

$$MT_GC_p + H_R\left(\frac{\pi D^2}{4}\Delta X\right) + Q\cdot(\pi D\cdot\Delta X) = MT_GC_p + \Delta X\frac{d}{dX}(MT_GC_p)$$

This equation reduces to

$$\frac{d}{dX}(MT_GC_p) = \left(Q + H_R\cdot\frac{D}{4}\right)\pi D$$

The rate of transmission of heat $Q$ from the furnace heaters across the walls to the gas flowing inside the tube can be approximated by the following expression

$$Q = h(T_W - T_G)$$

The value of the coefficient $h$ can be established by flowing an inert gas at a similar rate down the tube, measuring the rise in temperature, and matching the results by simulating the simplified version of the heat-balance equation

$$\frac{dT}{dX} = \frac{h \cdot \pi D}{MC_p} (T_W - T_G)$$

Since $h$ will vary for the packed and unpacked sections of the tube, it has to be defined as a function of the length $X$, that is, $h = f(X)$. Also, for each particular run, the wall temperature $T_W$ is defined as a function of length; that is, $T_W = f(X)$, as shown in Figure 9-19. The specific heat $C_p$ of the gas will vary with composition; that is,

$$C_p = Y_1 C_{p1} + Y_2 C_{p2} + Y_3 C_{p3}$$

where $Y_3$ is the composition of diluent gas.

The heat of reaction $H_R$ is defined in terms of the separate reactions as follows:

$$H_R = R_1 \cdot \Delta H_{R1} + R_2 \cdot \Delta H_{R2}$$

where $\Delta H_{R1}$ = heat of dehydrogenation of ethylene,
$\Delta H_{R2}$ = heat of acetylene decomposition.

The pressure of the gas as it travels through the packed section can be computed from the approximate expression that assumes a linear gradient.

$$\frac{dP}{dX} = \frac{P_0 - 1}{L}$$

NOTE. $P$ at $X = L$ is 1 atm.

where $P_0$ = inlet pressure (atm),
$L$ = length of tube (total).

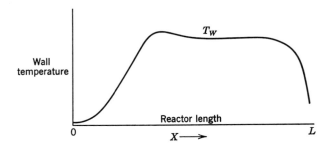

FIG. 9-19. Pipeline Reactor Wall Temperature Gradient

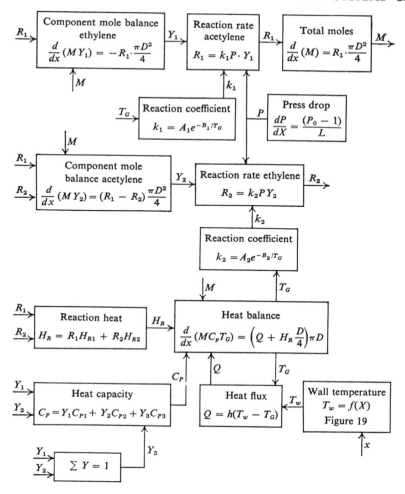

FIG. 9-20. Model for Pipeline Reactor

All of these equations are assembled in a model in Figure 9-20. Observation of how each equation is used will show that the arrangement is similar to the models discussed in preceding chapters. It is also apparent that in this case, because all the initial conditions to the differential equations are known at $X = 0$, the computation is not a "split boundary value problem" as in some of the other examples.

A computer study based on the mathematical model in Figure 9-20 would calculate the compositions resulting from various cases of input flow rate, and ambient temperature conditions by assuming reaction rate constants $A_1$, $B_1$, and $A_2$, $B_2$ and comparing the computed compositions with experimentally measured values; optimum values of $A$'s and $B$'s are obtained which best fit the data. Using these values, it is then possible to proceed on computer studies to scale-up the reactor to commercial size or to find the optimum design to obtain maximum yield (for a similar example see Reference 13).

## CONDENSATION

The particular phenomenon to be studied here is heat and mass transfer of the barrier-diffusion type, as exemplified by the process of total or partial condensation. In preceding chapters condensers have been defined by the simple assumption that the multicomponent vapor temperature is reduced to a specified value and that as it leaves the condenser it is in equilibrium with the condensate formed. Such a simplification is often quite adequate, especially for oversized condensers (a common situation). There are situations, however, in which this simplification cannot be made because the presence of a gas film barrier on the condensate surface prevents equilibrium from being achieved. For these a different approach is required, which involves a mass transfer mechanism across the gas film covering the condensate. The next three examples develop the approach starting from the simplest case to a general multicomponent system. All three examples assume steady-state condensation inside vertical tubes with coolant on the outside.

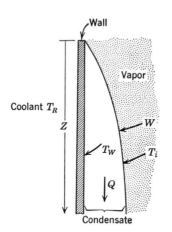

FIG. 9-21. Condensation on Vertical Wall

EXAMPLE 9-6. CONDENSATION OF A PURE VAPOR (6, 7)

The simplest condensation system is a single-component vapor condensing on a vertical wall, with a coolant on the other side of constant uniform temperature $T_R$ (Figure 9-21). As the vapor condenses on the wall, a condensate layer forms and flows down the wall. This layer decreases the overall heat transfer coefficient from vapor to coolant. To determine the effectiveness of the condensing surface a relationship between the distance $Z$ and the rate of condensation $W$ is required.

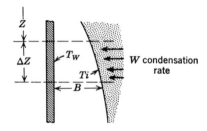

FIG. 9-22. $\Delta X$ Section of Condensation System

The system is assumed to be in steady state, and therefore at any point on the surface of the condenser the heat flux across the condensate layer equals the flux across the wall into the coolant. Now consider a thin slice $\Delta Z$, (Figure 9-22). Neglecting the sensible heat carried by the condensate, the heat flux across the condensate is

$$H = K\frac{(T_i - T_w)}{B} = \text{conductivity} \times \text{temperature gradient}$$

where $K$ is the thermal conductivity for the condensate fluid (PCU/(°C/ft)· time·length$^2$) and $B$ is the thickness of the condensate layer at that point in the condenser ($Z$). Also,

$T_i$ = temperature at the condensate/vapor interface
$T_w$ = temperature at the condensate/wall interface

The heat flux across the wall is

$$H = U_w(T_w - T_R)$$

where $U_w$ = wall heat transfer coefficient,
$T_R$ = coolant temperature.

The heat flux $H$ is the heat liberated by the condensing vapor at the surface of the condensate and so

$$H = W\lambda$$

where $W$ = condensation rate,
$\lambda$ = latent heat.

The condensate thickness $B$ at any level $Z$ can be obtained from the flow rate $Q$ of condensate at that level by the following relationship (1):

$$Q = \frac{\phi g B^3}{3\mu}$$

where $\mu$ = viscosity,
$\phi$ = density,
$g$ = 32 ft/sec².

The above expression for $Q$ is derived below.

NOTE. The student should already be familiar with derivations such as the following. It is included here as a reminder and to show how mathematics can be used to derive *simple* relationships that are a part of larger complex mathematical models.

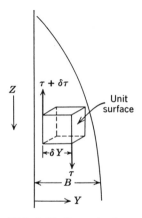

FIG. 9-23. Stress in Condensate Layer

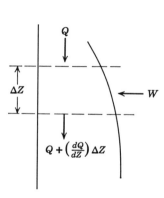

FIG. 9-24. Material Balance in Section $\Delta Z$

Consider a small elemental volume in the condensate film (Figure 9-23) with a unit surface and thickness $\delta y$. The stress on the unit surface nearest the film surface is $\tau$ and that closer to the wall is $(\tau + \delta\tau)$. Now

$$\delta\tau = -\frac{d\tau}{dy}\,\delta y$$

Because the fluid is in steady motion at this point and is not accelerating, the gravity force must be balanced by the inequality in the shear stress.

$$\phi g\,\delta y = -\frac{d\tau}{dy}\,\delta y$$

or

$$\frac{d\tau}{dy} = -\phi g$$

From the basic definition of viscosity, $\tau = \mu \cdot (du/dy)$, where $u = $ velocity,

$$\frac{d\tau}{dy} = \mu \frac{d^2u}{dy^2} = -\phi g$$

$$\therefore \frac{d^2u}{dy^2} = -\frac{\phi g}{\mu} \tag{1}$$

Integrating (1),

$$\frac{du}{dy} = -\frac{\phi g}{\mu} Y + C_1 \tag{2}$$

Now, at $y = B$,

$$\frac{du}{dy} = 0$$

$$\therefore C_1 = \frac{\phi g}{\mu} B$$

Integrating (2),

$$u = -\frac{\phi g}{\mu} \cdot \frac{Y^2}{2} + \frac{\phi g}{\mu} BY + C_2$$

at $y = 0$, $u = 0$, giving $C_2 = 0$,

$$\therefore u = \frac{\phi g}{\mu} \left( BY - \frac{Y^2}{2} \right) \tag{3}$$

The flow rate $Q$ at a distance $Z$ is given by integrating the velocity $u$ across the film from $y = 0$ to $y = B$; that is,

$$Q = \int_0^B u \, dy$$

Substituting (3) for $u$ and integrating, we obtain

$$Q = \frac{\phi g B^3}{3\mu}$$

Now $Q$ is the accumulation of condensate from 0 to $Z$, or, in differential form, the rate of increase in $Q$ with respect to distance $Z$ is the condensation rate $W$; that is, (Figure 9-24) by mass balance,

$$\text{in} = \text{out}$$

$$Q + W \cdot \Delta Z = Q + \Delta Z \cdot \frac{dQ}{dZ}$$

$$W = \frac{dQ}{dZ}$$

Because this is a single-component condensation, the surface temperature of the condensate $T_i$ is the boiling point of that component at the operating pressure $\pi$ of the condenser. The equations that have been developed for this system can now be listed:

$$H = W\lambda \qquad \text{heat flux = condensation rate}$$

$$\frac{dQ}{dZ} = W \qquad \text{change of condensate flow = condensation rate}$$

$$Q = \frac{B^3\phi g}{3\mu} \qquad \text{condensate layer thickness} = f(\text{condensate flow})$$

$$H = \frac{K(T_i - T_w)}{B} \qquad \text{heat flux across condensate film}$$

$$H = U_W(T_W - T_R) \qquad \text{heat flux across wall}$$

There are thus five equations to solve for the five unknowns $H$, $Q$, $B$, $T_W$, $W$.

The arrangement of these equations into a model requires an understanding of the fundamental mechanism of this process; namely, what establishes the condensation rate $W$? A moment's thought will reveal that the heat flux $H$ establishes the condensation rate $W$, and, in turn, the heat flux is established by the condensate film resistance between the known

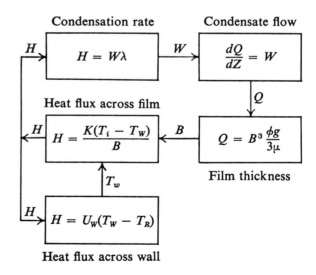

FIG. 9-25. Model for Condensation System

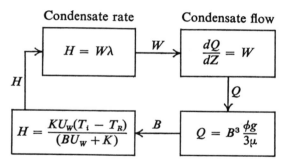

FIG. 9-26. Alternative Model for Condensation System

temperature $T_i$ and the coolant temperature $T_R$. The information flow block diagram for this model, which shows this cause and effect relationship, is illustrated in Figure 9-25, examination of which will reveal a potential computational difficulty. At $Z = 0$, $B = 0$, so that in the film heat flux equation the form $0/0$ is encountered. This could be made finite by assuming a small starting value for $B$ at $Z = 0$, but a better way is to eliminate $T_W$ from the pair of heat flux equations by defining an overall coefficient from $T_i$ to $T_r$. This leads to the form

$$H = \frac{KU_W}{BU_W + K}(T_i - T_R)$$

A further advantage of this form is that $H$ is obtained *explicitly* from $B$, with no minor loop iterations required. The complete model with this modification is shown in Figure 9-26. A computer solving these equations in this manner produces curves of condensation rate $W$ versus length $Z$ for various conditions of coolant temperature, vapor pressure (i.e., $T_i$), wall thickness, and so forth, from which a rating can be made.

EXAMPLE 9-7. CONDENSATION OF MULTICOMPONENT VAPORS (8, 9, 10)

The more complex case of condensation of mixed vapors can now be developed. There is a fundamental difference between the condensation of pure vapors and the condensation of vapors in the presence of an excess of less condensable vapors. For the pure vapor described in Example 9-6 the condensation rate is controlled by the rate at which the evolved latent heat of condensation can be transferred to the coolant, whereas for mixed vapors the condensation rate is determined largely by the rate of diffusion of the vapor to the condensing surface through a layer of either incondensable gas or more slowly condensing vapors. For equimolar counter-diffusion through a vapor film the rate of arrival of any component $j$ at

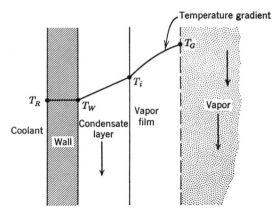

FIG. 9-27. Temperature Gradient from Vapor to
Coolant

the condensate surface is a diffusion factor $F_j$ (11) times the composition
gradient $dy_j/d\eta$, where $\eta$ is the normalized vapor film thickness (Figure
9-27), that is,

$$W_j = F_j \frac{dy_j}{d\eta}$$

After integrating this equation between limits $\eta = 0$ and $\eta = 1$, we obtain
the component condensation rate $W_j = F_j(y_j - y_{ij})$, where $y_j$ is the component composition in the vapor stream and $y_{ij}$ is the composition at the
interface in equilibrium with the condensate composition $X_{ij}$.

For the more general case of condensation, the rate of arrival of a component at the condensate surface is defined as $Wf_j$, where $W$ is the total
condensation rate at any point and $f_j$ is the mole fraction of any particular
component $j$ in the condensing stream. This rate will be the sum of the
diffusion rate $F_j \, dy_j/d\eta$, plus the bodily transport of matter across each
plane in the gas film $Wy_j$, that is,

$$Wf_j = F_j \frac{dy_j}{d\eta} + Wy_j$$

Integration of this equation between limits $\eta = 0$ and $\eta = 1$ results in

$$W = F_j \log_n \left( \frac{f_j - Y_{ij}}{f_j - Y_j} \right)$$

where $Y_{ij}$ = interface vapor composition,
$Y_j$ = vapor composition.

The latent heat liberated at the surface is

$$q\lambda = W \sum_j f_j\lambda_j$$

The sensible heat $q_s$ arriving at the surface arises from the heat flux due to the temperature gradient $(T_G - T_i)$, that is, gas temperature-interface temperature, plus the sensible heat carried to the interface by the condensing molecules:

$$\text{sensible heat} = \text{conduction} + \text{convection}$$

$$q_v = h_G \frac{dT_G}{d\eta} + WC_f(T_f - T_i)$$

After integrating this equation between limits $\eta = 0$ to $\eta = 1$, we reduce the sensible heat flux $q_s$ to

$$q_s = h_G(T_G - T_i)A_k$$

where $A_k$ = Akerman correction,

$$= -\frac{\alpha}{1 - e^{-\alpha}},$$

$$\alpha = \frac{WC_f}{h_G}.$$

The total heat arriving at the surface is

$$q_s + q_\lambda = h_G(T_G - T_i)A_k + W \sum_j f_j\lambda_j$$

Consider now the case of a condensable vapor and an inert gas flowing down a tube with coolant flowing countercurrently (Figure 9-28). The

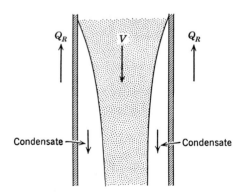

FIG. 9-28.  Build Up of Condensate Layer

surface temperature of the condensate $T_i$ is established by the partial pressure of the condensable vapor in equilibrium with the surface, that is, $T_i = f(PY_i)$, where $P$ is the pressure of the gas stream. Adapting the diffusion equation to this case ($f_j = 1$) gives the condensation rate

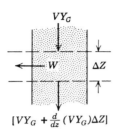

$[VY_G + \frac{d}{dz}(VY_G)\Delta Z]$

FIG. 9-29. Material Balance on Gas Phase Section $\Delta Z$

$$W = F \log_n \left(\frac{1 - Y_i}{1 - Y_G}\right)$$

where $Y_G$ is the net fraction of condensable vapor in the gas stream, and $F$ is the diffusion factor for the condensable component through the inert gas. A mass balance on a small slice $\Delta Z$ (Figure 9-29) of the vapor stream gives the following equation:

$$VY_G = -Wa\,\Delta Z + VY_G + \Delta Z \frac{d}{dZ}(VY_G)$$

which reduces to

$$\frac{d}{dZ}(VY_G) = -Wa$$

where $a$ = condensing area/unit length.

The first step toward arranging a model for this system is to list the equations and select the variable for which each equation is to be solved (Table 9-1). In this table one key variable remains undefined, namely, $T_i$,

**Table 9-1**

| | |
|---|---|
| $W = F \log_n [(1 - Y_i)/(1 - Y_G)]$ | diffusion equation, solves for condensation rate $W$ |
| $\dfrac{d}{dZ}(VY_G) = -Wa$ | component mass balance on gas solves for $Y_G$ |
| $\dfrac{d}{dZ}(Q) = +Wa$ | condensate mass balance solves for $Q$ |
| $\dfrac{d}{dZ}(V) = -Wa$ | gas mass balance solves for $V$ |
| $\dfrac{d}{dZ}(VC_G T_G) = -h_G(T_G - T_i)A_k'a$ | heat balance on gas, solves for $T_G$ |
| $A_k' = \alpha/(1 - e^\alpha) \qquad \alpha = WC_F/h_G$ | Akerman correction, solves for $A_k'$ |
| $H = q_s + q_\lambda$ | total heat flux, solves for $H$ |
| $q_s = h_G(T_G - T_i)A_k$ | sensible heat flux, solves for $q_s$ |
| $q_\lambda = W\lambda$ | latent heat flux, solves for $q_\lambda$ |
| $A_k = \alpha/(1 - e^{-\alpha})$ | Akerman correction (condensation) solves for $A_k$ |
| $\pi Y_i = f(T_i)$ | equilibrium at interface, solves for $Y_i$ |

which has to be obtained by a mechanism similar to that developed for the single-component condensation. Regardless of the heat carried by the condensate film (in most cases it is insignificant), $T_i$ has to be such that the gradient $T_i$ to coolant temperature $T_R$ through the condensate layer and wall conducts away from the heat flux $H$ liberated at the condensate surface. In equation form this is stated as

$$H = \frac{KU_w}{BU_w + K}(T_i - T_R) \rightarrow T_i$$

The coolant temperature $T_R$ is obtained by accumulating the heat flux $H$ in the refrigerant flow

$$\frac{d}{dZ}(-T_R Q_R C_R) = Ha$$

NOTE. The refrigerant flow $Q_R$ is countercurrent to the direction of $Z$, and thus the negative sign applies in the above equation.

These equations are arranged as an information flow block diagram in Figure 9-30.

EXAMPLE 9-8. MULTICOMPONENT CONDENSATION (10)

The general case of multicomponent condensation can now be considered. All the mass balance equations for each component in the gas stream are identical to that developed in Example 9-7, that is,

$$\frac{d}{dZ}(VY_{Gj}) = Wf_j a$$

$$\frac{d}{dZ}(V) = -Wa$$

The component mass balance on the condensate is

$$\frac{d}{dZ}(QX_j) = Wf_j a$$

and the total condensate flow is obtained by the total mass balance

$$\frac{d}{dZ}Q = Wa$$

The diffusion formula developed before

$$W = F_j \log_n\left(\frac{f_j - Y_{ij}}{f_j - Y_{Gj}}\right)$$

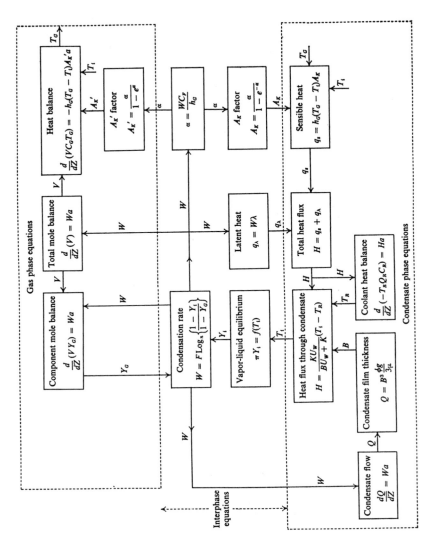

FIG. 9-30. Model for Condensation in the Presence of an Inert Gas

can be rearranged to the more convenient form

$$f_j = Y_{Gj} + \left(\frac{Y_{Gj} - Y_{ij}}{e^{W/F_j} - 1}\right)$$

NOTE. $f_j$ can be negative, indicating counterdiffusion.

This equation is used to establish $f_j$ for each component. The condensate surface temperature is obtained by an equilibrium balance:

$$\pi Y_{ij} = X_j f_j(T_i) \rightarrow Y_{ij}$$

$$\Sigma Y_{ij} = 1 \rightarrow T_i$$

The heat flux $H = q_s + q_\lambda$ is obtained from the heat flux across the condensate layer

$$H = \frac{KU_W}{BU_W + K}(T_i - T_R) \rightarrow H$$

because $T_i$ is obtained by equilibrium, and $T_R$ by coolant heat balance. The sensible heat flux $q_s$ is obtained as in Figure 9-30; therefore by difference $q_\lambda$ is computed; hence the total condensation rate is $W$. The model developed in Example 9-7 can be expanded to accommodate this multicomponent situation by incorporating these changes (Figure 9-31).

Variation in heat capacity, latent heat, and so forth can be readily calculated from the composition, as is explained in preceding chapters. The diffusion factors $F_j$ and gas heat transfer coefficient $h_g$ are complex functions of the Schmidt and Prandtl numbers, which in turn are functions of the Reynolds number and $J$ factor. They all depend on specific heat, viscosity, vapor velocity, and tube diameter. As a first approximation, average constant values for $F_j$ and $h_g$ can be inserted into the model. If a more precise result is required, these factors have to be computed. For details see References 6 and 12. All these equations are assembled in model form in Figure 9-31. If the flow of coolant is countercurrent, the computation becomes a split boundary value problem in which the iteration is on the exit refrigerant temperature.

Other complications in partial condensation can occur, some of which are the following:

1. Reaction in either vapor or condensate stream
2. Separate cooling sections
3. Diffusion resistance within the condensate layer (9)
4. Supercooling of the gas stream, and so forth (11)

These complications can, in many cases, be defined, thus adding more equations to the model and making solution by computer more laborious but not necessarily more difficult.

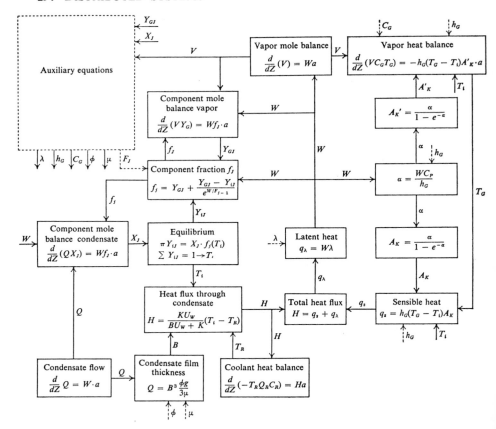

FIG. 9-31. Model for Multicomponent Condensation

It should be evident to the reader that the INT program developed in Chapters 2 and 3 for solving ordinary differential equations is not restricted to systems where time is the independent variable. This chapter deals with unit operations where the variables involved vary with distance but are invariant with time, that is, they are in steady state. The differential equations in Chapters 2 and 3 contain distance as the independent variable, and they are readily programmed with the INT program in the standard procedure described in Chapter 3. In this section several examples are described that demonstrate some typical control techniques frequently required for arriving at a solution.

EXAMPLE 9-9

The first programming example is the countercurrent heat exchanger described at the beginning of this chapter, where it will be assumed that the tube inlet temperature $T_{Bi}$ and the shell-side inlet temperature $T_{Si}$ are known

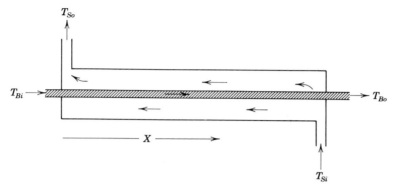

FIG. 9-32. Tabular Countercurrent Heat Exchanger

(Figure 9-32) and the shell-side exit temperature $T_{So}$ is specified. The problem is to find the length of the heat exchanger that will match the specified temperatures and to establish the exit-tube temperature for this calculated length. The equations are

1. Tube temperature

$$\frac{d(TB)}{dX} = \frac{Q}{FWCP}$$

where FWCP = tube fluid flow∗heat capacity = 8000 PCU/(°C min)
$\qquad$ X = distance (ft) from tube inlet

2. Shell temperature

$$\frac{d(TS)}{dX} = \frac{Q}{FBCP}$$

where FBCP = shell fluid flow ∗ heat capacity = 6000 PCU/(°C min)
3. Heat flux shell → tube:

$$Q = UA(TS - TB)$$

where UA = 200 PCU/(min °C ft)

4. Boundary conditions

$\qquad$ shell inlet temperature $T_{Si}$ $\quad$ = 100°C
$\qquad$ shell outlet temperature $T_{So}$ = 40°C
$\qquad$ tube inlet temperature $T_{Bi}$ $\quad$ = 15°C

This problem as specified may appear to be a split boundary value problem, but actually it is not, since both shell and tube temperature are specified at the tube inlet end. The procedure is simply to integrate the differential

```
 1*          DATA FWCP,FBCP,UA,TSI,TBI/8000.,6000.,200.,100.,15./
 2*          STP=2000.
 3*          TS=40.
 4*          TB=TBI
 5*          X=0.
 6*    C **DERIVATIVE SECTION**
 7*        6 Q=(TS-TB)*UA
 8*          DTB=Q/FBCP
 9*          DTS=Q/FWCP
10*          IF(TS.GE.TSI) STP=0.
11*          CALL PRNTF(10.,STP,NF,X,TS,TB,Q,DTB,DTS,0.,0.,0.,0.)
12*          GO TO (4,5),NF
13*    C **INTEGRATION SECTION**
14*        4 CALL INTI(X,5.,4)
15*          CALL INT(TS,DTS)
16*          CALL INT(TB,DTB)
17*          GO TO 6
18*        5 STOP
19*          END
```

FIG. 9-33.   Program for Tabular Heat Exchanger

equations and to stop when the shell temperature reaches the specified inlet temperature of 100°C.

The program for this procedure is shown in Figure 9-33, where the derivative and algebraic equations are grouped in the derivative section in a logical sequence, followed by the integration section where the integrations are performed.

In the preliminary section, the integration limit (STP), that is, the value of length X where the calculation must stop, is specified as a number (2000) well

| LENGTH | TS | TB | Q | DTB | DTS |
|---|---|---|---|---|---|
| .00000 | .40000+02 | .15000+02 | .50000+04 | .83333+00 | .62500+00 |
| .10000+02 | .45997+02 | .22996+02 | .46002+04 | .76670+00 | .57503+00 |
| .20000+02 | .51514+02 | .30352+02 | .42324+04 | .70540+00 | .52905+00 |
| .30000+02 | .56590+02 | .37120+02 | .38940+04 | .64900+00 | .48675+00 |
| .40000+02 | .61263+02 | .43347+02 | .35827+04 | .59711+00 | .44783+00 |
| .50000+02 | .65557+02 | .49076+02 | .32962+04 | .54937+00 | .41203+00 |
| .60000+02 | .69510+02 | .54347+02 | .30327+04 | .50544+00 | .37908+00 |
| .70000+02 | .73147+02 | .59196+02 | .27902+04 | .46503+00 | .34877+00 |
| .80000+02 | .76494+02 | .63658+02 | .25671+04 | .42785+00 | .32089+00 |
| .90000+02 | .79572+02 | .67763+02 | .23618+04 | .39364+00 | .29523+00 |
| .10000+03 | .82405+02 | .71540+02 | .21730+04 | .36217+00 | .27162+00 |
| .11000+03 | .85011+02 | .75015+02 | .19992+04 | .33321+00 | .24991+00 |
| .12000+03 | .87409+02 | .78212+02 | .18394+04 | .30657+00 | .22992+00 |
| .13000+03 | .89615+02 | .81153+02 | .16923+04 | .28205+00 | .21154+00 |
| .14000+03 | .91645+02 | .83860+02 | .15570+04 | .25950+00 | .19463+00 |
| .15000+03 | .93512+02 | .86350+02 | .14325+04 | .23875+00 | .17907+00 |
| .16000+03 | .95230+02 | .88640+02 | .13180+04 | .21966+00 | .16475+00 |
| .17000+03 | .96811+02 | .90748+02 | .12126+04 | .20210+00 | .15158+00 |
| .18000+03 | .98265+02 | .92687+02 | .11157+04 | .18594+00 | .13946+00 |
| .19000+03 | .99603+02 | .94471+02 | .10264+04 | .17107+00 | .12831+00 |
| .19500+03 | .10023+03 | .95309+02 | .98456+03 | .16409+00 | .12307+00 |

FIG. 9-34.   Numerical Results of Tabular Heat Exchanger Problem

beyond the expected length. Statement 10 in the derivative section checks TS against TSI, and when this is satisfied, STP is changed to 0. so that on the succeeding legitimate pass the print subroutine will automatically stop and print the final result at that value of length X.

The numerical results are shown in Figure 9-34, where it is seen that 195 ft are required for the exchanger length and that the exit tube temperature is 95.3°C.

EXAMPLE 9-10  SPLIT BOUNDARIES

By specifying the heat exchanger in the previous case as having a length of 200 ft and inlet shell and tube temperature of 100° and 150°C respectively, the calculation becomes a split boundary value problem that requires multiple passes in order to converge on the solution.

The calculation starts at the tube inlet end (Figure 9-32) with $T_{Bi}$, the known inlet tube temperature of 15., as the initial condition for $T_B$ and an estimated value of 35. for TSO, the exit shell temperature. The program is shown in Figure 9-35, where these substitutions are made on lines 2 and 3. The derivative and integration sections are the same as in the previous case, except that the computation finishes at X = 200 ft. At this point the calculation proceeds to the Exit Convergence section. Here an error term E is generated by comparing the shell temperature TS at X = 200 ft with the specified inlet temperature TSI. If this is less than a tolerance level of 0.1 (line 18), the calculation is complete, if not, a new value for the exit shell temperature TSO is calculated from the error value E using an arbitrary gain factor of 0.4 (line 19). The calculation returns to the initiation section at X = 0, and the run is repeated. This continues until the inlet shell temperature matches TSI within 0.1°C. The value of TSO for each trial for this

```
 1*          DATA FWCP,FBCP,UA,TSO,TSI,TBI/8000.,6000.,200.,35.,100.,15./
 2*        7 TS=TSO
 3*          TB=TBI
 4*          X=0.
 5*    C **DERIVATIVE SECTION**
 6*        6 Q=(TS-TB)*UA
 7*          DTB=Q/FBCP
 8*          DTS=Q/FWCP
 9*          CALL PRNTF(10.,200.,NF,X,TS,TB,Q,DTB,DTS,0.,0.,0.,0.)
10*          GO TO (4,5),NF
11*    C **INTEGRATION SECTION**
12*        4 CALL INTI(X,5.,4)
13*          CALL INT(TS,DTS)
14*          CALL INT(TB,DTB)
15*          GO TO 6
16*    C **EXIT CONVERGENCE**
17*        5 E=TS-TSI
18*          IF(ABS(E).LT..1) STOP
19*          TSO=TSO-E*.4
20*          GO TO 7
21*          END
```

FIG. 9-35.  Program for Split-Boundary Value Tabular Heat Exchanger

| LENGTH X | TS | TB | Q | DTB | DTS |
|---|---|---|---|---|---|
| .00000 | .39744+02 | .15000+02 | .49488+04 | .82480+00 | .61860+00 |
| .10000+02 | .45679+02 | .22914+02 | .45531+04 | .75886+00 | .56914+00 |
| .20000+02 | .51140+02 | .30195+02 | .41891+04 | .69818+00 | .52364+00 |
| .30000+02 | .56164+02 | .36894+02 | .38541+04 | .64236+00 | .48177+00 |
| .40000+02 | .60787+02 | .43057+02 | .35460+04 | .59100+00 | .44325+00 |
| .50000+02 | .65039+02 | .48727+02 | .32625+04 | .54374+00 | .40781+00 |
| .60000+02 | .68952+02 | .53944+02 | .30016+04 | .50027+00 | .37520+00 |
| .70000+02 | .72552+02 | .58744+02 | .27616+04 | .46027+00 | .34520+00 |
| .80000+02 | .75864+02 | .63160+02 | .25408+04 | .42347+00 | .31760+00 |
| .90000+02 | .78912+02 | .67223+02 | .23377+04 | .38961+00 | .29221+00 |
| .10000+03 | .81715+02 | .70961+02 | .21507+04 | .35846+00 | .26884+00 |
| .11000+03 | .84295+02 | .74401+02 | .19788+04 | .32980+00 | .24735+00 |
| .12000+03 | .86668+02 | .77565+02 | .18206+04 | .30343+00 | .22757+00 |
| .13000+03 | .88851+02 | .80476+02 | .16750+04 | .27917+00 | .20938+00 |
| .14000+03 | .90860+02 | .83155+02 | .15411+04 | .25685+00 | .19263+00 |
| .15000+03 | .92709+02 | .85619+02 | .14179+04 | .23631+00 | .17723+00 |
| .16000+03 | .94409+02 | .87887+02 | .13045+04 | .21742+00 | .16306+00 |
| .17000+03 | .95974+02 | .89973+02 | .12002+04 | .20003+00 | .15002+00 |
| .18000+03 | .97413+02 | .91892+02 | .11042+04 | .18404+00 | .13803+00 |
| .19000+03 | .98737+02 | .93658+02 | .10159+04 | .16932+00 | .12699+00 |
| .20000+03 | .99956+02 | .95282+02 | .93471+03 | .15579+00 | .11684+00 |

FIG. 9-36. Final Converged Values for Split Boundary Heat Exchanger

```
C **EXIT CONVERGENCE**
  5 E=TS-TSI
    TSI=TSO-E*.4
    CALL CONV(TSO,TSI,1,NC)
    GO TO (9,7),NC
  9 STOP
    END
```

FIG. 9-37. Use of CONV to Accelerate Convergence

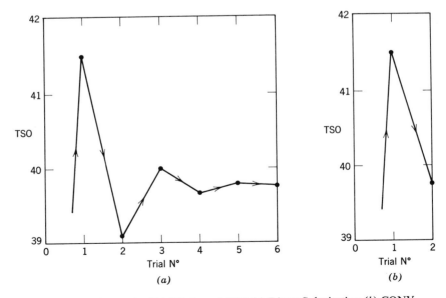

FIG. 9-38. Successive Trial Values of TSO (a) Direct Substitution (b) CONV

case is shown in Figure 9-38*a*, where it is seen that six trials are required to converge to the final value of 39.74. The final run is shown in Figure 9-36.

The gain factor, here 0.4, can be optimized to achieve an efficient convergence. However, a more general approach is to use the CONV subroutine, as shown in Figure 9-37, which by applying projection and interpolation can achieve convergence on the third pass, as shown in Figure 9-38*b*. Use of CONV will ensure rapid convergence, even for cases where the gain factor is larger than critical for stability when using direct substitution for TSO.

When there are several variables that have to be converged, the CONV routine is useful only in situations where the fluctuations of one of the variables predominates over the other, more sluggish variables. The CONV routine is applied to this variable, while the others are readjusted by direct substitution using a suitable gain factor.

## 9-10  STEADY-STATE HEAT EXCHANGER FOR DYFLO LIBRARY

There are several types of heat exchangers in such classifications as single or multipass, co- or countercurrent, gas or liquid. However, it is useful at this time to add a subroutine to the DYFLO library that will simulate the steady-state performance of one of the common types of heat exchangers, namely, the countercurrent single-pass unit discussed in the previous section. Chapter 10 develops the subroutine for the dynamic simulation of this exchanger.

In Chapter 5 a subroutine HTEXCH is developed (Section 5-6-1) that performs the function of adding a specified heat flux (positive or negative) to a particular stream, and then calculates the exit temperature. Here we are concerned with the general situation of specifying the state of two inlet flows (shell side ISI, and tube side ITI, Figure 9-39*a*) and the capacity of the exchanger in the form of the total heat transfer coefficient UA. The objective is to determine the exit temperature for both streams (ISO,ITO).

Most textbooks on heat transfer (Ref. 6) develop the well-known logmean temperature difference relationship. This relates the average temperature difference between shell and tube side to the temperature difference at the

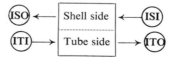

FIG. 9-39*a*.   Counter Current Heat Exchanger for DYFLO Library

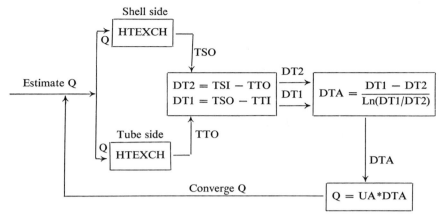

FIG. 9-39$b$.  Model for CSHE Subroutine

ends of the exchanger, that is,

$$DTA = (DT1 - DT2)/Ln(DT1/DT2)$$

where DT1 = TSO − TT1 temperature difference at tube inlet

DT2 = TSI − TTO temperature difference at shell inlet

The net total heat flux will then be

$$Q = UA*DTA$$

This Q will determine the total change in temperature on both the shell and tube sides by a simple heat balance. Since the exit temperatures have to be specified to determine Q, this procedure is implicit. Thus a convergence procedure is required. Starting with an estimated heat flux Q (Figure 9-39$b$) the exit temperatures are calculated by the HTEXCH subroutine, which in turn leads to the log-mean average temperature difference and a calculation for Q. This cycle is repeated to convergence on Q.

### 9-10-1  Subroutine CSHE

This subroutine embodies the procedure described in the previous section for a countercurrent steady-state heat exchanger. The listing is shown in Figure 9-39$c$, together with the definition of the argument list. Since internal accumulations are neglected, the compositions and flows are transferred from both input streams to the corresponding output streams (lines 3 → 8).

```
 1 *            SUBROUTINE CSHE(ISI,ISO,ITI,ITO,UA,IPS,IPT,NSF,NSL,NTF,NTL)
 2 *            COMMON/CD/STRM(300,24),DATA(20,10),RCT(22),NCF,NCL,LSTR
 3 *            DO 7 J=NSF,NSL
 4 *          7 STRM(ISO,J)=STRM(ISI,J)
 5 *            DO 8 J=NTF,NTL
 6 *          8 STRM(ITO,J)=STRM(ITI,J)
 7 *            STRM(ITO,21)=STRM(ITI,21)
 8 *            STRM(ISO,21)=STRM(ISI,21)
 9 *            Q=STRM(ISI,21)*(STRM(ISI,23)-STRM(ISO,23))
10 *          5 NCF=NSF
11 *            NCL=NSL
12 *            CALL HTEXCH(ISI,ISO,-Q,IPS)
13 *            NCF=NTF
14 *            NCL=NTL
15 *            CALL HTEXCH(ITI,ITO,Q,IPT)
16 *            DT1 = STRM(ISO,22)-STRM(ITI,22)
17 *            DT2=STRM(ISI,22)-STRM(ITO,22)
18 *            R=DT1/DT2
19 *            DTA=(DT1-DT2)/ALOG(ABS(R))
20 *            QC=UA*DTA
21 *            CALL CPS(Q,QC,R/(1.+R),NC)
22 *            GO TO (6,5), NC
23 *          6 RETURN
24 *            END
```

(c)

FIG. 9-39c.  Listing for Subroutine CSHE
Argument List

| | | | |
|---|---|---|---|
| ISI | shell-side feed stream number | IPS | PHASE (shell) (vap = 0, liq = 3) |
| ISO | shell-side exit stream number | IPT | PHASE (tube) (vap = 0, liq = 3) |
| ITI | tube-side feed stream number | NSF | first component number (shell) |
| ITO | tube-side exit stream number | NSL | last component number (shell) |
| UA | total capacity (PCU/(C min)) | NTF | first component number (tube) |
| | | NTL | last component number (tube) |

For the sake of efficiency, the composition ranges for each side are specified. Since usually one side will be a single component, such as air or cooling water, line 9 estimates the heat flux Q, so that for the initial pass an estimate of the exit enthalpy (STRM(ISO,23)) must be provided from the main program. The subroutine HTEXCH is called for each side after specifying the component limits NCF,NCL on lines 10 to 15, after which the exit (lines 16, 17) and average temperature differences (line 19) are calculated, followed by the heat flux QC. The ratio of the end temperature differences (R line 18) is always positive. However, it is possible that while cycling to convergence it could become temporarily negative. To prevent an error termination on taking the log, the absolute value of R is used. This sometimes occurs when one of the flow streams becomes small with respect to the other, that is, DT1 $\gg$ DT2 or vice versa. It was found that the best convergence procedure in this case is the partial substitution subroutine CPS (line 21) with a variable substitution factor based on the value of R, that is, $R/(1 + R)$. This will provide convergence for R ranging from 0.02 to 50.

A numerical case was solved with this subroutine as follows (units, moles,

°C, min, PCU):

| Component | Liquid Enthalpy Coefficients | | Feed Composition | |
|-----------|:---:|:---:|:---:|---|
| Tube side | | | | |
| 1 | 18. | 0.01 | 0.2 | Feed flow = 50 mole/min |
| 2 | 14. | 0.02 | 0.5 | Feed temperature = 20°C |
| 3 | 16 | 0.001 | 0.3 | Capacity UA = 540 |
| Shell side | | | | |
| 4 | 15. | 0.003 | 1. | Feed flow = 30 mole/min |
| | | | | Feed temp. = 120°C |

The routine calculated the exit temperatures as

$$\text{shell} = 73.25°C$$

$$\text{tube} = 69.67°C$$

EXAMPLE 9-11   TUBULAR REACTOR

This example will demonstrate the use of variable step and print size in a simulation exhibiting very sharp gradients. A tubular reactor is surrounded by a jacket containing a boiling fluid whose temperature is a constant level of 150°C. A gas flows down the pipe, consisting at the feed end of the following component flows:

Component 1    8 moles/min
Component 2    32 moles/min

The gas phase reaction follows the stoichiometric relation $A + B \rightarrow C$. This reaction is highly exothermic, having a heat of reaction of 38000 PCU/mole of $A$ reacting. The rate relationship is simply

$$R_a = k(T_G)(P \cdot Y_1)(P \cdot Y_2)(\text{moles } A/\text{min ft}^3)$$

that is, proportional to the partial pressures of the two reacting components. The gas pressure is $P$ (psi $A$); $Y_1$ = mole fraction of $A$; $Y_2$ = mole fraction of $B$. The reaction rate coefficient is $k(T_G) = Ae^{-B/TK}$ where $TK$ = temperature °$K$, $A = 17,500$, and $B = 7377$. The basic equations are

1. Component mole balance $A$:

$$\frac{d}{dl}(N \cdot Y_1) = -R_a \left(\frac{\pi D^2}{4}\right)$$

where $D$ = pipe diameter
      $N$ = total moles

2. Component mole balance $B$:

$$\frac{d}{dl}(N \cdot Y_2) = -R_a\left(\frac{\pi D^2}{4}\right)$$

3. Total mole balance:

$$\frac{d}{dl}(N) = -R_a\left(\frac{\pi D^2}{4}\right)$$

$N$ = moles of gas/min

4. Gas pressure $P$ (turbulent pressure drop)

$$\frac{dP}{dl} = -\frac{k_f M N^2(TK)}{pD^5}$$

where the molecular weight $M = M_1 Y_1 + M_2 Y_2 + M_3 Y_3$
  component $C$ composition $Y_3 = 1 - Y_1 - Y_2$
  tube diameter $D = 0.5$ ft
  friction factor $k_f = 0.66 \times 10^{-8}$
component molecular weights $M_1, M_2, M_3 = 42, 74, 116$, respectively.

5. Heat balance:

gas sensible heat change = reaction heat − heat loss to jacket

$$\frac{d}{dl}(NC_pT_G) = R_a h_R \frac{(\pi D^2)}{4} - \pi DU(T_G - T_j)$$

where $C_p = 25$ PCU/mole °C gas specific heat
  $\pi DU = 100$ PCU/min °C ft
  $h_R = 38,000$ PCU/mole heat of reaction

All the above equations are shown in a signal flow model in Figure 9-40, where there is a relatively high ratio of differential equations to be solved. This simplifies the ordering procedure in the FORTRAN program shown in Figure 9-41. The symbology in the program is almost identical to that of the model with the following exceptions:

1. For efficiency, the pipe area term ($\pi D^2/4 = 0.19$) has been included in the reaction rate equation (line 12).
2. The gas sensible heat $NC_pT_G$ is TGN in the program
3. Inlet gas temperature = 0°C
4. Inlet gas pressure $P = 90$ psi $A$

Since the initial values of all the integrated variables are known at the feed end of the tube, only one pass is required to obtain a complete solution for a total tube length of 50 ft. This solution will consist of the temperature and

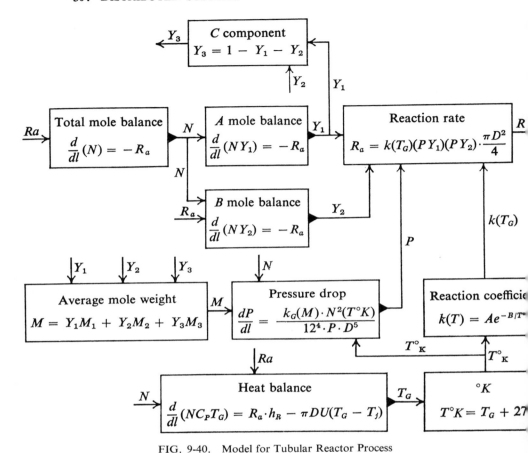

FIG. 9-40.   Model for Tubular Reactor Process

composition profile down the tube, and for this case is shown in Figure 9-42. It will be seen that an almost vertical slope is exhibited around $L = 30$ ft, because the exothermic reaction produces heat energy faster than its dissipation rate to the jacket. This causes a rise in gas temperature that, via the exponential rate coefficient $k(T_G)$, increases the reaction rate. This sequence results in a runaway reaction, rapidly increasing the temperature, stops increasing only when all the $A$ component has been reacted. This is depicted by its composition $Y_1$ in Figure 9-47. For such a situation, it is necessary to decrease the integration step size to very small values in the high reaction region. It will be recalled from the discussion in Chapter 3 that the INT system was intentionally constructed to accept a variable step size. This was

```
 1*          REAL KT,M,L,NY1,NY2,N
 2*          DATA NY1,NY2,N,P,TGN/8.,32.,40.,90.,0./
 3*   C **INITIATION SECTION**
 4*          TS=150.
 5*   C ***DERIVATIVE SECTION***
 6*        6 TG=TGN/(N*25.)
 7*          TK = TG + 273.
 8*          KT = 17500.*EXP(-7377./TK)
 9*          Y1 = NY1/N
10*          Y2 = NY2/N
11*          Y3 = 1.-Y1-Y2
12*          RA = KT*Y1*Y2*P**2*.196
13*          M = Y1*42.+Y2*74.+Y3*116.
14*          DTGN= (RA*38000.-100.*(TG-TS))
15*          DP = -.66E-8 * M*N**2*TK/(P*.031)
16*          DL=.2/(1.+ABS(DTGN)/2000.)
17*          TPR=50.*DL
18*          CALL PRNTF(TPR,50.,NF,L,P,TG,Y1,Y2,Y3,RA,DTGN,DL,TS)
19*          GO TO (4,5),NF
20*   C ***INTEGRATION SECTION***
21*        4 CALL INTI(L,DL,1)
22*          CALL INT(TGN,DTGN)
23*          CALL INT(NY1,-RA)
24*          CALL INT(NY2,-RA)
25*          CALL INT(N,-RA)
26*          CALL INT(P,DP)
27*          GO TO 6
28*        5 STOP
29*          END
```

FIG. 9-41.  Program for Tubular Reactor Process

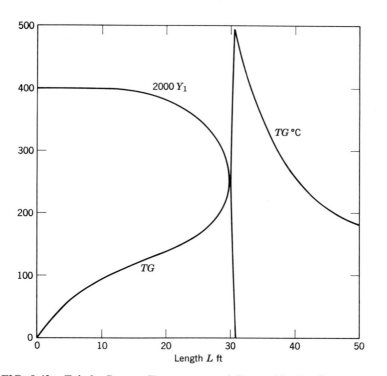

FIG. 9-42.  Tubular Reactor Temperature and Composition Profiles

accomplished by specifying the interval size DTD in the INTI subroutine's argument list. The step size can now be varied by calculating a new value on each pass through the derivative section. A convenient form is (line 16, Figure 9-40)

$$DL = \frac{0.2}{1. + |DTGN|/G}$$

where $G$ = gain factor

This has the characteristic of having a maximum value of 0.2 when the rate of change of sensible heat DTGN is low, to a much smaller value when DTGN is high. The gain factor G is found by trial, in this case 2000.

The same comments apply to the print interval (TPR), which should be very small in the high rate region in order to display the complete profile of the variables. Since this would require functional relationship similar to dl, it is made a multiple of dl, that is, TPR = 50. *DL (line 17).

EXAMPLE 9-12   SERIAL INTEGRATION

Example 9-10 was a case of a split boundary value problem, where the boundary conditions are specified at different boundaries. Most problems of this class can be solved in the manner described by estimating the unknown values at one boundary, solving the equations to the other boundary, comparing calculated with known variables, reestimating the initial values, and continuing the cycle to achieve a match to all boundaries. This procedure invariably involves integrating one or more differential equations in the direction contrary to an actual physical flow. In the case of the shell and tube heat exchanger, starting the integration at the tube inlet end, the shell temperature is solved by integrating backward against the direction of shell-side flow. Fortunately, the coefficients relating to the shell and tube temperatures were of such a magnitude as to produce a stable solution. Other similar situations can have coefficients that result in wildly unstable solutions when integrating against the physical direction of flow. Figure 9-43 shows a typical

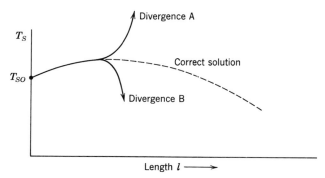

FIG. 9-43.   Typical Divergent Solutions

case, where if the integration is started with the correct value $TSO$, the computation rapidly becomes unstable, resulting in a divergent solution. An infinitesimal change in $TSO$ produces a divergent solution in the other direction. This effect can be traced to the presence of strong, positive feedback in the differential equation.

Clearly, for these problems the solution method described breaks down, so an alternative approach is used. This alternative is generally termed "serial integration" and consists of arranging the calculation sequence so as always to solve the equations in the direction of physical flow.

The previous case of a pipeline reactor (Example 9-11) can become an unstable situation by assuming that a coolant is flowing in a countercurrent direction to the tube flow on the shell side (Figure 9-44). The shell-side temperature now varies with length and will be

$$\frac{d(TS)}{dl} = -0.2\,(TG - TS)$$

where the coefficient 0.2 is the usual combination of heat transfer and shell-side flow capacity. If the inlet shell-side temperature is 0°C, the shell temperature profile has to end at $1 = L$ (50 ft) with $TS = 0$°C. If the integration for $TS$ is not started with a value very close to the correct one, the solution for all the problem variables (i.e., including the tube variables) deteriorates rapidly. A successful method is to store an assumed profile of $TS(1)$ in an array ATS and solve the tube equations using $TS$ obtained from this profile by a function generator. The calculated gas temperature $TG$ is stored in an array (ATG). When the tube solution is completed at $1 = L$, a return integration is made solving only the shell-side equation, using the profile for tube-gas temperature $TG$ previously calculated and stored. The new calculated profile for $TS$ is re-stored in the array ATS. The tube equations are again solved using the new profile for $TS$. This cycling procedure is repeated until the profiles for both temperatures $TG$ and $TS$ settle down to steady-state values.

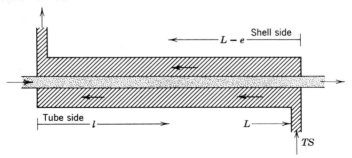

FIG. 9-44.  Tubular Reactor with Countercurrent Coolant flow

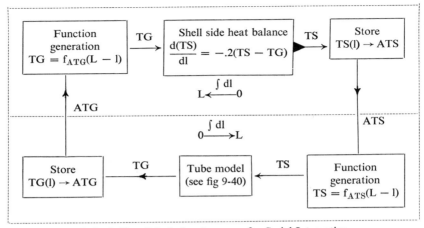

FIG. 9-45.  Calculation Sequence for Serial Integration

Figure 9-45 shows a pictorial schematic of this procedure, while Figure 9.46 shows the computer program embodying the procedure for this case. This program should be largely self-explanatory, since most of the equations have already been described in Example 9-11. The test for convergence is made on line 41, where successive values of tube-gas temperature TG are stored (line 42) and compared with a tolerance value (line 40). The temperature profiles are stored by the subroutine STOR (Figure 9-47) where both the temperature and length coordinates are stored at the print interval, the storing function being triggered by the print index NPR set in the print subroutine PRNTF (Chapter 3). The STOR subroutine also counts the number of points stored (NG) on each pass, since this information is required by the function generator FUN1. Note that a variable step and print interval are also used on the return integration (lines 48, 49).

The shell-side temperature profile initially is set as three points in a data statement (lines 3 and 4). As the cycles are repeated, the temperature TS will approach a stable limit. The profile for the first, sixth, twelfth, and eighteenth passes are shown in Figure 9-48 where the approach to convergence is apparent. The final gas temperature TG for the eighteenth pass is also shown.

### Exercise Problems

1. A water-purifying process that operates on the principle of counter-osmotic pressure is to be designed. A brine solution flows down a tube at high pressure. The inlet brine concentration is $c$ lb salt/ft$^3$, and the tube wall is permeable only to water. The permeation coefficient is $K$ (lb water/ft$^2$, psi

```
 1*            REAL KT,M,L,NY1,NY2,N
 2*            DIMENSION ALS(100),ALG(100),ATS(100),ATG(100)
 3*            DATA(ALS(NS),ATS(NS),NS=1,3)/0.,0.,42.,642.,50.,200./
 4*            NS=3
 5*     C **INITIATION SECTION**
 6*         7 NY1=8.
 7*            NY2=32.
 8*            N=40.
 9*            P=90.
10*            TGN=0.
11*            L=0.
12*            NG = 0
13*     C ***DERIVATIVE SECTION***
14*         6 TG=TGN/(N*25.)
15*            TK = TG + 273.
16*            KT = 17500.*EXP(-7377./TK)
17*            Y1 = NY1/N
18*            Y2 = NY2/N
19*            Y3 = 1.-Y1-Y2
20*            RA = KT*Y1*Y2*P**2*.196
21*            M = Y1*42.+Y2*74.+Y3*116.
22*            TS=FUN1(50.-L,NS,ALS,ATS)
23*            DTGN= (RA*38000.-100.*(TG-TS))
24*            DP = -.66E-8 * M*N**2*TK/(P*.031)
25*            DL=.2/(1.+ABS(DTGN)/2000.)
26*            TPR=50.*DL
27*            CALL PRNTF(TPR,50.,NF,L,P,TG,Y1,Y2,Y3,RA,DTGN,DL,TS)
28*            CALL STOR(TG,L,ALG,ATG,NG)
29*            GO TO (4,5),NF
30*     C ***INTEGRATION SECTION***
31*         4 CALL INTI(L,DL,1)
32*            CALL INT(TGN,DTGN)
33*            CALL INT(NY1,-RA)
34*            CALL INT(NY2,-RA)
35*            CALL INT(N,-RA)
36*            CALL INT(P,DP)
37*            GO TO 6
38*     C **SHELL SIDE RETURN**
39*         5 CONTINUE
40*            E=TG-TGO
41*            IF(ABS(E).LT..05) STOP
42*            TGO=TG
43*            TS = 0.
44*            NS=0
45*            L=0.
46*         9 TG=FUN1(50.-L,NG,ALG,ATG)
47*            DTS=-.20*(TS-TG)
48*            DL=.2/(1.+ABS(DTS)/2000.)
49*            DPR=10.*DL
50*            CALL PRNTF(DPR,50.,NF,L,TS,TG,DTS,0.,0.,0.,0.,0.,0.)
51*            CALL STOR(TS,L,ALS,ATS,NS)
52*            GO TO (10,7),NF
53*        10 CALL INTI(L,DL,1)
54*            CALL INT(TS,DTS)
55*            GO TO 9
56*            END
```

FIG. 9-46.   Computer Program for Serial Integration of Tubular Reactor

```
1*          SUBROUTINE STOR(TG,X,ALG,ATS,NG)
2*          DIMENSION ALG(100),ATS(100)
3*          COMMON/CPR/NPR
4*          IF(NPR.EQ.1) GO TO 5
5*          RETURN
6*        5 NG=NG+1
7*          ALG(NG)=X
8*          ATS(NG)=TG
9*          RETURN
10*         END
```

FIG. 9-47.   Subroutine STOR

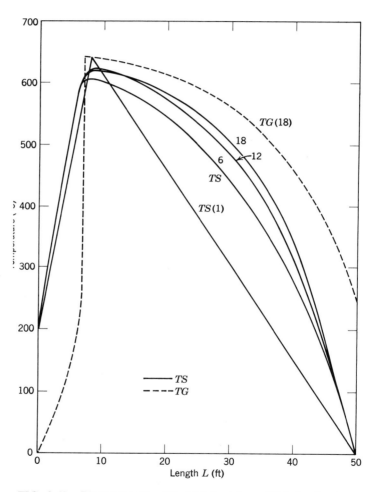

FIG. 9-48.   Plotted Results of Serial Integration of Tabular Reactor

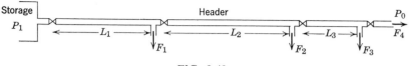

FIG. 9-49.

min) and the driving force (psi) across the wall of the tube is the pressure above the reverse osmotic pressure. The osmotic pressure is directly proportional to the salt content of the brine.

Construct a mathematical model that will determine the yield of fresh water as a function of tube length, diameter, inlet flow rate, and inlet pressure, assuming significant pressure drop caused by friction down the tube.

2. A gaseous fuel is to be supplied to three locations from a central storage area by flowing down a main header in which are located the four valves shown in Figure 9-49. Flows $F_1$, $F_2$, and $F_3$ are withdrawn upstream and adjacent to the three header valves. The pressure downstream of the fourth valve is $P_0$.

Construct a model that establishes the supply pressures for flows $F_1$, $F_2$, and $F_3$ and the flow through the fourth valve $F_4$, assuming that the valves $Cv^s$ are known.

3. A catalytic reactor consists of a fixed bed of catalyst pellets through which flows a gas containing reagents $A$ and $B$, products $R$ and $S$, and inerts $I$. The reaction is

$$A + B \rightarrow R + S$$

and the reaction rate is

$$r = \frac{k(P_A \cdot P_B - P_R \cdot P_S/K)}{(1 + K_A \cdot P_A + K_B \cdot P_B + K_R \cdot P_R + K_S \cdot P_S + K_i \cdot P_i)^2}$$

where $r$ = (lb mole/min lb catalyst)
$P_i$ = partial pressure (component $i$)
$K$ = equilibrium constant (temperature dependent)
$k$ = rate constant (temperature dependent)

The reaction is exothermic and the reactor is adiabatic. Also, the pressure drop through the bed is proportional to the square of the mass flow rate, the length of the bed, and the inverse of the fourth power of the diameter. Construct a model that defines how the exit compositions vary with inlet temperature, total bed volume, length and diameter of bed, inlet composition, and so on.

4. The fluid phase reaction

$$A + B \rightleftharpoons C + D$$

takes place in a heated pipeline reactor. The reaction is endothermic, and additional heat is supplied from the jacket to raise the temperature to the boiling point and to promote vaporization. Only components $C$ and $D$ vaporize but in a ratio favorable to component $C$. This enriches the fluid phase with component $D$, and tends to drive the reaction to completion. At several locations in the reactor the vapor phase is withdrawn, and from experimental data the following correlation for the pressure drop is obtained:

$$\frac{dP}{dl} = k_1 W_L + K_2 V^2$$

where $W_L$ = lb/min liquid phase
$\quad\quad\;\; V$ = ft³/min vapor phase

The heat transfer coefficient $U$ across the pipe wall is also correlated by the relationship

$$U = \frac{K_3 + K_4 W_L}{1 + K_5 V} \text{ PCU/min, C, ft}^2$$

Assuming equilibrium between liquid and vaporizing stream and reaction coefficient temperature dependency, construct a model that relates the composition as a function of length, positions of vapor draw-off, pipe diameter, inlet pressure, and jacket temperature.

5. A spherical polymer particle (diameter $D$), suspended in a fluid medium, is bombarded with radicals. The rate of entry of radicals through the particle surface is $R$ (moles/cm² sec), and the rate of diffusion within the particle is proportional to the concentration gradient. The rate of radical termination within the particle is $k_T \cdot C^2$, where $C$ is the radical concentration (moles rad/cm³ particle).

Construct a model and a method of computer solution to define the radical concentration within the particle.

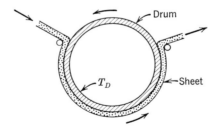

6. A sheet of thickness $t_f$ is to be cooled by allowing it to touch a rotating metal drum (thickness $t_d$) cooled on the inside by a coolant (temperature $T_d$). The temperature fluctuations of the drum surface and sheet are to be computed for various sheet thicknesses, drum speed, and contact length.

Develop a model for the system, assuming negligible temperature drop through the sheet and drum and resistance to heat flow only between the coolant/drum and the drum/sheet interface.

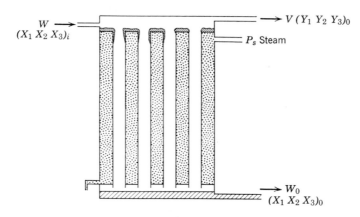

7. A falling film evaporator consists of a bundle of vertical tubes enclosed in a shell. A constant steam pressure is maintained in the shell, which transfers heat to the film of the liquid flowing down on the inside of the tubes, thus causing vaporization. The vapor is drawn off at the top of the evaporator. The feed consists of three components, all of which vaporize to some degree, and the pressure in the evaporator is constant.

Construct a model of the system to define the outlet flows and compositions as a function of the system parameters, assuming equilibrium at all points between the liquid phase and the vaporizing stream (i.e., interface equilibrium).

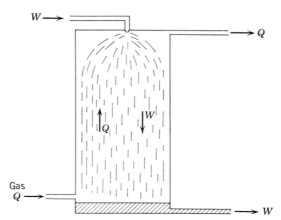

8. A spray tower of volume $V$ humidifies a single component gas (flow rate $Q$) with a vapor from a single-component liquid flowing counter-current to the gas at a rate $W$. The conventional two-film theory postulates that equilibrium is achieved at the interface between the gas and the liquid droplets (i.e., the temperature on each side of the two phases is the same) and that the gas is saturated with vapor at the interface.

Construct a model of the process, assuming that the sensible heat of the vaporized liquid is negligible and that the quantity of vaporized liquid is negligible compared with the flow of liquid down the tower and the upflow of gas.

9. How would the program for the heat exchanger described in Example 9-10 be modified to calculate the shell-side flow rate required to achieve a specified cooling load, assuming the tube and shell-side feed temperatures and tube-side flow rate?

10. Suppose the reaction in the tabular reactor described in 9-11 includes an additional degradation reaction involving the product $C$, such as

$$C + C \rightarrow D$$

The reaction rate is proportional to the square of the partial pressure of $C$ and the rate constant is $k_c = 1050 * \mathrm{EXP}(-9000/\mathrm{TK})$. How would the model shown in Figure 9-40 be modified to include this reaction? Modify the program to include this reaction, and determine the optimum shell temperature $TS$ that will maximize the production rate of product $C$.

## REFERENCES

1. "Membrane Separation Process," N. N. Li, R. B. Long, and E. J. Henley, *Ind. Eng. Chem.*, **57**, No. 3, 1965.
2. "Horizontal Pipeline Flow of Equal Density Oil/Water Mixtures," M. E. Charles, G. W. Govier, and G. W. Hoggson, *Can. J. Chem. Eng.*, February 1961.
3. "A Flow for Two Phase Slug Flow in Horizontal Tubes," E. S. Koroyban, *Journal of Basic Eng.*, December 1961.
4. "Apply Analog Techniques to Equipment Design," R. G. Franks, *Chem. Eng.*, April 1963.
5. "Gas Liquid Flow in Horizontal Pipes," C. J. Hoogendorn, *Chem. Eng. Sci.*, **9**, 1959.
6. *Process Heat Transfer*, D. Q. Kern, McGraw-Hill, New York, 1950.
7. "Transient Film Condensation," E. M. Sparrow and R. Siegel, *ASME*, 58-A-13, 1958.
8. "Application of a General Purpose Analog Computer to the Design of a Cooler Condenser, R. G. Franks and N. G. O'Brien, *Chem. Engr. Progr. Symp. Ser.*, **56**, No. 31, 1960.
9. "Calculation of the Performance of a Mixed-Vapor Condenser by Analog Computation," N. G. O'Brien, R. G. Franks, and J. K. Munson, *Chem. Engr. Symp. Ser.*, **55**, No. 29, 1957.
10. A. P. Coburn and O. A. Hougen, *Ind. Eng. Chem.*, **26**, 1178, 1934.

11. "A Generalized Correlation of Diffusion Coefficients," J. R. Fair and B. J. Lerner, *A.I.Ch.E. J.*, **2**, No. 1, 1956.
12. "Predicting Physical Properties of Gases and Gas Mixtures," J. T. Holmes and M. G. Burns, *Chem. Eng.*, May 1965.
13. "Steady State Simulation of an Ammonia Synthesis Converter," R. F. Baddour, P. L. T. Brian, B. A. Longeais, and J. P. Emery, *Chem. Eng. Sci.*, **20**, 1965.
14. "Analog Computer Method for Designing a Cooler Condenser with Fog Formation," D. R. Coughanowr and E. O. Stensholt, *Ind. Eng. Chem.*, **3**, No. 4, October 1964.
15. "On-Line Computer Control of Thermal Cracking," S. M. Roberts and C. G. Laspe, *Ind. Eng. Chem.*, May 1961.

## NOMENCLATURE

| | |
|---|---|
| $a$ | constant or tube area/unit length (Example 9-7) |
| $A_k$ | Ackerman correction factor |
| $B$ | condensate thickness |
| $C$ | heat capacity |
| $C_p$ | specific heat |
| $D$ | diameter |
| $F$ | flow rate |
| $F_j$ | diffusion factor component $j$ |
| $f_j$ | mole fraction in condensing stream component $j$ |
| $G$ | gain factor |
| $H$ | heat flux |
| $H_R$ | heat of reaction |
| $\Delta H_R$ | molal heat of reaction |
| $h$ | conduction coefficient |
| $h_g$ | heat transfer coefficient (Example 9-7) |
| $K$ | thermal conductivity |
| $K_t$ | friction factor |
| $k_f$ | friction factor |
| $k$ | reaction rate coefficient |
| $L$ | total length or liquid flow rate (Example 9-4) |
| $l$ | length |
| $M$ | mole flow rate |
| $P$ | pressure |
| $Q$ | heat flux (Example 9-2), gas flow rate (Example 9-3), condensate flow (Example 9-6) |
| $q_\lambda$ | latent heat liberated |
| $q_s$ | sensible heat liberated |
| $R$ | gas constant or reaction rate |
| $S$ | shell side |
| $T$ | temperature or tube side |
| $t$ | wall thickness |

| | |
|---|---|
| $U$ | overall heat transfer coefficient |
| $u$ | velocity |
| $V$ | volumetric flow rate, or mole flow rate (Example 9-4) |
| $v$ | vaporization rate |
| $W$ | mass flow (Example 9-4); condensation rate (Example 9-6) |
| $X$ | distance |
| $Y$ | gas mole fraction |
| $y$ | distance through condensate film |
| $Z$ | distance |
| $\phi$ | density |
| $\alpha$ | radiation constant |
| $\beta$ | permeation constant |
| $\mu$ | viscosity |
| $\lambda$ | latent heat |
| $\tau$ | stress on unit surface |
| $\gamma$ | activity coefficient |

# CHAPTER X

# PARTIAL DIFFERENTIAL EQUATIONS

The area of partial differential equations is the last and most complex phase in the development of mathematical model building. The first phase dealt with systems described by ordinary, nonlinear differential and algebraic equations with time as the independent variable. The second phase was an extension of the first phase to multistage systems, and the third phase was a study of systems described by ordinary nonlinear differential equations, in which either distance or geometric length was the independent variable. For this last phase, steady-state operation was assumed in order to eliminate time as an independent variable. The most general situation, however, is that in which variables such as temperature, flow rate, pressure, and composition not only are distributed in space dimensions, but also vary with time. Some very elementary examples of this classification lend themselves to analytical solutions, but those are rarely of practical interest.

The most common approach to partial differential equations that is suited to computer solution is that of converting the partial differential equations into multiple sets of simultaneous, ordinary differential equations, for which there is only one independent variable, usually time. This is accomplished by "finite differencing" the space variables, which merely means that the space dimensions are divided into a number of "cells" and the usual mass and energy balance equations are defined for each cell in turn. The number of cells that is adequate for any situation is discussed at great length in the literature (Refs. 1, 2, 3) because if a large number of cells are defined, the mathematical model will approach the physical situation very closely. However, this requires a large amount of computer running time per solution. Fortunately, in most cases, an engineering solution having adequate accuracy can be obtained from a surprisingly small number of cells. The distribution of the cells through the distributed system should be

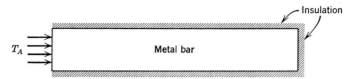

FIG. 10-1.  Insulated Metal Bar

determined by the analyst; that is, since the cells need not all be the same size, the accuracy of the approximation can be improved by crowding more cells in those areas where higher spatial gradients will occur during a transient and by approximating the low-gradient areas with fewer cells. This aspect can be demonstrated in some of the following examples.

EXAMPLE 10-1  (Refs. 4, 5)

This first example is one of the elementary cases often described in texts on heat transfer and for which an analytical solution exists. It is used here to demonstrate the simplicity of the finite differencing approach combined with the model building technique. The system consists of a metal bar insulated on all sides and at one end (Figure 10-1). At time $O$, there is a uniform temperature through the bar (say $O°C$), and a temperature of $100°C$ is applied and maintained at the uninsulated end. Figure 10-2 shows the temperature gradient along the bar at successive times $\theta_i$. As is apparent, the temperature $T$ varies with both distance $X$ and time $\theta$. This is usually expressed in notation form as $T(\theta, X)$, and the partial differential equation that relates the temperature to time and distance is

$$\frac{\partial^2 T}{\partial X^2} = h^2 \frac{\partial T}{\partial \theta}$$

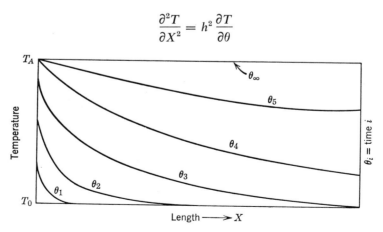

FIG. 10-2.  Temperature Gradients in Bar at Time $\theta_1$

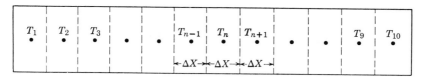

FIG. 10-3.   Sectionalized Bar

where $h$ is a property of the material, that is,

$$h = \frac{\text{density} \cdot \text{specific heat}}{\text{conductivity}}$$

The finite difference approach to the solution of this problem is to imagine that the bar is divided into a number of sections and that the temperature at the center of each section is the average temperature of that section (Figure 10-3).

Assuming that each section is the same size and that the distance between the centers is $\Delta X$, then the rate of heat transfer from section $n - 1$ to section $n$ (see Figure 10-4) is

$$Q_{n-1} = \text{conductivity} \times \text{area} \times \text{temperature gradient}$$

$$Q_{n-1} = k \cdot A \cdot \frac{(T_{n-1} - T_n)}{\Delta X}$$

where $A$ is the cross-sectional area of the bar.

Similarly, the rate of heat conduction from section $n + 1$ to section $n$ is

$$Q_{n+1} = k \cdot A \cdot \frac{(T_{n+1} - T_n)}{\Delta X}$$

A heat balance made on section $n$ results in the following differential equation:

rate of heat accumulation = sum of all heat fluxes to the section

$$\frac{d}{d\theta} [C\phi(A \, \Delta X)T_n] = k \cdot A \cdot \frac{(T_{n-1} - T_n)}{\Delta X} + k \cdot A \cdot \frac{(T_{n+1} - T_n)}{\Delta X}$$

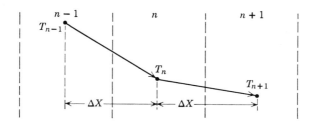

FIG. 10-4.   Gradients on Section $n$

$$T_{n-1} \rightarrow \boxed{\frac{dT_n}{d\theta} = \frac{k}{\phi C (\Delta X)^2} [(T_{n-1} - T_n) + (T_{n+1} - T_n)]} \rightarrow T_n$$

FIG. 10-5.   Model for Section $n$

This equation can be reduced by simple manipulations to

$$\frac{d}{d\theta} T_n = \frac{k}{\phi C(\Delta X)^2} [(T_{n-1} - T_n) + (T_{n+1} - T_n)]$$

As part of an information flow diagram, this equation should be arranged as in Figure 10-5. It can also be developed by formal methods of finite differencing the original partial differential equation; however, the approach developed above is both easier to visualize and straightforward to implement. The mathematical model for the entire bar is merely one equation block for each section; the assembly is illustrated in Figure 10-6 for five sections.

The mathematical accuracy of the representation can be greatly improved by redistributing the sections without increasing the number of sections (Figure 10-7). Because the sharpest temperature gradients occur at the end of the bar where the forcing function is applied (see Figure 10-1), that end should be divided into a finer grid than the insulated end where at all times the gradients are low. Solution of the resulting set of simultaneous, ordinary differential equations by machine computation is relatively simple; all that is required is the initial value of $T$ for each section and the value of the forcing function, $T_A$. Although analytic solutions exist for specific forcing functions such as sinusoids or step functions, it is more convenient, even in this elementary case, to program the equations for a computer, thus allowing solution to a variety of arbitrary functions. The computation yields specific values of $T$ for each section $n$, as a continuous function of time or as the temperature gradient versus length $X$ at specific values of time $\theta_i$.

A problem that develops in sectionalizing a distributed system is calculating the number of sections required to achieve various degrees of accuracy. There are many references in the literature dealing with this

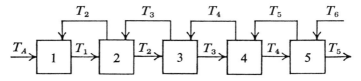

FIG. 10-6.   Signal Flow for Five Sections of Bar

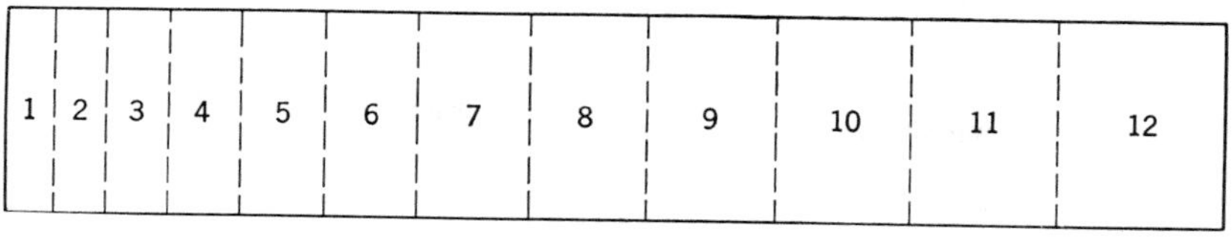

FIG. 10-7.  Improved Distribution of Sections

subject (1, 2, 3), but a useful rule is to assign the sections according to the number of points required to define the temperature gradients through the distributed medium.

### 10-1-1  Computer Program for Insulated Metal Bar

The finite difference approach to distributed systems leads to mathematical models that are ideally suited to digital computer programming. The elementary example described in the previous section will be programmed as a set of ordinary differential equations, but, in order to capitalize on the advantages offered by the computer, it will be programmed in a most general form. The various distances and dimensions for each section will be calculated by the computer; the only specification required for this is the total length of the bar $L$, the number of sections $N$, and the ratio of the length of each succeeding section (cross sectional area will be assumed to be unity). If the length of section $J$ is $\Delta L_J$ (Figure 10-8a) and the length of the $J + 1$ section is $\Delta L_{J+1}$, then the ratio $R = \Delta L_{J+1}/\Delta L_J$. If this ratio is maintained along the length of the bar and if the length of the first section is $DL$, then the length of any section $J = DL(R)^{J-1}$. The total length

$$L = \sum_{J=1}^{N} DL(R)^{J-1}$$

This presupposes that the sections become progressively larger, as shown in Figure 10-7. Therefore, by specifying $L$, $R$, and the total number of sections $N$, the base length

$$DL = \frac{L}{\sum\limits_{J=1}^{N} (R)^{J-1}}$$

The program (Figure 10-8b) shows this calculation performed on lines 6 to 9.

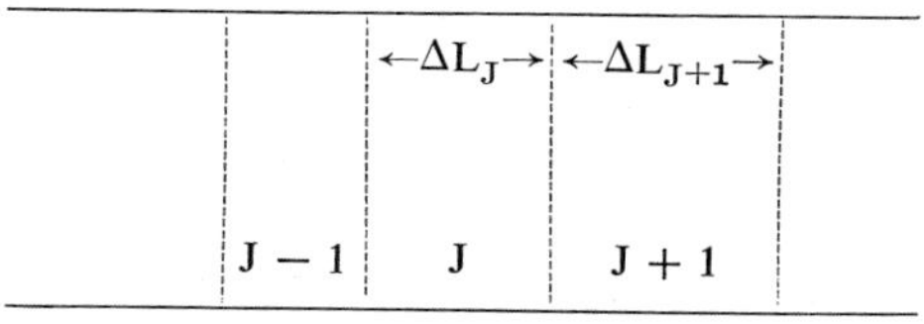

FIG. 10-8a.  Variable Bar Sections

```
 1*          DIMENSION XC(30),T(30),DTC(30),A(30)
 2*          REAL L,K
 3*          DATA L/10./K/1.44/RO/160./C/.222/N/10/TA/100./R/1.2/NM1/9/
 4*      100 FORMAT(10E12.4)
 5*   C ** INITIATION SECTION **
 6*          DO 5 J=1,N
 7*          T(J) = 0.
 8*        5 SR = SR + R**(J-1)
 9*          DL = L/SR
10*          XC(1) = DL/2.
11*          DO 6 J=1,N
12*          A(J) = K*2./(RO*C*DL**2)
13*          XC(J+1) = XC(J) + DL/2.*(1.+R)
14*        6 DL = DL * R
15*          PRINT 100 (XC(J),J=1,N)
16*   C ** DERIVATIVE SECTION **
17*        9 DTC(1) = A(1)*((TA-T(1)) + (T(2)-T(1))/(1.+R))
18*          DO 7 J=2,NM1
19*        7 DTC(J) = A(J)*((T(J-1)-T(J))/(1.+1./R)+(T(J+1)-T(J))/(1.+R))
20*          DTC(N) = A(N)*(T(N-1)-T(N))/(1.+1./R)
21*          CALL PRNTF(100.,1000.,NF,TIM,T(1),T(2),T(3),T(4),T(5),T(6),T(N-2),
22*         1T(N-1),T(N))
23*          IF (NF .EQ. 2) STOP
24*   C ** INTEGRATION SECTION **
25*          CALL INTI(TIM,.1,4)
26*          DO 8 J=1,N
27*        8 CALL INT(T(J),DTC(J))
28*          GO TO 9
29*          END
```

FIG. 10-8b.  Program for Insulated Metal Bar

Adopting the differential equation developed previously to the general situation of gradually increasing sections we have

$$\frac{dT_n}{d\theta} = \frac{k}{\phi C \, \Delta L_n}\left[\frac{(T_{n-1} - T_n)}{\frac{1}{2}(\Delta L_{n-1} + \Delta L_n)} + \frac{(T_{n+1} - T_n)}{\frac{1}{2}(\Delta L_{n+1} + \Delta L_n)}\right]$$

Substituting $\Delta L_{n-1} = \Delta L_n/R$ and $\Delta L_{n+1} = \Delta L_n R$, we have

$$\frac{dT_n}{d\theta} = \frac{2k \cdot}{\phi C (\Delta L_n)^2}\left[\frac{T_{n-1} - T_n}{(1/R + 1)} + \frac{T_{n+1} - T_n}{(1 + R)}\right]$$

The coefficient $A = 2k/(\phi C (\Delta L_n)^2)$ is calculated on line 12 as $A(J) = K * 2./(RO * C * DL ** 2)$, and DL is calculated on line 14 for each section. The location of the center of each section as a coordinate distance from the end of the bar is calculated from the expression

$$XC_{J+1} = XC_J + \frac{DL}{2}\sum_{J-1}(1 + R)$$

where $XC_1 = DL/2$. This is performed as part of the DO loop on line 13. The derivative of the temperature for the general section is calculated on line 19 for all sections except the first and last. These two exceptions are calculated separately on lines 17 and 20. It will be recalled that the last section has only one heat flux, since the end is insulated, while the first section has its first

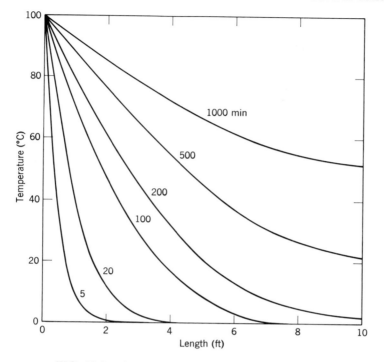

FIG. 10-8c.   Plotted Results of Program in Figure 10-8a

flux from the boundary temperature TA to the first temperature T(1) across a distance of ½DL. The expressions resulting from these factors are as shown in the program. The integration of the derivatives is done in the DO loop on line 27.

The power of digital programming for this type of problem should now be apparent. The 10 differential equations require only one line of coding, achieved by using subscripts and the DO loop. For example, if the number of sections have to be changed it is only necessary to change the index limits of the DO loops N and NM1 specified in the data statement on line 3.

A run was made using the parameter values specified in the data statement, which are for an aluminum bar. The numerical results are plotted in Figure 10-8c as a series of temperature gradients down the length of the bar at different times up to 1000 min, at which time the bar is approaching stabilization at the hot-face temperature.

The next examples of partial differential equations are more complex than the preceding cases but are typical of the degree of complexity common to industrial applications. The elementary cases already considered demonstrate the principle of finite differencing down to a set of ordinary differential equations and their solution using a FORTRAN program. The next example is a more complex situation, but it does not involve any principle beyond those already covered.

EXAMPLE 10-2 (Refs. 6, 7)

The liquid-liquid countercurrent double-pipe heat exchanger is studied in Chapter 9. Equations are developed that define the temperature gradient as a function of location in the exchanger only, and in order to eliminate time as an additional independent variable, steady-state operation is assumed. The more general case of unsteady-state operation can now be assumed so that the *dependent* temperature variables are functions of the *independent* variables, time and distance; the exchanger is shown in Figure 10-9. Since there is more than one independent variable, the relationship between temperature, time, and distance can be stated as a partial differential equation. However, since the objective is to reduce the mathematical model to a set of simultaneous *ordinary* differential equations (i.e., only *one* independent variable, time), the system will be "finite differenced" or "lumped" at the outset.

The exchanger is divided into 10 sections (Figure 10-10) with each section on the tube side corresponding to a section on the shell side; a typical section $n$ appears in Figure 10-11. Fluid on the shell side is flowing from section $(n + 1)$ to $(n - 1)$, whereas on the tube side, it is flowing from section $(n - 1)$ to section $(n + 1)$. Section $n$ is fixed in space, so that fluid flows in and out of section $n$ continuously. Heat is also transmitted from inside the tube $(T)$ to the shell side $(S)$. If the temperature at the center of each section is specified, as $T_T$ and $T_S$, respectively, the rate of heat transfer is

$$H_n = U(\pi D \, \Delta l)(T_{Tn} - T_{Sn})$$

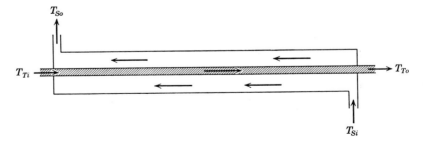

FIG. 10-9. Shell and Tube Heat Exchanger

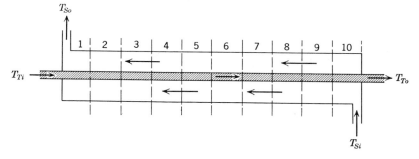

FIG. 10-10.  Sectionalized Heat Exchanger

where $D$ is the mean diameter of the tube and $\Delta l$ is the length of the section.

In order to perform a heat balance on the shell and tube sections, the temperatures of the flowing fluid entering and leaving each section have to be defined in terms of the temperatures at the center of each section. As a first approximation it can be assumed that the temperatures at the boundaries of the sections are the arithmetic mean of the temperatures at the center of the sections. The temperature of the fluid entering section $S_n$ can be expressed as $\frac{1}{2}(T_{Sn} + R_{Sn+1})$, and, similarly, the temperature of the fluid on the tube side entering section $n$ from $n - 1$ will be $\frac{1}{2}(T_{Tn-1} + T_{Tn})$. A heat balance equation on each section can now be written as follows:

rate of heat accumulation = heat flow in − heat flow out ± heat transferred

*Section n* shell side

$$\frac{d}{d\theta}(C\phi V_S T_{Sn}) = CF_S[(T_{Sn} + T_{Sn+1})\tfrac{1}{2} - (T_{Sn} + T_{Sn-1})\tfrac{1}{2}] + H_n$$

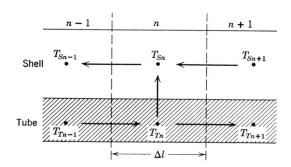

FIG. 10-11.  Section $n$ of Heat Exchanger

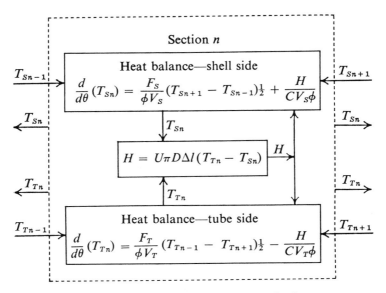

FIG. 10-12.   Model for Heat Exchanger Section $n$

*Section n* tube side

$$\frac{d}{d\theta}(C\phi V_T T_{Tn}) = CF_T[(T_{Tn} + T_{Tn-1})^{\frac{1}{2}} - (T_{Tn} + T_{Tn+1})^{\frac{1}{2}}] - H_n$$

where $V_S$ and $V_T$ are the volumes of fluid in each section $n$, shell and tube, respectively, and are simply

$$V_S = \Delta l\pi(D_S{}^2 - D_T{}^2) \qquad V_T = \Delta l\pi D_T{}^2$$

where $D_S$ and $D_T$ are the diameters of the two pipes.

These equations for section $n$ can be simplified and arranged in model form (Figure 10-12); the other sections would then be similar to section $n$, and all the sections interconnected, as shown in Figure 10-13. The end sections, 1

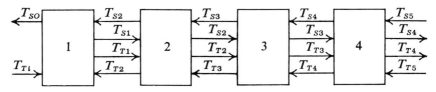

FIG. 10-13.   Information Flow Between Sections of Heat Exchanger

and $N$, require special treatment. Figure 10-14 shows the location of the boundary and temperature points for the shell side flow. The temperature of the fluid entering section 1 is the average between $T_{S2}$ and $T_{S1}$, but the temperature leaving section 1, $T_{S0}$, can either be assumed to be $T_{S1}$ (i.e., this section has a uniform temperature), or the gradient $T_{S2} \rightarrow T_{S1}$ can be projected to establish $T_{S0}$; that is,

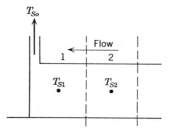

FIG. 10-14. End Section (Boundary) on Shell Side

$$T_{S0} = T_{S1} - \tfrac{1}{2}(T_{S2} - T_{S1})$$

The first assumption, $T_{S0} = T_{S1}$ is often satisfactory, especially if the exchanger is divided into many sections. A similar approach is used to approximate the tube side exit temperature in the last section $N$. Suppose the inlet temperatures to the exchanger, $T_{Si}$ and $T_{Ti}$, vary arbitrarily as a function of time, simultaneously with a variation in the flows $F_S$ and $F_T$. The exit temperatures, $T_{S0}$ and $T_{T0}$, commonly termed the system response resulting from the variation of input conditions, can be established by solving the equations shown in the model.

## 10-2-1 DYFLO Subroutine DHE for Dynamic Heat Exchanger

The occurrence of a heat exchanger in a chemical process is sufficiently frequent to warrant the inclusion of a subroutine simulating its dynamic response in the general DYFLO library. Two subroutines, HTEXCH and CSHE, have already been developed to simulate the steady-state performance of heat exchangers. The DHE subroutine simulates the dynamic temperature and composition responses to input disturbances. As demonstrated in the previous section, the heat exchanger is divided into N sections, judiciously selected by the programmer based on the relative importance of the heat exchanger dynamics with respect to the total system to be simulated. Each section in the shell and in the tube side will be assumed to be a backmix section simulated by the HLDP subroutine (Figure 10-15).

The HDLP subroutine will calculate the exit temperatures from each section, as well as compositions, but when calculating the heat fluxes QS, QT (Figure 10-15), the average temperature across each section is used; that is, TAS=(STRM(JS,22)+STRM(JS+1,22))/2. The temperature of the tube wall separating the shell and tube sections is stored in position 20 of the shell-side STRM array, that is, STRM(JS,20). The equation for the tube-wall temperature in each section will be

$$\frac{d(\mathrm{TW})}{d\theta} = -\frac{\mathrm{QS} + \mathrm{QT}}{\mathrm{WHC}/N}$$

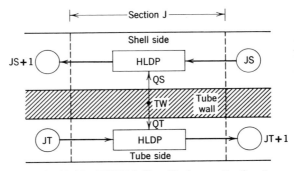

FIG. 10-15. DYFLO Heat Exchanger Section J

```
 1 *       SUBROUTINE DHE(IS,IT,N,UA,WHC,IPS,IPT,NSF,NSL,NTF,NTL,HS,HT,NC)
 2 *       COMMON/CD/STRM(300,24),DATA(20,10),RCT(22),NCF,NCL,LSTR
 3 *       LOGICAL LSTR
 4 *       DIMENSION HE(10,4)
 5 *       DO 5 J=1,N
 6 *       STRM(IS+J,21)=STRM(IS,21)
 7 *     5 STRM(IT+J,21)=STRM(IT,21)
 8 *       IF(LSTR) GO TO 8
 9 *     9 NCF=NSF
10 *       NCL=NSL
11 *       DO 6 J=1,N
12 *       JT=IT+J
13 *       JS=IS+N-J+1
14 *       QS=(STRM(JS,20)-(STRM(JS,22)+STRM(JS-1,22))/2.)*HE(NC,1)
15 *       STRM(JT,20)=QS
16 *     6 CALL HLDP(JS-1,JS,IPS,HE(NC,3),QS)
17 *       NCF=NTF
18 *       NCL=NTL
19 *       DO 7 J=1,N
20 *       JT=IT+N-J+1
21 *       QT=(STRM(JS+J-1,20)-(STRM(JT,22)+STRM(JT-1,22))/2.)*HE(NC,1)
22 *       CALL HLDP(JT-1,JT,IPT,HE(NC,4),QT)
23 *       IF(LSTR) RETURN
24 *       DTW=-(STRM(JT,20)+QT)/HE(NC,2)
25 *     7 CALL INT(STRM(IS+J-1,20),DTW)
26 *       RETURN
27 *     8 HE(NC,1)=UA/N*2.
28 *       HE(NC,2)=WHC/N
29 *       HE(NC,3)=HS/N
30 *       HE(NC,4)=HT/N
31 *       CALL CSHE(IS,IS+N,IT,IT+N,UA,IPS,IPT,NSF,NSL,NTF.NTL)
32 *       CALL ICHE(IS,IT,N,UA,IPS,IPT,NSF,NSL,NTF,NTL)
33 *       GO TO 9
34 *       END
```

(a)

FIG. 10-16a. Listing for Subroutine DHE
Argument list

| | | | |
|---|---|---|---|
| IS | line number feed shell | NSF | first component number shell |
| IT | line number feed tube | NSL | last component number shell |
| N | number of sections | NTF | first component number tube |
| UA | total heat transfer | NTL | last component number tube |
| WHC | total wall-heat capacity | HS | total mole holdup shell |
| IPS | phase shell (vap = 0, liq = 3) | HT | total mole holdup tube |
| IPT | phase tube (vap = 0, liq = 3) | NC | call number |

**328**

where

$$QS = UA_J(TW - TAS)$$

$$QT = UA_J(TW - TAT)$$

$$WHC = \text{total wall heat capacity.}$$

The listing for the subroutine is shown in Figure 10-16a, together with a definition of the argument list. Only the input stream numbers and the number of sections desired are specified, since all the other internal and exit streams will be internally numbered by the DO loops. Consequently, if the inlet stream numbers are IS and IT, the exit stream numbers will be IS+N and IT+N. The first function performed by the subroutine is to transfer the input flow rate to each internal and exit stream (lines 5 → 7). The shell-side heat flux QS is calculated on line 14 and is stored temporarily in the tube-stream array STRM(JT,20). This heat flux is then supplied to the HLDP subroutine (line 16) that calculates the exit temperature from each section. Note that the calling sequence for the sections is in a direction opposite to the shell-side flow direction, a basic requirement when using the INT program. Proceeding to the tube side, the heat flux QT is calculated on line 21 and passed to the tube side HLDP subroutine on line 22. The two heat fluxes QT and QS, the latter stored in the proper section array STRM(JT,20), are used to calculate the derivative of the tube-wall temperature DTW on line 25. Again, note that the calling sequence is opposite to the tube-side flow direction.

On the initial pass (LSTR=.TRUE.) through the routine, the calculation is diverted (line 8) to an initiation section (lines 27 → 32) where the section parameters are calculated from overall values supplied in the argument list. These are then stored in a special internal array HE in a location specified by a call number NC from the argument list, thus allowing the routine to be used several times in a program. This initiation section also conveniently calculates the initial conditions of temperature and composition for each section (based on steady-state operation) by first calling the steady-state routine CSHE (see Chapter 9) that calculates the temperature and enthalpy in the exit streams, then calling subroutine *ICHE*(*I*nitial *C*ondition *H*eat *E*xchanger) described in the following section.

### 10-2-2 Subroutine ICHE

After the exit temperatures are obtained from the steady-state subroutine CSHE, the next step is to determine the initial temperatures in each of the internal sections (Figure 10-16b). The procedure is to define a ratio factor

$$R = Ln(DT1/DT2) * (1/N)$$

as on line 11 (Figure 10-16c), where DT1 and DT2 are the temperature differences at the ends of the exchanger. It can be shown that for each of the

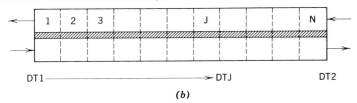

*(b)*

FIG. 10-16*b*.  Internal Sections for ICHE

N equal sections in the exchanger the ratio of the temperature difference at one end of the exchanger to the temperature difference at the boundary of each section J will follow the relationship (line 14)

$$DTJ = DT1/EXP(J * R)$$

The average temperature difference from the end of the exchanger to section J (line 15) will be

$$DTA = (DT1-DT2)/Ln(DT1/DT2) = (DT1-DT2)/(J * R)$$

```
 1 *          SUBROUTINE ICHE(IS,IT,N,UA,IPS,IPT,NSF,NSL,NTF,NTL)
 2 *          COMMON/CD/STRM(300,24),DATA(20,10),RCT(22),NCF,NCL,LSTR
 3 *          DO 6 J=1,N
 4 *          DO 6 JC = 1,20
 5 *          STRM(IS+J,JC) = STRM(IS,JC)
 6 *        6 STRM(IT+J,JC) = STRM(IT,JC)
 7 *          NCF=NSF
 8 *          NCL=NSL
 9 *          DT1=STRM(IS,22)-STRM(IT+N,22)
10 *          DT2=STRM(IS+N,22)-STRM(IT,22)
11 *          R=ALOG(DT1/DT2)/N
12 *          UAN=UA/N
13 *          DO 5 J=1,N
14 *          DTJ=DT1/EXP(J*R)
15 *          DTA=(DT1-DTJ)/(J*R)
16 *          Q=UAN*J*DTA
17 *          CALL HTEXCH(IS,IS+J,-Q,IPS)
18 *          STRM(IT+N-J,22)=STRM(IS+J,22)-DTJ
19 *          STRM(IS+J,20)=(STRM(IS+J-1,22)+STRM(IS+J,22)+STRM(IT+N-J,22)+
20 *         1STRM(IT+N-J+1,22))/4.
21 *        5 CONTINUE
22 *          NCF=NTF
23 *          NCL=NTL
24 *          DO 7 J=1,N
25 *          IF(IPT.EQ.3) CALL ENTHL(IT+J)
26 *        7 IF(IPT.EQ.0) CALL ENTHV(IT+J)
27 *          RETURN
28 *          END
```

*(c)*

FIG. 10-16*c*.  Listing for Subroutine ICHE
Argument list (supplied internally by DHE)

| | | | |
|---|---|---|---|
| IS | feed number shell | NSF | first component shell |
| IT | feed number tube | NSL | last component shell |
| N | number of sections | NTF | first component tube |
| UA | total heat transfer | NTL | last component tube |
| IPS | phase shell | | |
| IPT | phase tube | | |

Line 16 then calculates Q, the heat flux for each section. Q is supplied to the HTEXCH subroutine (line 17) to calculate the shell temperature entering section J. The tube temperature at the corresponding boundary is calculated by subtracting the temperature difference DTJ (line 18) from the shell temperature. The remaining function is to calculate the wall temperature in each section by assuming it to be the average of the inlet and outlet temperatures on the shell and tube side (lines 19, 20). This entire sequence is repeated for each section (DO loop lines 13–21), after which the enthalpy of each stream on the tube side corresponding to the temperature that was calculated on line 18 is determined according to its phase (lines 24–26). The subroutine also initializes the composition of each section to equal the feed composition (lines 3 → 6).

It should be noted that each section uses the arithmetic average temperature for calculating the heat fluxes; however the initial temperatures are based on log-mean temperature differences. This minor discrepancy decreases as the number of sections (N) is increased. The only penalty is a slight readjustment during the initial part of the dynamic simulation.

EXAMPLE 10-3

A feasible method for the terminal disposition of low-energy atomic waste material is to bury the material in a spherical cavity, deep underground. The radioactive waste continues to generate heat but at an exponentially decaying rate for a number of years, as shown in Figure 10-17. The heat generated is dissipated into the surrounding ground formation, and the temperature of the contents in the cavity gradually rise and eventually subside as the rate of heat generation diminishes. The feasibility of this method of storage depends on the maximum temperature reached in the cavity. Above a certain temperature, the walls of the cavity will fail and allow some of the radioactive

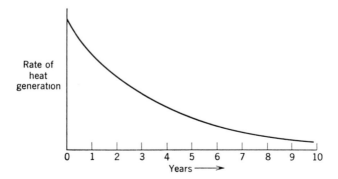

FIG. 10-17.  Rate of Heat Generation of Radioactive Waste

waste to seep out. A mathematical model is required that can predict the temperature over a future period of years, assuming that the entire cavity, or volume $V$, is entirely filled at time $\theta = 0$ and that the cavity is sufficiently deep (Figure 10-18) that heat loss from ground surface may be neglected. Clearly, heat will be conducted away from the cavity by a temperature gradient in the surrounding ground, but at all times the cavity will have the highest temperature. It becomes necessary then to compute the temperature gradient as a time translent.

This problem has four independent variables, time and three space dimensions, but because of symmetry around the cavity, the three space dimensions can be reduced to a single independent variable, namely, the radial distance from the cavity. In other words, the temperature at a particular radial distance is the same in all directions. The first step in constructing the mathematical model is to sectionalize the surrounding ground into appropriate sections. A typical temperature/distance gradient is shown in Figure 10-19, in which it can be seen that a high change of gradient exists near the cavity, becoming progressively less steep as the radial distance increases. This indicates that the sections should become progressively larger as the radial distance increases. Figure 10-20 shows a two-dimensional representation of the sections which actually are shells concentric with the cavity. If we consider a typical section, $n$, and define the temperature $T_n$ at the midpoint of the section (Figure 10-21), the heat balance equation for section $n$ is

$$
\begin{array}{ccccc}
\text{rate of change of} & & & & \\
\text{heat content} & = & \text{heat flux from} & + & \text{heat flux from} \\
\text{section } n & & \text{section } n - 1 & & \text{section } n + 1
\end{array}
$$

$$
\frac{d}{d\theta}(C\phi V_n T_n) = \frac{(T_{n-1} - T_n)}{\frac{1}{2}(\Delta R_{n-1} + \Delta R_n)} \cdot A_{n-1}k + \frac{(T_{n+1} - T_n)}{\frac{1}{2}(\Delta R_{n+1} + \Delta R_n)} \cdot A_n k
$$

Ground surface

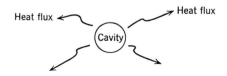

FIG. 10-18.   Underground Spherical Cavity

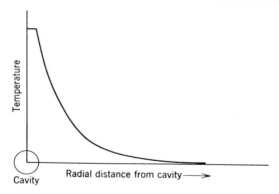

FIG. 10-19. Typical Temperature Gradient as a Function of Distance

where $V_n$ = volume of section $n$,
    $A$ = log mean area between midpoints of sections,
    $C$ = heat capacity of ground,
    $k$ = thermal conductivity,
    $\phi$ = density of the ground.

There are two end sections that constitute the boundaries for the system and

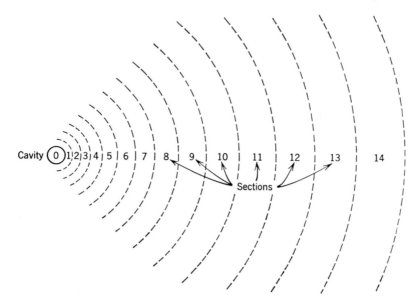

FIG. 10-20.  Sections of Spherical Shells

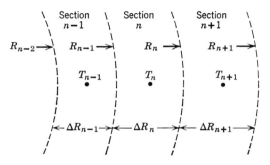

FIG. 10-21. Section $n$

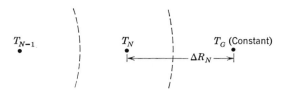

FIG. 10-22. End Section $N$

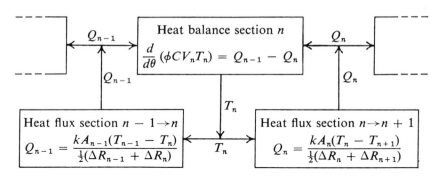

FIG. 10-23. Model for Section $n$

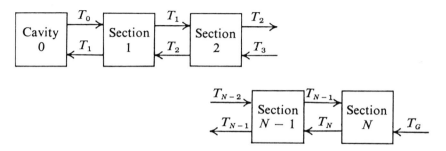

FIG. 10-24. Information Flow between Sections

require special definition. The cavity boundary is simply a heat balance equation on the material in the cavity, that is,

$$\frac{d}{d\theta}(C_W \phi_W V_0 T_0) = V_0 H - \frac{T_0 - T_1}{\frac{1}{2}\Delta R_1} k \cdot A_0$$

In this equation, the input heat flux is the heat generated by the radioactive material and is provided as basic data in the form $H$ Btu/unit vol unit time. The other boundary section is located sufficiently distant from the cavity for two safe assumptions to be made: that the temperature beyond it is the base temperature of the ground and remains constant at all times. This is sometimes called an "infinite sink." The equation for this last section $N$ would thus be (Figure 10-22)

$$\frac{d}{d\theta}(C\phi V_N T_N) = kA_{N-1} \cdot \frac{T_{N-1} - T_N}{\frac{1}{2}(\Delta R_N + \Delta R_{N-1})} + kA_N \frac{(T_G - T_N)}{\Delta R_N}$$

where $T_G$ is the ground temperature located $\Delta R_N$ from $T_N$.

The model for section $n$ is shown in Figure 10-23; the complete model, in signal flow form, in Figure 10-24. Solving these equations results in the temperature of the cavity, and the temperatures at specific radial locations, as functions of time. The results can be displayed either as the temperature at specific locations $n$ (center of each section), as functions of time (Figure 10-25a) or as temperature gradients as functions of radial distance at specific times, $\theta_i$ (Figure 10-25b). From these results, it is possible to determine whether the heat conduction and capacity of the ground is high enough to absorb the heat generated in the cavity and thus prevent the maximum temperature from rising to a dangerous level.

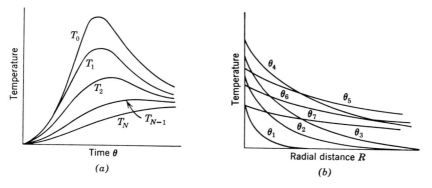

FIG. 10-25.  Results: Temperature versus Time

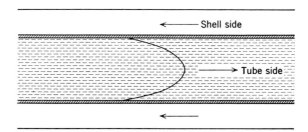

FIG. 10-26. Shell and Tube Heat Exchanger with Laminar Flow in the Inner Tube

EXAMPLE 10-4

The following example is a steady-state situation described by a partial differential equation with two independent variables, both of which are space dimensions. The system consists of a concentric pipe heat exchanger that transfers heat from a fluid flowing in the center pipe to a fluid flowing countercurrent on the shell side (Figure 10-26). The flow in the center pipe is laminar, but is turbulent on the shell side. The radial temperature gradient on the shell side can be considered uniform, but on the tube side, there is a gradient from the center to the tube wall. If a velocity profile is assumed for the laminar flow, a heat conduction model is required to establish the exchange of heat or the capacity of the unit for a given length.

The first step is to divide the inner tube into a suitable number of concentric annuli, 10 are sufficient in this case. Because the steepest gradients are near the wall, it is best to divide the total area into annuli having equal areas rather than equal radial increments. In this way the annuli near the wall will have smaller radial increments than the annuli near the center of the tube (Figure 10-27). Since the velocity distribution is known, the average flow rate in each annulus can be readily calculated from the total flow rate. A set of ordinary differential equations can now be written to define the temperature at the center of each annulus as a function of the distance $Z$ down the

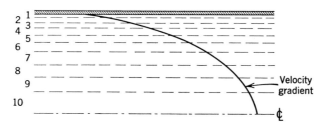

FIG. 10-27. Sectionalization of Inner Tube Flow

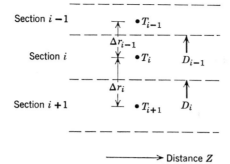

Section $i-1$    $\bullet T_{i-1}$

$\Delta r_{i-1}$

Section $i$    $\bullet T_i$    $D_{i-1}$

$\Delta r_i$

Section $i+1$    $\bullet T_{i+1}$    $D_i$

$\longrightarrow$ Distance $Z$

FIG. 10-28.  Section $i$

pipe. Analyzing the $i$th annulus (Figure 10-28) a heat-balance equation would require the heat transmitted from the adjacent annuli; using the average temperature gradients, these are (Figure 10-28)

$$kA_{i-1}\frac{T_{i-1} - T_i}{\Delta r_{i-1}} \quad \text{and} \quad kA_i\frac{T_{i+1} - T_i}{\Delta r_i}$$

where $A_i = \pi D_i$ and $D_i$ is the log mean diameter between points $T_i$ and $T_{i+1}$. The same holds for the areas $A_{i+1}$ and $A_{i-1}$.

The heat-balance equation for annulus $i$ is as follows:

$$\text{rate of change of heat content with length} = \text{sum of the heat fluxes}$$

$$\frac{d}{dZ}(v_i a_i T_i C) = k\left[A_{i-1}\frac{(T_{i-1} - T_i)}{\Delta r_{i-1}} + A_i\frac{(V_{i+1} - T_i)}{\Delta r_i}\right]$$

where $v_i$ = average fluid velocity in section $i$,
     $a_i$ = cross-section area of annulus section $i$,
     $A_i = \pi D_i$,
     $C$ = heat capacity.

The flow rate on the shell side is $F_S$, and its temperature $T_S$ is obtained by the heat balance

$$\frac{d}{dZ}(-F_S C T_S) = Q_S$$

The negative sign in the above equation is due to the countercurrent flow $F$ ($X$ is defined as positive with the direction of flow in the inner pipe). The flux from the shell side to the first or outermost concentric annulus is as follows:

$$Q_S = \pi D U_W(T_1 - T_S)$$

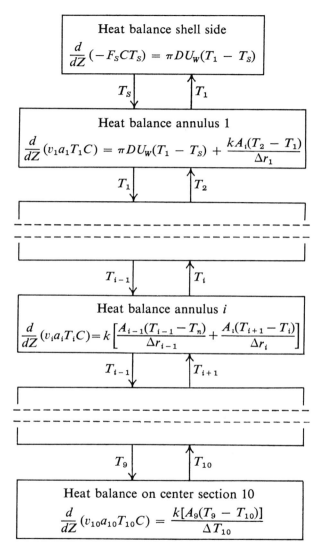

FIG. 10-29.   Model for Laminar Flow Heat Exchanger

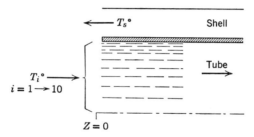

FIG. 10-30. Boundary Conditions for Heat Exchanger

where $U_W$ is the overall conductivity from the shell side through the wall and half way through the first annulus. Figure 10-29 shows the complete model; note the presence of only one flux term in the last section, that is, the center section. A computer mechanization of this model will result in a typical split boundary problem. Because the flows on the shell and tube sides are countercurrent, there will be one unknown initial condition at $Z = 0$ (inner tube inlet end), that of the shell side exit temperature $T_S^\circ$ (Figure 10-30). The starting values for the temperature in each annulus of the inner tube are assumed to be a uniform value. The value for the shell side temperature $T_S^\circ$ is estimated, and, after integrating down the length of the tube, it is checked with the known inlet temperature. If they differ, then the computation is repeated with an improved estimate of $T_S^\circ$; the cycle is repeated until a satisfactory match is obtained. This procedure is usually implemented on modern computers by automatic techniques. The advantage of starting the integration at the tube inlet end is obvious, since of the eleven initial conditions required, ten are known. The reverse would be true if the integration started at $Z = L$ and proceeded to $Z = 0$, for the number of integration cycles required to match the boundaries of that situation would be far greater.

EXAMPLE 10-5   DYNAMIC SIMULATION OF A FIXED-BED REACTOR

The hydrogenation reaction studied in Chapter 6, Section 4 as an experimental batch reaction is conducted commercially in a large fixed-bed reactor. By operating at very high pressures, the gas phase is compressed into solution in the liquid phase, so that only a liquid phase containing the reactants pass through the bed of catalyst pellets (Figure 10-31).

The reactor has no provision for removing the heat of reaction, such as a jacket, so it operates as an adiabatic process. The heat generated by the reaction, therefore, heats up the fluid as it flows through the reactor; the

FIG. 10-31. Fixed-Bed Reactor

fluid then exits at a higher temperature than the inlet feed temperature (Figure 10-32).

By a suitable combination of flow rate and feed temperature the reaction is not permitted to proceed to completion, thus avoiding excessive exit temperatures that produce undesirable by-products. An unwelcome characteristic of this type of operation is the erratic fluctuation of the outlet temperature and composition, caused by small changes in the inlet temperature. Various control strategies are proposed to reduce these fluctuations to acceptable limits, which generates the need to develop a computer simulation of the dynamic behavior of the reactor that can be used to test the strategies. A brief review of the kinetics of the reaction (already covered in Chapter 6) follows.

The reaction is a two-step hydrogenation of an organic $A$ to form a product $C$ via an intermediate compound $B$, that is,

$$A + H_2 \rightarrow B$$
$$B + H_2 \rightarrow C$$

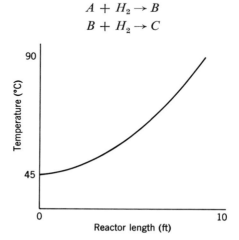

FIG. 10-32. Temperature Gradient through Reactor

The reaction takes place on the surface of the catalyst, which consists of spherical pellets sufficiently small that they sustain a negligible internal temperature gradient. Since there is no external heat sink or source for the reactor, conditions of temperature and composition in the radial direction are uniform. The only gradients are in the axial direction; thus it is necessary to finite difference this axis into a number of back-mix sections. This will approximate the physical mixing occurring because of the turbulence of the flow stream. Figure 10-33 shows one section, consisting of a mass of static catalyst particles with the liquid flowing through the interstices.

This situation can be reduced to an equivalent section, consisting of a solid phase and a liquid phase. The difference between the temperature of the liquid $(T_L)$ and solid $(T_C)$ creates a gradient that transfers heat $(QTR)$ between the phases under dynamic conditions. If the reactor were able to operate at steady-state conditions there would be no heat flux between the phases. This heat flux $QTR$ is calculated from the temperature gradient as

$$QTR = (T_L - T_C)W * UA$$

where $W$ = weight of catalyst in the section (lb)
$\quad UA$ = heat transfer coefficient of the catalyst surface
$\quad\quad$ (PCU/lb °C/min)

The rate of change of catalyst temperature in the section is defined by the heat balance

$$\frac{dT_C}{d\theta} = \frac{QTR}{W * C}$$

where $C$ = heat capacity of catalyst (PCU/lb °C).

It will be assumed that even though the reaction occurs on the catalyst surface, the heat of reaction is generated in the fluid phase. The model for the reaction is described in Chapter 6, and a subroutine for the calculation of the reaction rates is shown in Figure 6-12. The only modification that is required is to specify the mass of catalyst as an input constant in the argument

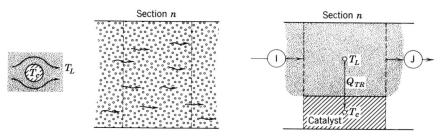

FIG. 10-33. Reactor Section

```
 1*          SUBROUTINE REACT(W,J)
 2*          COMMON/CD/STRM(300,24),DATA(20,10),RCT(22),NCF,NCL,LSTR
 3*          REAL K1,KG,KB,KC,K5
 4*          TR=(STRM(J,22)+273.)*1.98
 5*          K1=13.E4*EXP(-11130./TR)
 6*          K5=2.E4*EXP(-12200./TR)
 7*          KB=1.5E-3*EXP(4468./TR)
 8*          KC=8.*EXP(-1930./TR)
 9*          KG=.0144*EXP(-1130./TR)
10*          DEN=1.+KB*STRM(J,2)+KC*STRM(J,3)+KG*STRM(J,4)
11*          RA=W*K1*STRM(J,1)/DEN
12*          RB=W**2*K5*STRM(J,2)*STRM(J,4)/DEN**2
13*          RCT(1)=-RA
14*          RCT(2)=RA-RB
15*          RCT(3)=RB
16*          RCT(4)=-RA-RB
17*          RCT(21)=-RA-RB
18*          RCT(22)=RA*14700.+RB*15600.
19*          RETURN
20*          END
```

FIG. 10-34.   React Subroutine

list. This modified subroutine is shown in Figure 10-34. The subprogram will calculate the two reaction rates $R_A$ and $R_B$ (lines 11, 12) as a function of fluid temperature and composition. These quantities are listed in the J STRM array, where J is specified as the second item in the argument list. The reaction rates are transferred to the RCT array on lines 13 to 18.

The REACT subroutine is called from another subroutine, SECT, (Figure 10-35) that calculates the transfer of mass and energy in one section of the reactor. Each section will consist of a back-mixed fluid phase with an inlet stream I and an outlet stream J (Figure 10-33). There will be a constant liquid holdup HL in the section that is maintained by specifying the exit flow rate STRM (J,21) from an instantaneous mole balance (line 6). Since the change in mole flow is due to the reaction rate, a quantity that appears in RCT (21) is included in the mass balance. This balance is made after a call on

```
 1*          SUBROUTINE SECT(I,J,TC,W,HL)
 2*          COMMON/CD/STRM(300,24),DATA(20,10),RCT(22),NCF,NCL,LSTR
 3*          LOGICAL LSTR
 4*          DATA UA/.30/C/.1/
 5*          CALL REACT(W,J)
 6*          STRM(J,21)=STRM(I,21)+RCT(21)
 7*          QTR=(STRM(J,22)-TC)*W*UA
 8*          CALL HLDP(I,J,3,HL,QTR)
 9*          DTC=QTR/(W*C)
10*          IF(LSTR)RETURN
11*          CALL INT(TC,DTC)
12*          RETURN
13*          END
```

FIG. 10-35.   Subroutine SECT

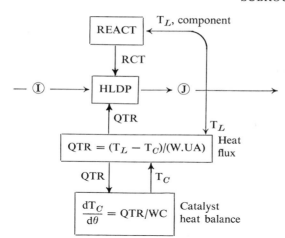

FIG. 10-36.   Information Flow for SECT

the REACT subroutine. The component and heat balances for the fluid phase is simulated by the HLDP subroutine (Chapter 5), where the external heat flux from the fluid phase for the catalyst QTR, calculated on line 7, is entered into the HLDP argument list. The remaining function of SECT is to calculate the catalyst temperature TC, which is identified for each section as the third item in the argument list to SECT. The derivative and integration are done on lines 9 and 11, respectively. A simplified schematic of SECT is shown in Figure 10-36.

The main program for the reactor simulation is shown in Figure 10-37a.

```
 1*              COMMON/CD/STRM(300,24),DATA(20,10),RCT(22),NCF,NCL,LSTR
 2*              DIMENSION TC(10)
 3*              LOGICAL NF,LSTR
 4*              DATA NCF/1/NCL/5/W/500./HL/2./LSTR/.TRUE./
 5*              DATA TC/49.3,50.6,52.5,54.6,57.1,60.1,63.8,68.8,75.8,87.1/
 6*              CALL DAT
 7*      C **DERIVATIVE SECTION**
 8*            6 STRM(1,22)=48.+.5*SIN(5*TIM)
 9*              CALL ENTHL(1)
10*              DO 5 J=10,1,-1
11*            5 CALL SECT(J,J+1,TC(J),W,HL)
12*              CALL PRL(.1,2.,NF,1,2,3,4,5,6,7,8,9,10,11,0)
13*              IF(NF) STOP
14*      C **INTEGRATION SECTION**
15*              CALL INTI(TIM,.005,1)
16*              GO TO 6
17*              END
```

(a)

FIG. 10-37a.   Main Program for Reactor

```
 1*            SUBROUTINE DAT
 2*            COMMON/CD/STRM(300,24),DATA(20,10),RCT(22),NCF,NCL,LSTR
 3*            DATA(DATA(1,J),J=1,8)/10.5,-2955.,273.,16.,0.,5805.,24.,0./
 4*            DATA(DATA(2,J),J=1,8)/10.5,-3010.,273.,19.,.1,8060.,30.,.03/
 5*            DATA(DATA(3,J),J=1,8)/11.2,-3400.,273.,22.,.2,9726.,33.,.02/
 6*            DATA(DATA(4,J),J=1,8)/10.5,-2750.,273.,6.9,.05,2600.,22.,.1/
 7*            DATA(DATA(5,J),J=1,8)/13.,-4600.,273.,33.,.06,12174.,41.,0./
 8*            DATA(STRM(1,J),J=1,22)/.15,0.,0.,.45,.4,15*0.,100.,40./
 9*            DATA(STRM(J,1),J=2,11)/.148,.146,.143,.141,.138,.134,.131,.126,
10*           1                      .120,.11/
11*            DATA(STRM(J,2),J=2,11)/.002,.004,.006,.008,.01,.012,.014,.015,
12*           1                      .016,.017/
13*            DATA(STRM(J,3),J=2,11)/.2E-3,.8E-3,.16E-2,.29E-2,.47E-2,.72E-2,
14*           1                      .01,.015,.023,.036/
15*            DATA(STRM(J,4),J=2,11)/6*.445,4*.42/
16*            DATA(STRM(J,5),J=2,11)/6*.4,4*.42/
17*            DATA(STRM(J,22),J=2,11)/49.3,50.8,52.5,54.6,57.1,60.1,63.8,68.7,
18*           1                      75.8,87.1/
19*            DATA(STRM(J,21),J=2,11)/99.7,99.4,99.,98.5,98.,97.4,96.6,95.5,
20*           1                      93.8,90.5/
21*            RETURN
22*            END
```

*(b)*

FIG. 10-37b.  Subroutine DAT for Reactor Simulation

There are 10 sections for the reactor; thus the streams are numbered 1 to 11, and the subroutine SECT is called in the DO loop (line 11) in the reverse order (line 10), that is, starting at the exit end and proceeding to the inlet end. It will be recalled that this procedure obeys a basic rule for proper sequencing of DYFLO subroutines.

The first run was made with a constant input-fluid temperature (STRM 1,22) of 48°C. When equilibrium was achieved, the values of catalyst and fluid temperature and fluid compositions in all 10 sections were read out. These formed the starting values for the dynamic test and are listed in subroutine DAT (Figure 10-37b) with the component physical property data listed in the DATA array.

The dynamic run on the program was made with the feed temperature fluctuating $\pm 0.5°C$ as a sine wave (line 8). It is then necessary to establish the corresponding enthalpy of this feed stream by calling the subroutine ENTHL (line 9). The "sensitivity" of the simulated reactor to this input temperature fluctuation is shown in Figure 10-37c as a line plot of the exit-fluid temperature (plot B), the catalyst temperature in section 10 (plot C), and the feed temperature (plot E). It will be seen that the amplitude of the exit temperature is about 14°C, which is an amplification factor of 14 over the 1° feed amplitude variation. There is also a phase lag of 0.25 min between the input and output wave.

This simulation is now ready to be used for testing control schemes. This will be done as a problem exercise in Chapter 11 (Problem 8).

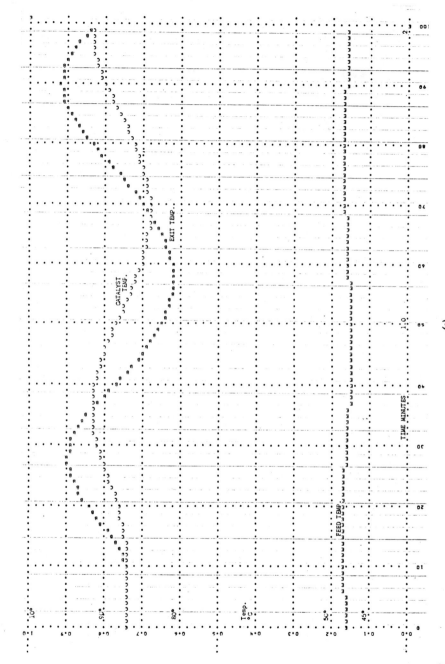

FIG. 10-37c. Plotted Results from Fixed-Bed Reactor Program

(c)

345

EXAMPLE 10-6  POLYMER KINETICS (REFS 9, 10)

One of the difficulties in an analytical approach to polymer kinetics is the relatively immense number of molecular species (i.e., different molecular weights or chain lengths) involved; consequently, successful methods have been developed for treating the chain length as a continuous variable rather than as a large number of discrete lengths (9). When a batch polymerization reaction is considered, there are two independent variables involved, time $\theta$ and chain length $N$, leading, of course, to a system of partial differential equations. A very simple example is considered here, involving only radical initiation and addition polymerization, with coupling termination. The reactions are as follows:

$$I \text{ (initiator)} \xrightarrow{k_D} 2R \text{ (initiating radicals)}$$

$$R + M \text{ (monomer)} \xrightarrow{k_A} C_0 \text{ (first chain radical)}$$

$$C_0 + M \xrightarrow{k_R} C_1 \text{ (radical 1)}$$

$$C_n + M \xrightarrow{k_R} C_{n+1} \text{ (radical } n + 1)$$

$$C_X + C_Y \xrightarrow{k_P} P_{X+Y} \text{ (coupling termination polymer chain)}$$

Assuming a vessel contains a batch of monomer and initiating agent $I$, the above reaction sequence then occurs and the distribution of molecular weights of the radicals proceeds as shown in Figure 10-38. Because the molecular weight distribution of the polymer is one of the factors that influence the physical properties of the polymer, it is often possible to control

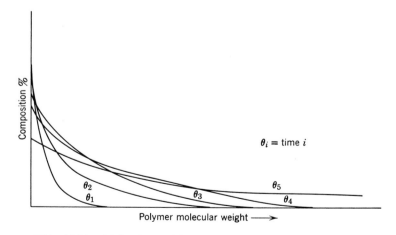

FIG. 10-38.   Molecular Weight Distributions as a Function of Time

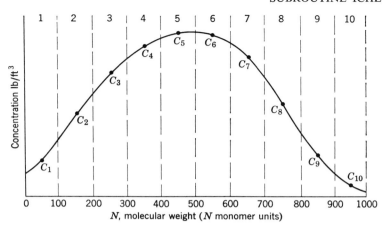

FIG. 10-39.  Sectionalized Molecular Distribution

the reacting environment thus to guide the distribution to one yielding favorable characteristics.

The molecular weight can increase up to very large values, and therefore the distribution is divided into sections (i.e., finite differenced), and ordinary differential equations written for the average molecular weight in each section. Suppose a typical distribution shown in Figure 10-39 is divided into 10 equal sections spaced so that the sections encompass 95% of the predicted distribution (Figure 10-39). The center of each section has a specific molecular weight $n$. For example, Section 6 (Figure 10-39) has an average molecular weight of 550 monomer units. If the total weight concentration of each section $n$ is $C_n$ (lb/ft³), then the average mole concentration will be $Cn/Nn \cdot \Delta n$ (moles/ft³), where $Nn$ is the average molecular weight of section $n$, and $\Delta n$ is the number of species in the region. The rate equation defining radical chain growth is

$$\text{rate moles/ft}^3 = k_R \left(\frac{C_m}{N_m}\right) \left(\frac{C_i}{N_i \cdot \Delta n}\right)$$

where $Cm$ = lb/ft³ monomer,
  $C_i$ = lb/ft³ chain radical of $i$ mol. wt.,
  $N_m$ = mol. wt. monomer,
  $N_i$ = mol. wt. chain radical $i$,
  $k_r$ = rate constant.

If the molecular weight at each boundary of section $n$ is $i$ or $j$, respectively

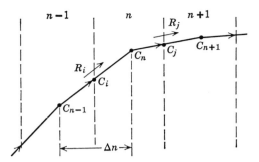

FIG. 10-40.  Rate of Growth through Section $n$

(Figure 10-40), then the mole concentration of the radical species $i$ will be

$$C_i = \left(\frac{C_{n-1}}{N_{n-1}} + \frac{C_n}{N_n}\right) \cdot \frac{1}{2} \cdot \frac{1}{\Delta n}$$

where $\Delta n$ is the number of species in the region bounded by $n$ and $n-1$. The rate of chain radical growth into and out of section $n$ can be defined as follows:

$$R_i = k_r \cdot \left(\frac{C_m}{N_m}\right) C_i \quad \text{and} \quad R_j = k_r \cdot \left(\frac{C_m}{N_m}\right) C_j$$

where $C_i$ and $C_j$ are the mean concentrations between sections $n-1$, and $n$ and $n$ and $n+1$, respectively.

Chain radical $C_n$ can also be consumed by coupling with each of the other radicals and of course with itself. The expression for this rate is

$$Rc = k_P \cdot \left(\frac{C_n}{N_n}\right) \sum_{n=1}^{n=10} \frac{C_n}{N_n}$$

With $C_n$ the total weight fraction of section $n$, the distribution can be considered to be concentrated at the points specified as $n$. An overall mole balance on chain radical section $n$ can now be assembled as follows:

$$\frac{d}{d\theta}\left[\frac{C_n}{N_n}\right] = \text{(rate of growth across boundary } n/n-1) - \text{(rate of growth across boundary } n/n+1) - \text{(rate of coupling termination)}$$

$$= k_R \frac{C_m}{N_m} [C_i - C_j] - k_P \frac{C_n}{N_n} \sum_1^{10} \frac{C_n}{N_n}$$

The weight concentration of polymer can be obtained by a similar method, for if 10 specific molecular weight radicals are defined, there will be 20 polymer species resulting from coupling terminations. For example, any particular polymer species, say, $P5$, can be formed by a combination of several pairs, that is, $(C1 + C4)$, $(C3 + C2)$, and so forth, and the number of pairs increases with the order of the species. This can be summarized in the statement that $P_n$ is formed from the combination $\sum_a^b C_i \cdot C_{n-i}$, where the conditions on the lower and upper limit are as follows:

$$\text{lower limit } a = 1 \text{ for } n < 10$$
$$a = n - 10 \text{ for even } n > 11$$
$$a = n - 9 \text{ for odd } n > 10$$
$$\text{upper limit } b = n/2 \text{ for even } n$$
$$b = (n - 1)/2 \text{ for odd } n$$

A mass-balance equation for each polymer species $Pn$ would be

$$\frac{d}{d\theta}\left(\frac{P_n}{N_n}\right) = k_P\left(\sum_a^b \frac{C_i}{N_i} \cdot \frac{C_{n-i}}{N_{n-i}}\right)$$

Because $C_n$ and $P_n$ are the weight concentrations of the radical and polymer species, the monomer concentration can be obtained by difference; that is,

$$C_m = M^0 - \left(\sum_{n=1}^{10} C_n - \sum_{n=1}^{20} P_n\right)\Delta n$$

The rate of formation of initiating radicals is

$$\frac{d}{d\theta} \cdot C_R = 2 \cdot k_D \cdot C_i - k_A \cdot C_R \cdot C_m$$

The mass balance on the initiator is

$$\frac{d}{d\theta} C_I = -k_D \cdot C_I$$

The term $k_A \cdot C_B \cdot C_m$ is the rate of growth of chain radicals into Section 1.

These equations are arranged in model form as shown in Figure 10-41. Basically, what has been achieved here is the reduction of a kinetic polymer system sometimes involving chains containing over 100,000 monomer units into a manageable set of approximately 30 equations. All the usual complexities can now be included such as heat of formation and reaction rates as functions of temperature. This approach can also be extended to include chain branching, and chain transfer as well as other complexities.

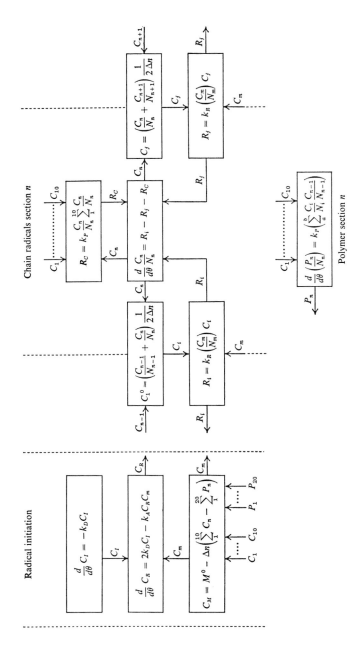

FIG. 10-41. Model for Batch Polymerization

### Exercise Problems

1. A large reaction vessel contains a bolted flange around its perimeter. The temperature of the inside surface of the reactor $T(t)$ cycles through violent fluctuations, and it is feared that the thermal gradients in the flange may cause stress failure.

Assuming that the outside surface of the flange is insulated and that $T(t)$ is known, construct a model that establishes the temperature variations at locations $A$, $B$, $C$ (Figure 10-42).

2. A pipeline reactor contains a highly viscous fluid that flows in the laminar region with a known radial velocity distribution $V(R)$. The reaction within the pipe is

$$A + B \underset{k_2}{\overset{k_1}{\rightleftharpoons}} C + D$$

and the forward and reverse reaction rate coefficients are known.

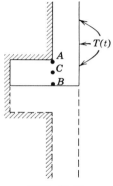

FIG. 10-42

Characterize the reactor with a mathematical model that will establish the approach to equilibrium for a range of reactor lengths, assuming no component radial diffusion.

3. Suppose the reactants in the above problem are "carried" by a neutral but highly viscous fluid. How would the mathematical model, developed in the preceding problem, be modified to include the carrier fluid and the effect of radial diffusion of the reacting components, assuming equimolar counter diffusion of the carrier fluid?

4. In order to damp out temperature oscillations in a fluid stream a fixed bed of steel balls of uniform diameter are placed in a vessel, and the fluid is passed through the bed. The heat transfer coefficient from the liquid to the solid phase is known, and the heat capacity of the fluid stream and the thermal diffusivity of the steel balls are also known.

Construct a model that will establish the total weight of steel required to damp out sine wave oscillations with an inlet amplitude $A(°C)$ and a frequency $\omega$ (cycles/min) for a particular flow of fluid $F$ lb/min, assuming plug flow through the bed.

How would the model be modified if axial mixing was significant?

5. A water phase contains a suspension of polymer particles and a catalyst system that generates radicals according to the relationship

$$A + B \xrightarrow{k_R} R\cdot$$

The radicals can terminate in the water phase as follows:

$$R\cdot + R\cdot \xrightarrow{k_T} C$$

The water phase radical population rises to a maximum, then falls to zero over a period of time. The radicals diffuse into the particles (assume spherical shape, diameter $D$) and terminate at a rate $kC^2$ where $C$ is the radical concentration in the particle (moles radical/cm³). The rate of entry of radicals into the particle surface is proportional to the water phase radical concentration, that is,

$$R_S = k_E C_w$$

where $C_w$ is the water phase radical concentration and $R_S$ is the radical entry rate (moles/cm² min). The radical diffusion rate within the particle is proportional to the radical concentration gradient.

Construct a model that defines how the radical population within the particle varies with time and radical position.

6. Redefine the model in Problem 9-6, Chapter 9, for a significant temperature gradient existing in both the metal and the sheet.

7. A simplified mechanism for the ion exchange process can be described as follows:

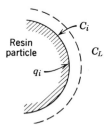

If a resin particle is immersed in a fluid containing a solute that can be absorbed by the particle, a concentration gradient between the solvent concentration $C_L$ and the center of the particle will ensue that will drive the absorbable solute into the particle. A linear gradient is assumed across the fluid film, and the surface concentration on the fluid side of the interface $C_i$

bears a constant relationship to the surface concentration on the solid side $q_i$; that is, $C_i = f(q_i)$. In addition, $D_L$ is the film side coefficient of diffusion and $D_r$ is the resin side of diffusion coefficient, $R = $ Radius of the particle and $f_v$ is the void fraction for a bed of particles.

Construct a model that determines the behaviour of an ion exchange process in the form of a fixed bed of particles length $Z$, diameter $D$, through which flows a solution with a flow rate $S$, and an inlet composition $C_0$. The model should define the outlet concentration of solute as a function of time.

The following problems will require the use of a computer.

8. *Cooling fin design:* Determine the optimum ratio of base width to height $(w/h)$ of an aluminum fin that will provide the maximum rate of heat loss to a circulating cooling gas. Assume:

(a) maximum cross section area $= 1$ in.$^2$ $(= wh/2)$
(b) base temperature $= 200°C$
(c) gas temperature $= 15°C$
(d) surface heat flux to gas is proportional to the temperature difference at a rate of $0.01$ PCU/(sec, °C, in.$^2$)
(e) conductivity of aluminum is $0.024$ PCU/(ft$^2$, sec, °C/ft)
(f) the width of the fin tip is $\frac{1}{10}$ the width of base.

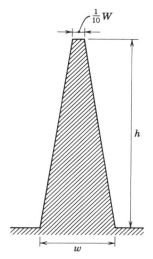

$\frac{1}{10}W$

$h$

$w$

9. The countercurrent heat exchanger described in Section 9-10 has a total liquid holdup of 300 moles on the shell side and 100 moles on the tube side. Determine the minimum number of sections required to provide a dynamic model that will simulate an exit temperature response to a 10°C step change in inlet shell-side feed temperature with less than 1°C error. The total tube-wall heat capacity is 10,000 PCU/°C.

10. Obtain a numerical solution for Problem 1 using the following data:

(a) conductivity of metal 0.05 PCU/(ft² sec °C/ft.)
(b) heat capacity of metal 0.1 PCU/(lb °C)
(c) wall thickness 3 in.
(d) flange thickness 4 in.
(e) flange width (top) 6 in.
(f) step change of inside surface temperature 400°C → 100°C.

## NOMENCLATURE

$A$     cross-sectional area
$a$     cross-sectional area of annulus
$C$     heat capacity
$C_G$   heat capacity of gas
$C_i$    chain radical concentration
$D$     diameter or diffusion coefficient (Example 10-5)
$F$     flow rate
$h$     thermal diffusivity
$H$     heat flux
$I$     initiator
$k$     conductivity, or catalyst activity (Example 10-5)
$k_1$    reaction rate constant
$l$     length
$M$    monomer
$N$     molecular weight
$P$     polymer
$Q$     heat flux
$R$     radius, or radicals (Example 10-6), or reaction rate
$r$     radius
$T$     temperature
$V$     volume
$X$     distance
$\theta$     time
$\phi$     density
$\pi$     3.1412

## REFERENCES

1. "Higher Order Differences in the Analog Solution of Partial Differential Equations," M. E. Fisher, *J. Assoc. Comp. Mach.*, **3**, 1956.
2. "The Solution of Partial Differential Equations by Difference Methods," R. M. Howe and V. S. Haneman, *Proc. I.R.E.*, 1953.
3. "Application of a Frequency Response Error Criterion to Distributed System Simulation Models," W. F. Hillyard, *A.I.Ch.E.*, Buffalo, 1963.

4. "Heat Flow and Simulation," A. V. Branker, *Petroleum*, July 1956.
5. "Computer Solves Heat Flow Problem," C. A. Levine and A. Opler, *Petroleum*, July 1956.
6. "Process Dynamics and Analog Computer Simulation of Shell and Tube Heat Exchanger," L. H. Fricke, H. J. Morris, R. C. Otto, and T. J. Williams, *Chem. Eng. Progr. Symp. Ser.*, **56**, No. 3, 1960.
7. "Simulation of the Dynamic Characteristics of a Superheater," D. Groupe and A. S. Aldred, *J. Mech. Sci.*, **5**, 1963.
8. "Hybrid Computation Applied to Counter Current Process," A. Frank and L. Lapidus, *Chem. Eng. Progr.*, January 1964.
9. "Continuous Polymerization Models," R. J. Zeman and N. R. Amundsen, *Chem. Eng. Sci.*, **20**, 1965.
10. "Effect of Reaction Paths and Initial Distribution on Molecular Weight Distribution of Irreversible Condensation Polymers," H. Kilkson, *Ind. Eng. Chem.*, **3**, No. 4, 1964.
11. "DSS—Distributed System Simulator," M. G. Zellner and W. E. Schiesser, *Summer Comp. Simulation Conf.*, Denver, 1970.
12. *Applied Numerical Methods*, B. Carnahan, Luther Wilkes. Wiley, New York, 1970.

## References for Additional Study

### Hydrology

13. "Flood Simulator for the River Kitakami," K. Otoba, K. Shibatami, and H. Kuwata, *Simulation*, **4**, No. 2, 1965.
14. "Analog Simulation of Underground Water Flow in the Los Angeles Coastal Plain," D. A. Darms and H. N. Tyson, *W.J.C.C.*, 1961.
15. "Progress in Ground Water Studies with the Electronic Analog Model," R. H. Brown, *J.A.W.W.A.*, August 1962.
16. *The Analog Computer and Estuarine Pollution Problems*, D. J. O'Connor, L. L. Falk, and R. G. Franks, 18th Ind. Water & Wastes Conf., 1963.

### Steel Industry

17. *Analog Computer Analysis of a Direct Iron Ore Reduction Process*, N. F. Simcic and J. C. Buker, ISA Paper-71-NY60
18. "An analog Study of the Transient Behaviour of a Batch Type Fluidized Bed for Metallurgical Heat Transfer," C. O. Pedersen, *ASME*, 61-WA-214

### Fixed Bed Systems

19. *A Computer Model for the Regenerative Bed*, P. K. Leung and D. Quon, C.I.C. Chem. Eng. Conf., Hamilton, Ont., 1964.
20. "Transient Behaviour of an Ammonia Synthesis Reactor," P. L. T. Brian, R. F. Baddour, and J. P. Eymery, *Chem. Eng. Sci.*, **20**, 1965.
21. "Packed Catalytic Reactors" *I.&E.C.* **5**, No. 7, July 1970.

# CHAPTER XI

# PROCESS CONTROL

The field of process control has been one of the most important stimuli to the development of computer simulation, and today the more complex problems in this area are solved by computer techniques. A brief outline of the basic elements in typical control loops is described in this chapter, together with the method of computer simulation. For a more detailed coverage of this subject, the reader is referred to texts dealing exclusively with process control technology (Refs. 1, 2, 3). One point should be observed, that is, that control engineering, if practiced without the aid of a computer, requires a thorough grasp of the basic principles involved together with a knowledge of the techniques available for the approximate solution of problems in process dynamics. A computerized approach to problems allows a more general solution that retains all the nonlinear complexities, as well as an optimization of control configurations by rapid trial and error procedures. Experts in advanced control techniques have predicted "Computer hardware and software are approaching the point where man/machine symbiosis appears feasible. The man tackles a problem, thinks it out, calls on the computer for instant calculational help, then continues to think and let the machine calculate until the two converge on an answer" (Richard Bellman, 1964). It appears that synthesis of complex controls for processes will rely to an ever increasing extent in the future on computer technology.

## 11-1  BASIC CONTROL CONFIGURATION

Figure 11-1 shows a typical scheme for a single process control loop. The process itself viewed from the control loop consists of an output variable such as temperature, pressure, and composition that is directly influenced

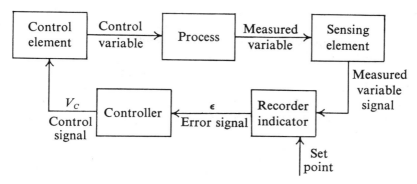

FIG. 11-1.   Basic Arrangement of a Process Control Loop

by some input control variable such as flow rate, pump speed, heating rate. The relationship between the input "force" and the output variable is described by a mathematical model that can range from a single equation to a whole set of complex interacting nonlinear expressions. Most of the previous chapters in this text are directed towards developing the technique of model building for a variety of process systems; this aspect, therefore, is not repeated in this chapter other than at the end, with the example of a batch kettle reactor. There is no doubt that formulating the equations for the process of the loop is by far the most burdensome task, and by comparison, formulating the analytical definition for the other elements in the loop is almost trivial. One reason for this is that except for unusual applications, the relationships as "transfer functions" are well known.

## 11-2   SENSING ELEMENT: FIRST-ORDER TRANSFER FUNCTION TFNI

The majority of sensing elements fall into one of three groups, thermal elements, pressure or flow sensors. Generally, thermal elements such as thermowells or resistance bulbs are the more complex to describe because of the inherent dynamic lag. For example, consider a thermocouple enclosed in a thermowell as shown in Figure 11-2. An equation relating the temperature $T_c$ "sensed" by the thermocouple and the actual temperature $T$ of the fluid into which the thermowell is immersed can be developed as follows.

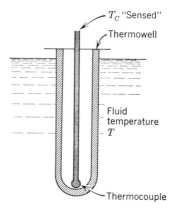

FIG. 11-2.   Thermocouple Enclosed in a Thermowell

The thermal gradient across the wall of the thermowell is $(T - T_c)$. If $h$ is the overall heat transfer coefficient, then the rate of heat transfer into the thermowell will be $h(T - T_c)$. It is assumed that the thermocouple measures the temperature on the inside of the thermowell wall and also that the thermocouple itself has negligible heat capacity.

The heat transferred into the thermowell is stored in the heat capacity of the walls, and if it is assumed that nearly all the resistance is on the outside skin, then the average temperature of the thermowell wall is almost the same as the temperature of the thermocouple. On this assumption, the following heat balance can be made

$$\frac{d}{dt}(WCT_c) = h(T - T_c)$$

where $WC$ is the heat capacity of thermowell plus thermocouple. Because $WC$ is constant, it follows that

$$\frac{d}{dt}(T_c) = \left(\frac{1}{WC/h}\right)(T - T_c)$$

The expression $WC/h$ has units of time and is generally termed the "time constant." It is denoted by the symbol $\tau$. The above differential equation can be converted to the Laplace form by substituting $d/dt$ by the symbol $S$, and treating $S$ as an algebraic quantity. Making these substitutions and rearranging the equation, the following "transfer function" is obtained:

$$\frac{T_c}{T} = \frac{1}{\tau S + 1}$$

In block diagram notation this can be shown as

$$T \longrightarrow \boxed{\frac{1}{(\tau S + 1)}} \xrightarrow{T_c}$$

that is, $T$ is "acted upon" by the transfer function $1/(\tau S + 1)$ to give $T_c$. The Laplace notation for the dynamic characteristics of the control elements is common in control engineering and is presented here for that reason, although the original equation could also be used, that is,

$$T \longrightarrow \boxed{\frac{d}{dt}T_c = \frac{1}{\tau}(T - T_c)} \xrightarrow{T_c}$$

```
1  *        SUBROUTINE TFN1(OUT,XIN,TC,GAIN)
2  *        DTF=(GAIN*XIN-OUT)/TC
3  *        CALL INT(OUT,DTF)
4  *        RETURN
5  *        END
```

*(a)*

FIG. 11-3a.  Listing for First-Order Transfer Function
Argument list:
OUT = output signal
IN = input signal
TC = time constant
GAIN = gain

A more general transfer function would include a gain factor G, that is,

$$\xrightarrow{\text{XIN}} \boxed{\frac{\text{OUT}}{\text{XIN}} = \frac{G}{\tau S + 1}} \xrightarrow{\text{OUT}}$$

The equivalent differential equation for this transfer function is:

$$\frac{d(\text{OUT})}{dt} = \frac{(G * \text{XIN} - \text{OUT})}{\tau}$$

Using the INT program developed in this text, the equation above can be readily programmed as two statements comprising the subroutine TFN1 shown in Figure 11-3a. Only four items are required for the argument list, the output, the input (XIN), the time constant (TC), and the gain. The derivative to be integrated is coded on line 2, followed by a CALL on the subroutine INT to integrate this derivative. All TFN1 subroutine calls must be listed in the integration section of the main program, observing the basic rule that the input to TFN1 (XIN) must never be the output from a preceding routine listed in the integration section.

## 11-3  SECOND-ORDER TRANSFER FUNCTION

The general form of a second-order transfer function is as follows:

$$\xrightarrow{\text{XIN}} \boxed{\frac{\text{OUT}}{\text{XIN}} = \frac{G}{\tau^S S^2 + \zeta \tau S + 1}} \xrightarrow{\text{OUT}}$$

This expression defines the dynamic relationship between the input and output signal passing through a double-capacity linear system. The dynamic characteristics of such a second-order system are characterised by $\tau$, the time constant and $\zeta$, the damping ratio. If $\zeta$ is greater than unity, the system is

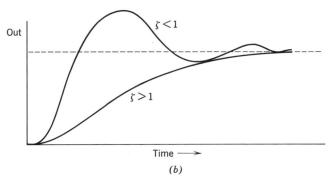

FIG. 11-3*b*. Effect of Damping Factors on Second-Order response

over damped and will respond to a step disturbance as shown in Figure 11-3*b*. If the damping factor is less than unity, an oscillatory response will follow a step disturbance.

The transfer function can be converted to its equivalent differential equation and will have the following form:

$$\frac{d^2}{dt^2}(\text{OUT}) = (\text{GAIN} \cdot \text{XIN} - \text{OUT}) \cdot A_1 - A_2 \frac{d}{dt}(\text{OUT})$$

where

$$A_1 = 1/\tau^2$$
$$A_2 = \zeta/\tau$$

The double integration involved in this case is implemented in the subroutine TFN2 shown in Figure 11-3*c*, by extending the method used in

```
1 *          SUBROUTINE TFN2(OUT,XIN,A1,A2,GAIN)
2 *          COMMON/CINT/T,DT,JS,JN,DXA(500),XA(500),IO,JS4
3 *          JN=JN+1
4 *          J1=JN
5 *          DTF=(GAIN*XIN-OUT)*A1-XA(J1)*A2
6 *          CALL INT(OUT,XA(J1))
7 *          CALL INT(XA(J1),DTF)
8 *          RETURN
9 *          END
```
(c)

FIG. 11-3*c* Listing for Second-Order Transfer Function
Argument list:
OUT = output signal
IN = input signal
$A_1 = 1/2$
$A_2 = \delta/\tau = \delta\tau/\tau$
GAIN = gain

TFN1. There are five entries in the argument list, the input and output variables, the two coefficients $A_1$ and $A_2$, and the gain. The double integration requires two calls on the subroutine INT, but the successive integrations are listed in the reverse order, as recommended previously, since the output of the first integration is the derivative of the second integration. At the start of a run, the initial condition for the first integration must always be zero. It is necessary, then, that this variable be cleared to zero at the end of each run to be ready for the succeeding run. This is done by storing this variable in the array XA in position J1 obtained from the INT call counter JN (lines 3 and 4); at the end of each run, the PRNTF subroutine (Chapter 3) clears the entire XA array.

The CALL for this subroutine must be listed in the integration section, again taking care that the input XIN is not the output of a preceding subroutine listed in the integration section.

## 11-4 CONTROLLERS

Nearly all control studies of industrial processes involve one or more control loops. In order to avoid having to program these controllers repetitiously, subroutines are developed next that simulate the action of the three basic modes of industrial controllers. These three modes are proportional, automatic reset or integral, and rate actions. The three most common combinations of these modes are:

1. Single mode: proportional mode
2. Two mode: proportional, plus automatic reset
3. Three mode: proportional, reset, and rate action

A separate subroutine is developed for each of these controllers.

### 11-4-1 Proportional Controller CONTR1

The functions of a controller can be grouped as shown in Figure 11-4. The measuring section converts the input signal, such as temperature, pressure, or flow, to a normalized signal based on the zero and range of the instrument.

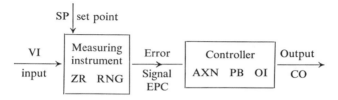

FIG. 11-4. Instrument Controller Combination

In the subroutine to be developed, this normalized signal will range from 0. to +100. For example, if the input signal is 260°C and the range of the instrument is 150° to 350°C, the normalized signal will be 55. The instrument section also contains a set point made either by an external manual setting or, as in cascade systems, by the output of another controller. This set point is the level desired for the controlled variable and must be normalized on the instrument's zero and range. The signal fed to the controller section will be an error signal that will be the difference between the normalized input and set-point signals and, consequently, will vary from −100 to +100. If, for example, the instrument just mentioned had a set point of 250°, the error signal passed to the control section would be +5.

The control section, in the case of a proportional controller, has three parameters that define its action, namely, the gain, which is 100/proportional band; the action, that is, either direct or reverse acting; and the output setting, termed the manual reset. The expression that relates these parameters represents a proportional controller and is

$$CO = (EPC * 100/PB) * AXN + OI$$

where  $CO$ = controller output ranging from 0 to +100,
$EPC$ = normalized error,
$PB$ = proportional band,
$AXN$ = action, +1. for direct, −1. for reverse,
$OI$ = manual reset.

The manual reset is usually set to a value that will reduce the error EPC close to zero under normal steady-state conditions. Since it is possible under certain conditions for the controller equation to calculate a theoretical output CO that is beyond the 0 or 100 limit, it is necessary to limit the output signal to these maximum and minimum values. Similarly, it is necessary to limit the input signal VI to the top and bottom limit of the instrument range if it is beyond these limits.

The subroutine CONTRI for a proportional controller is shown in Figure 11-5a. The first two variables in the argument list are the input (VI) and the output (CO). These are followed by the zero (ZR) and range (RNG) values of the instrument, the set point (SP), numerically equal to a value between ZR and RNG, and the action (AXN), either +1 or −1. The last two items are the proportional band PB and the manual reset OI. Since the input is limited to a travel between ZR and RNG, the input is converted to a dummy symbol VIM to prevent the value of VI from being changed by the controller on lines 4 and 5. The limiting procedure for the output is performed on lines 8 and 9.

```
 1  *          SUBROUTINE CONTR1 (VI,CO,ZR,RNG,SP,AXN,PB,OI)
 2  *          SPAN = ABS(RNG-ZR)
 3  *          VIM = VI
 4  *          IF (VI .GT. RNG) VIM=RNG
 5  *          IF (VI .LT. ZR) VIM= ZR
 6  *          EPC = 100. * AXN * (VIM-SP)/SPAN
 7  *          CO = 100./PB * EPC + OI
 8  *          IF(CO.LT.0.)CO=0.
 9  *          IF(CO.GT.100.)CO=100.
10  *          RETURN
11  *          END
```

*(a)*

FIG. 11-5a. Subroutine Listing for Proportional Controller CONTRI

| | |
|---|---|
| VI = input | ANX = action: +1. direct, −1. reverse |
| CO = output | PB = proportional band |
| ZR = zero | OI = manual reset |
| RNG = range | |
| SP = set point | |

## 11-4-2 Two-Mode Controller

This controller consists of proportional plus reset (i.e., integral) action. The equation describing its action is

$$CO = \frac{100.}{PB}\left(EPC + RPT\int EPC \cdot dt\right) * AXN$$

This can be transformed to a more convenient form:

$$CO = \frac{100}{PB} \cdot AXN \cdot EPC + OI$$

$$OI = \frac{100}{PB} \cdot AXN \cdot RPT\int EPC \cdot dt$$

RPT is the reset control setting and usually has units of repeats per minute. Before discussing the subroutine for this controller, an important aspect has to be understood. Examination of the controller equation above shows that the action of the two-mode controller consists of two parallel paths, the proportional action, which is algebraic, and the reset action OI, which involves an integration. In order not to violate the rule stated earlier concerning the segregation of algebraic equations and the CALL statements for the integral equation into separate sections, it will not be possible to perform the integration for this controller within the subroutine. An attempt to do this would be incorrect. Consequently, the last item DI to be brought out in the argument list will be the value to be integrated, namely, EPC * AXN * RPT. The integrated value OI (Output of Integrator) also serves as the initial manual setting, since it is necessary to supply an initial value at the

```
 1 *          SUBROUTINE CONTR2(VI,CO,ZR,RNG,SP,AXN,PB,RPT,OI,DI)
 2 *          SPAN = ABS(RNG-ZR)
 3 *          VIM = VI
 4 *          IF (VI .GT. RNG) VIM= RNG
 5 *          IF (VI .LT. ZR) VIM= ZR
 6 *          EPC = 100. * AXN * (VIM-SP)/SPAN
 7 *          DI = EPC * RPT * 100./PB
 8 *          IF(OI.GT.100.) OI=100.
 9 *          IF(OI.LT.0.) OI=0.
10 *          CO = 100./PB * EPC + OI
11 *          IF (CO .GE. 100.) CO = 100.
12 *          IF (CO .LT. 0.) CO = 0.
13 *          RETURN
14 *          END
```

(b)

FIG. 11-5b. Subroutine Listing for a Two-Mode Controller

| | |
|---|---|
| VI = input | ANX = action +1 direct, −1 reverse |
| CO = output | PB = proportional band |
| ZR = zero | RPT = repeats/unit time |
| RNG = range | OI = reset output |
| SP = set point | DI = reset derivative |

NOTE. This subroutine must have CALL INT (OI, DI) in the integration section

start of the problem. Coupled with this Subroutine CONTR2, then, has to be a CALL INT (OI,DI) statement listed in the integration section.

The subroutine is shown in Figure 11-5b and the argument list is similar to that for CONTR1. An additional precaution is required here with respect to the saturation characteristics. This term defines the behavior of the controller's reset action when the output CO is at its maximum and the input is still off the set point. The degree of saturation varies for each controller; thus in order to simplify the action, the subroutine is programmed so that the integrator is reset to the limit when the output OI tends to go beyond the limit (lines 8 and 9). The item RPT in the argument list is the integration constant in repeats per unit time. Its numerical value must be in terms of the time units of the independent variable in the problem differential equations. If, for example, the digital program is integrating with respect to seconds (i.e., INTI (T,DT,IO) where T is in seconds), a controller setting of 30 repeats/min would be specified in the subroutine argument list as RPT = 0.5 repeats/sec.

## 11-4-3  Three-Mode Controller CONTR3

This is the most complex of the three controllers discussed in this section. Perhaps the best way to understand its function is to regard it as being merely a two-mode controller with rate action added. This rate action can be placed in either of two locations, namely, acting on the error signal or on the input variable. If it is on the error signal, it will apply rate action to any set-point changes as well as to the input signal. Since this is not desirable

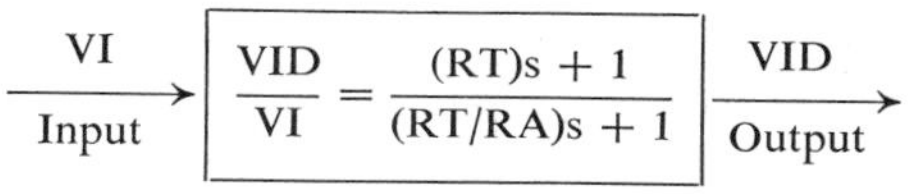

$$\xrightarrow[\text{Input}]{\text{VI}} \boxed{\dfrac{\text{VID}}{\text{VI}} = \dfrac{(\text{RT})s + 1}{(\text{RT/RA})s + 1}} \xrightarrow[\text{Output}]{\text{VID}}$$

FIG. 11-6. Transfer Function for Derivative Section

except for cascade situations, the alternate form where the rate action acts on the input variable will be programmed.

The purpose of the rate action is to provide an increment of change in the controller output, which is proportional to the *rate* of change of the input variable. The characteristics of a controller's rate section can be elegantly summarized by the transfer function shown in Figure 11-6. The rate action is provided by the numerator $[(\text{RT})S + 1]$ having the time constant RT, termed "rate time." The denominator $((\text{RT/RA})s + 1)$ provides the damping necessary to prevent high frequency, low amplitude noise in the input signal from becoming greatly amplified. The parameter RA is termed the "rate amplitude" and varies between manufacturers but generally has a value between 8. and 20.

The transfer function shown in Figure 11-6 can be transformed to the following integral equation:

$$VID = RA\left(VI + \frac{1}{RT}\int (VI - VID)\, dt\right)$$

At steady state, $VID = VI$, and since $RA > 1$, the integral term would have to be negative. A suitable subroutine for a three-mode controller is shown in Figure 11-7a. The argument list now contains the extra parameters for the rate section, namely, RT and RA, plus ODI and DODI for the integration involved in the rate section. This will also require an additional CALL INT (ODI, DODI) statement in the integration section to integrate DODI. On the first pass through the routine, the initialization of the rate section integrator is made (lines 6, 7) in order to start off with VID = VI. This avoids having to supply a suitable value for ODI. The remainder of the coding is similar to the CONTR2 routine.

Although this routine can be used for a two-mode proportional and rate controller by setting RPT = 0., it cannot be used as a proportional plus reset controller by making RT = 0. In fact, the smaller the value of RT, the more hazardous the use of this routine becomes. It is necessary for the programmer to make sure that the integration interval DT is reduced as RT becomes smaller, otherwise instability will result.

## 11-4-4 Dead Time or Transport Lag

Control difficulties are sometimes due to the presence within the control loop of a transport lag. A typical example would be a long pipe conducting

```
 1 •          SUBROUTINE CONTR3(VI,CO,ZR,RNG,SP,AXN,PB,RPT,RT,RA,OI,DI,ODI,DODI)
 2 •          COMMON/CINT/T,DT,JS,JN,DXA(500),XA(500),IO,JS4
 3 •          VIM = 100.*(VI-ZR)/ABS(RNG-ZR)
 4 •          IF(VIM.LT.0.) VIM = 0.
 5 •          IF(VIM.GT.100.) VIM = 100.
 6 •          IF(T.GT.0.) GO TO 2
 7 •          ODI = VIM*(1./RA-1.)
 8 •        2 VID = RA*(VIM+ODI)
 9 •          DODI=(VIM-VID)/RT
10 •          SPM=100.*(SP-ZR)/ABS(RNG-ZR)
11 •          EPC = (VID-SPM)*AXN*100./PB
12 •          IF(OI.GT.100.) OI=100.
13 •          IF(OI.LT.0.) OI=0.
14 •          DI = EPC*RPT
15 •          CO = EPC + OI
16 •          IF (CO .GE. 100.) CO = 100.
17 •          IF (CO .LT. 0.) CO = 0.
18 •          RETURN
19 •          END
```

*(a)*

FIG. 11-7*a*.  Subroutine Listing for a Three Mode Controller

| | |
|---|---|
| VI = input | RPT = repeats/unit time |
| CO = output | RT = rate time |
| ZR = zero | RA = rate amplitude |
| RNG = range | OI = output reset section |
| SP = set point | DI = derivative of OI |
| AXN = action +1 or −1 | ODI = output of rate section |
| PB = proportional band | DODI = derivative of OI |

NOTE.  This subroutine must have CALL INT (OI, DI)
CALL INT (ODI, DODI) in the integration section

fluid from one location to another. The flow rate of the fluid would be the same at both ends of the pipe (if incompressible), but the composition and temperature would be different under transient conditions. The situation can be characterized by specifying the inlet stream as I and the outlet stream as J (Figure 11-7*b*).

Both I and J streams then have the same flow rate, but the temperature in J will be "delayed" from the temperature in stream I by the time to travel down the pipe length L. Such a dead time can be simulated on a computer by a "time delay" subroutine. This can be constructed in several ways, the most direct of which would be the "bucket brigade" approach, a self-explanatory term. A more efficient method follows.

### 11-4-5  Subroutine TDL

The operating principle of this routine is to allocate N spaces to the delay channel, where N = time delay/DT and where DT is the integration interval.

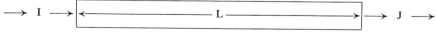

FIG. 11-7*b*. Pipeline Transport Lag

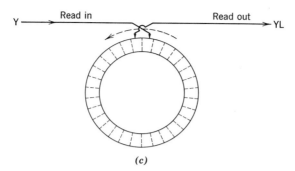

FIG. 11-7c. Schematic for Subroutine TDL

Instead of feeding the input signal (temperature, composition, etc.) at the front end, moving the value in each space one position toward the exit end, and reading the value in the last space as the exit value ("bucket brigade"), the value in each space remains in that location and the readin and readout move from space to space. This can be better understood if the spaces are visualized as being arranged in a circle (Figure 11-7c) where the readin and readout move around the circle.

Each CALL on the subroutine requires only two operations, readin and readout, versus N operations for the bucket brigade. This procedure is readily accomplished by the subroutine TDL, (time delay) shown in Figure 11-7d. The argument list contains the following items:

YA: the name of the array that is used to store the input function Y(t), that is, the delay channel. In order to conserve space in the computer, this will be dimensioned in the main program as YA(25,M), where 25 = number of channels required, M = time delay/integration interval DT.
    Y: input variable
  YL: output variable (delayed)
NTL: time delay/integration interval DT
  JC: subroutine call number

On the first pass, the entire array is loaded with the output variable Y (lines 7 and 8), and the output YL is made equal to Y (line 12). This is done since at the beginning of the run, no previous history of Y(T) exists. Subsequently the output YL is read out of the array (line 16) from the location specified by the call number JC and the time location counter NCJ. This time location

```
 1 *              SUBROUTINE TDL(YA,Y,YL,NTL,JC)
 2 *              COMMON/CINT/ T,DT,JS,JN,DXA(500),XA(500),IO,JS4
 3 *              DIMENSION YA(25,1),NC(25)
 4 *              IF (T .LE. 0.) GO TO 3
 5 *              GO TO (5,6,6,8),IO
 6 *            6 GO TO (5,7),JS
 7 *            3 DO 4 K = 1,NTL
 8 *            4 YA(JC,K) = Y
 9 *            5 NC(JC) = NC(JC)+1
10 *              IF (NC(JC) .GT. NTL) NC(JC)=1
11 *            7 NCJ=NC(JC)
12 *              YL = YA(JC,NCJ)
13 *              IF (JS .EQ. 2) YA(JC,NCJ)=Y
14 *           10 RETURN
15 *            8 GO TO (9,10,5,11),JS4
16 *            9 NCJ=NC(JC)
17 *              NC1=NCJ+1
18 *              IF (NC1 .GT. NTL) NC1 = 1
19 *              YL = (YA(JC,NCJ)+YA(JC,NC1))/2.
20 *              RETURN
21 *           11 NCJ=NC(JC)-1
22 *              IF(NCJ.EQ.0) NCJ=NTL
23 *              YA(JC,NCJ)=Y
24 *              RETURN
25 *              END
```

*(d)*

FIG. 11-7d.   Listing for Time Delay Subroutine TDL
Argument list:
YA = array (to be dimensioned in main program YA (25, NTL))
 Y = input signal
YL = output delayed signal
NTL = time delay/integration step size DT (integer)
JC = subroutine call number

counter is specific for each time delay call; thus it also has to be stored in the special array (NC) in a location specified by the call number JC. The routine is intended for use with any of the three integration methods, and the value read out YC will always correspond to the value of the independent variable T from COMMON. For the fourth-order method the value of YL read out on the two intermediate dummy passes (JS4=1&2), when T is at the midpoint of the integration interval, will be the average of the two values stored on either end of the corresponding time increment (line 19). For the second- and fourth-order methods, the new value of Y is stored only on the appropriate legitimate pass. (See line 15 for fourth order and line 6 for second order.)

This routine must be located in the derivative section of a main program. It should also be realized that the integration interval must be constant at all times when using this routine. A time-delay routine that did not have this restriction would require considerably greater sophistication and would be much slower in execution.

```
 1 *          SUBROUTINE LDL(YA,I,IO,NTL,JN)
 2 *          COMMON/CD/STRM(300,24),DATA(20,10),RCT(22),NCF,NCL,LSTR
 3 *          DIMENSION YA(25,1)
 4 *          JNT=(NCL-NCF+3)*(JN-1)
 5 *          M=0
 6 *          DO 10 N=NCF,NCL
 7 *          M=M+1
 8 *       10 CALL TDL(YA,STRM(I,N),STRM(IO,N),NTL,JNT+M)
 9 *          STRM(IO,21)=STRM(I,21)
10 *          CALL TDL(YA,STRM(I,22),STRM(IO,22),NTL,JNT+M+1)
11 *          CALL TDL(YA,STRM(I,23),STRM(IO,23),NTL,JNT+M+2)
12 *          STRM(IO,24)=STRM(I,24)
13 *          RETURN
14 *          END
```

(e)

FIG. 11-7e.  Listing for Line Delay Subroutine LDL
Argument list:
YA = array (to be dimensioned in the main program)
I = input line number
IO = output line number
NTL = time delay/integration step size (integer)
JN = subroutine call number

## 11-4-6  Stream or Pipeline Delay, Subroutine LDL. (DYFLO)

Generally, a chemical process is constructed so that the unit operations are located close together. Consequently, the time delay due to transport lag through the pipes is small compared with the vessel holdup times and can be neglected. In a few situations the pipeline delay is not insignificant and must be included in the simulation. For incompressible fluids the flow rate and pressure are transmitted to the exit end of the pipe without delay. The compositions, enthalpy, and temperature will suffer a time delay equal to the volume of the pipe/volumetric flow rate. If the simplification is made of assuming a constant time delay based on the average flow rate, a simple subroutine LDL can be used to simulate the line delay. This is shown in Figure 11-7e. The argument list requires dimensioning the storage array YA in the main program as for TDL (see Section 11-4-5) and a feed stream number I at the inlet to the pipe and an exit stream number IO at the exit to the pipe. The next item is NTL, which is the time delay/integration step size DT, and the last item is JC, the subroutine call number. The internal structure for LDL is simply a series of calls on TDL, the single channel delay described in the previous section. Each call is appropriately indexed (line 7) and will correspond to the compositions NCF → NCL. When this is completed, the temperature and enthalpy are delayed on lines 10 and 11.

## 11-5  CONTROL ELEMENTS

The most common control elements are diaphragm or motor-actuated control valves that receive the signal from the controller and convert this

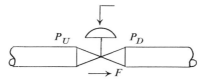

FIG. 11-8. Control Valve

into a stem position that in turn controls the port areas throttling the flow of fluid to the process (Figure 11-8). The variation of flow of liquids and gases through valves and restrictions has already been covered in Chapter 7 so we will concern ourselves here with the relation between the stem position and the port area. The flow through the valve can be expressed in its most general form as

$$F = A \cdot C_v \cdot F(P)$$

The constant $C_v$ is the valve capacity and is a function of the valve size and its service. It usually refers to the flow through the valve when the port area $A$ is wide open ($A = 1$) and when the pressure drop is 1 psi. It, therefore, has units of mass flow rate/pressure drop, and care should be exercised to convert $C_v$ to its proper value corresponding to the units used in the flow equation. The pressure function $F(P)$ in the equation depends on the phase of the flow and sometimes on the pressure levels. Valve manufacturer catalogs will usually specify the appropriate flow equations for each valve in various service conditions. Typically, these have the following form:

1. Liquids      $F = A \cdot C_v \sqrt{P_u - P_d}$

2. Gases      $F = A \cdot C_v \sqrt{\bar{P}(P_u - P_d)}$    subcritical flow

             $F = A \dfrac{C_v}{\sqrt{2}} P_u \, 0.85$        critical flow

where $A$ = fractional valve opening,
       $C_v$ = valve capacity,
       $P_u$ = upstream pressure,
       $P_d$ = downstream pressure,
       $\bar{P}$ = average pressure $(P_u + P_d)/2$.

Chapter 7 provided a more universal expression that covered flow in both the critical and subcritical flow region.

### 11-5-1 Valve Stem Position Versus Port Area

Valve port characteristics are generally expressed in graphical form, usually plotted from experimental data. With a constant pressure drop, the flow rate is rarely proportional to the port opening because of the variation of pressure loss, turbulence, and so on; thus the port area should be expressed as the equivalent flow area rather than the actual geometric area. The relationship between this area and stem position depends on the type of plug design, such as parabolic, v-port, bevel or rectangular port, but usually they can be classified in two general groups, the linear and equal-percentage characteristic. The flow area-stem position relationship for a linear valve is

$$A = A_o + (1 - A_o)P$$

where  $A$ = Fractional valve opening
  $A_o$ = Valve opening with fully closed stem position
  $P$ = Stem position $o - 1$

Because of fabrication difficulties, such as providing for clearances, it is not practical to shut off a control valve completely, so a minimum opening $A_o$ remains when $P = o$. The value of $A_o$ is obtained from the rangeability specification, which is the number of times the minimum flow may be increased before maximum flow is obtained, that is

$$A_o = \frac{1}{\text{rangeability}}$$

The turndown of the valve is also determined by the minimum controllable flow but is based on the nominal maximum instead of the ultimate maximum with $A = 1$.

Perhaps the most common valve used in process control is the equal-percentage relationship which is provided by a semilogarithmic port characteristic (Figure 11-9). This could be simulated in a computer by using an

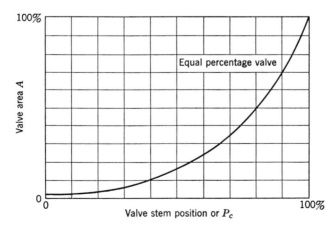

FIG. 11-9. Nonlinear Valve Areas versus Valve Stem Position

arbitrary function generator with about 10 coordinate points, but a more convenient method is to use the equivalent analytical expression:

$$A = A_o \, \text{EXP} \, (P/K)$$

where
$$A_o = 1/\text{rangeability},$$
$$K = -100/\text{Ln} \, (A_o) = 100/\text{Ln} \, (\text{rangeability}).$$

Either the linear or equal-percentage valve expressions described above are adequate for most computer simulations of control problems. In situations where it is judged that the valve-flow characteristics are a critical factor, then it is advisable to use any actual data available from laboratory or field measurements. A subroutine VALVE shown in Figure 11-10 simulates both linear or equal-percentage port characteristics.

EXAMPLE 11-6 BATCH KETTLE REACTOR CONTROL

This example incorporates a number of typical controls used for controlling processes; Figure 11-11 shows a batch kettle reactor that consists of a heavy-walled vessel containing reactants under a high pressure. The reaction can be summarized by

$$A \xrightarrow{k} B$$

```
 1 *        SUBROUTINE VALVE(F,SP,PU,PD,LV,KV,CV,AO)
 2 *        ADP=ABS(PU-PD)
 3 *        IF((ADP.LE.0.).AND.(ADP.GE.0.)) GO TO 6
 4 *        GO TO (1,2),KV
 5 *      1 A=SP/100.
 6 *        GO TO 3
 7 *      2 A=AO*EXP(-SP*ALOG(AO)/100.)
 8 *      3 IF(LV.EQ.0) GO TO 4
 9 *        F=A*CV*SQRT(ADP)*(PU-PD)/ADP
10 *        RETURN
11 *      4 IF(PD.LT. .53*PU) GO TO 5
12 *        F=A*CV* SQRT(PU*ADP)*(PU-PD)/ADP
13 *        RETURN
14 *      5 F=A*CV*PU*.85
15 *        RETURN
16 *      6 F=0.
17 *        RETURN
18 *        END
```

FIG. 11-10.   Listing for VALVE subroutine
Argument list:

F = flow rate
SP = stem position $0 \rightarrow 100\%$
PU = upstream pressure PSI
PD = downstream pressure PSI
LV = liquid or vapor (0 = vapor)
KV = port characteristics, 1 = linear, 2 = equal %
CV = valve capacity
AO = rangeability

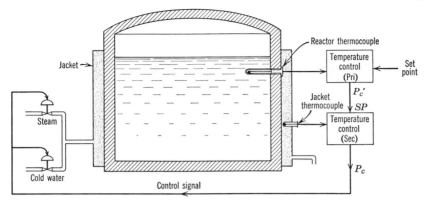

FIG. 11-11. Batch Kettle Reactor Process

where $k$, the reaction coefficient, is an exponential function of temperature, that is, $k = f(T)$. The mass-balance equation is

$$\frac{d}{dt}(MX_A) = -kX_A$$

where $kX_a$ is the reaction rate. This reaction is exothermic, and thus the rate of heat generation due to the reaction is $k \cdot X_a \cdot h_R$, where $h_R$ is the heat of reaction (PCU/mole). A heat-balance equation can now be formulated to include the heat of reaction as follows:

$$\frac{d}{dt}(MCT) = Q + H_R$$

where $H_R = kX_A \cdot h_R$.

In this equation $Q$ is the heat flux from the jacket and is defined as

$$Q = (2UA)(T_w - T)$$

where $UA$ is the overall heat transfer coefficient. The average temperature of the vessel wall, $T_w$, is obtained from a heat balance on the wall as follows:

$$\frac{d}{dt}(wcT_w) = Q_j - Q$$

where $wc$ is the heat capacity of wall.

The heat flux from the jacket is $Q_j$ and is defined as

$$Q_j = (2UA)(T_j - T_w)$$

The jacket is always filled with water, and the temperature can be changed either by flowing in cold water or sparging steam. Two valves control the water and steam from a single-control signal such that just as the steam valve closes, the water valve begins to open (Figure 11-12). The temperature of the reactor contents is measured with a thermal bulb having a dynamic time constant of $\tau$ sec.

$$\frac{dT_c}{dt} = \frac{1}{\tau}(T - T_c)$$

The signal from the thermal bulb is transmitted to a primary three-mode controller whose output directs the set point of a secondary cascaded controller. This controller in turn, receives a measured signal from a thermal bulb located in the jacket. The output of the secondary controller operates directly the valves that control the steam and cooling water flow. A heat balance on the contents of the jacket (assuming the contents are well agitated) is as follows:

$$\frac{d}{dt}(W_j T_j) = F_W T_W + F_S H_S - Q_j - (F_S + F_W)T_j$$

where  $W_j$ = jacket heat capacity (i.e., mass of water, since $c = 1.0$)
  $F_W$ = flow of cold water
  $F_S$ = steam flow
  $H_S$ = enthalpy of steam
  $T_W$ = inlet water temperature.

The difficulty experienced with this process is that initially the mass must be heated in order to initiate the reaction. Once this happens, the heat of reaction causes a rapid temperature rise unless cooling is applied; however, under no circumstances may the temperature rise beyond a specified level. Because of safety considerations, therefore, a computer study is planned

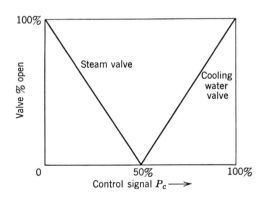

FIG. 11-12.  Tandem Control Valve Operation

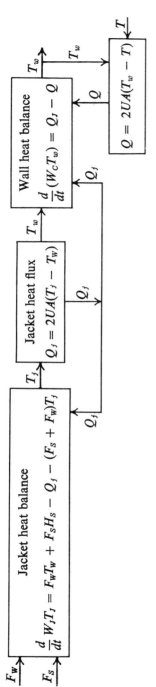

FIG. 11-13.   Model for Heat Balance on Wall and Jacket

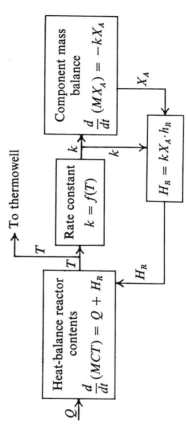

FIG. 11-14.   Model for Reactor Heat Balance

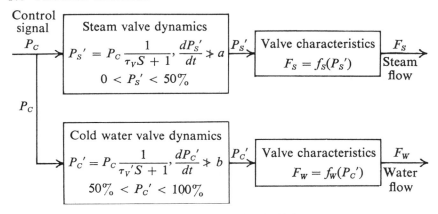

FIG. 11-15.  Model for the Valves

in order to establish the adequacy of the proposed controls to direct the reaction along the desired path.

The first step is to assemble the model based on the equations already developed. In this case the process has two inputs, steam or cooling water flow, and also two measured outputs, the jacket and reactor contents temperatures. The mathematical model for the process must connect the two inputs with the two outputs.

Combining the jacket and wall heat balances gives the partial model shown in Figure 11-13; the reactor content equations are assembled as in Figure 11-14. The control loops in this instance consist of controller equations, a first-order transfer function for the thermowell dynamics, and two first-order transfer functions, one for each valve. Each of these "drives" a function derived from Figure 11-12, representing the valve opening. It is assumed that the pressure drop across the valves for both the steam and cooling water valves is constant, so that the flow of both fluids is a direct function of stem position (Figure 11-15). The control loop is assembled together with the process equations (shown here as a single block) in Figure 11-16. Various computer runs can now be made with this model, using trial values of controller settings to find the optimum settings possible within the limitations of the system. If no settings can be found to meet the process specifications, then greater sophistication has to be added to the control system, and further computer runs made to establish the predicted behavior. Eventually, by this process of trial and error, a satisfactory control method is synthesized.

EXAMPLE 11-7  CENTRIFUGAL COMPRESSOR SURGE CONTROL

This example demonstrates the feasibility of studying control problems by digital simulation. Admittedly, batch processing of computer runs with a

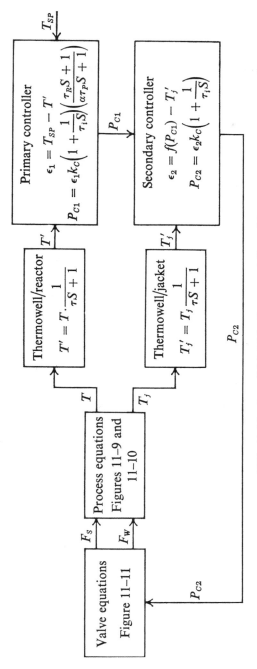

FIG. 11-16. Overall Model for Process and Control Loop

377

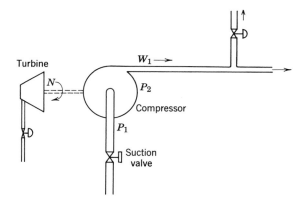

FIG. 11-17. Turbine Compressor System

$\frac{1}{2}$–1 hr turnaround time would make such a study rather cumbersome to execute. However, the imminent availability of time-shared computer terminals with graphic capabilities will largely overcome this drawback. Proceeding on this assumption, this example demonstrates how a control problem may be programmed using some of the subroutines developed in an earlier section.

### 11-7-1  Development of a Mathematical Model

The process to be analyzed is shown in Figure 11-17 and consists of a compressor driven by a steam turbine, compressing a gas to a large distribution manifold, which supplies gas to a process. A control system is to be designed that will maintain the manifold pressure at a desired level, within limits, for various changes in demand load. Several control schemes seem feasible, but since the compressor is highly nonlinear with the attendant possibility of surging, the most effective way of determining the best control scheme is by computer simulation.

### 11-7-2  Compressors

The most complex unit in this problem is the compressor. There are four variables associated with the compressor, namely, the shaft speed $N$, the gas inlet pressure $P_1$, the outlet pressure $P_2$, and the flow rate $W_1$ (Figure 11-17). The speed and pressures are classified as imposed conditions, while the flow rate is a resulting effect. The numerical relationships between these variables for a particular compressor is accurately expressed by a series of charts or compressor maps showing the relation between discharge pressure $P_2$ (psia) and flow rate ($W_1$ lb/hr) at various constant speeds $N$ for a range of suction pressures $P_1$. These relationships can all be combined into a single curve

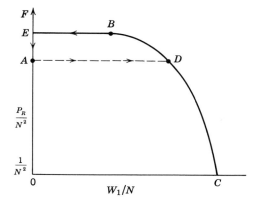

FIG. 11-18. Characteristic Curve for Compressor

called the "characteristic curve" for the compressor, which relates two parameters, $W_1/N$ and $P_R/N^2$, where $P_R$ is the pressure ratio $P_2/P_1$. Figure 11-18 shows the curve for the compressor in this example, with $W_1/N$ on the abscissa and $P_R/N^2$ as the ordinate. It consists of two sections $E$ to $B$, which is the maximum point, and $B$ to $C$. In order to understand the significance of this curve, it is convenient to imagine the compressor running at constant $N_o$. The point $C$, then, represents the maximum flow rate through the compressor with unity pressure ratio. If the output is now restricted so that the pressure ratio increases, the flow rate $W_1$ will decrease ($C$ to $B$, Figure 11-18). As the critical pressure ratio for that speed $N_o$ is approached, (point $B$) the flow rate decreases rapidly. If the pressure ratio increases beyond the critical, the compressor "surges"; that is, the flow rate $W_1$ drops to zero and will remain at zero until the value of $P_R/N^2$ drops below point $A$. When this happens, the flow will again resume at a value corresponding to point $D$ on the curve $C$ to $B$.

The suction pressure $P_1$ results from a pressure drop across a suction valve. Assuming that the gas is originally at atmospheric pressure $P_o$, the suction pressure $P_1$ will be

$$P_1 = P_o - \left(\frac{W_1}{C_{vs} \cdot A_{vs}}\right)^2$$

where $C_{vs}$ is the suction valve coefficient and $A_{vs}$ is the valve opening. Assuming that the compressor speed $N$ and the outlet pressure $P_2$ are available from other sections of the mathematical model, an information flow diagram for the compressor and suction valve is shown in Figure 11-19. There are two loops shown with the flow rate $W_1$ as the common point to both loops. The upper loop passes through the compressor curve, entering as $W_1/N$ and

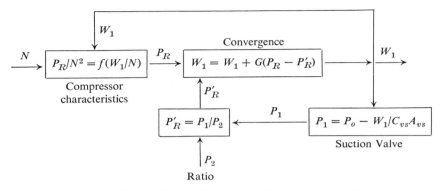

FIG. 11-19. Information Flow and Compressor Model

emerging as $P_R/N^2$. Supplying $N$ permits the necessary conversion to $W_1/N$ and $P_R$. The lower loop passes $W_1$ through the suction valve equation to produce $P_1$ from which the pressure ratio $P'_R$ is calculated by dividing by $P_2$. The two values of $P_R$ and $P'_R$ are "matched" by iterating $W_1$ in the converge equation:

$$W_1 = W_1 + G(P_R - P'_R)$$

A suitable value of $G$ was obtained by testing for sensitivity through each loop and was found to be 2000. Convergence can then be accelerated by feeding the calculated value of $W_1$ to the CONV subroutine.

### 11-7-3  Compressor Program

The program for the compressor and suction valve is shown in Figure 11-20 from lines 19 to 33. Line 19 is the first line of the derivative section. Starting values for $W_1$, N, and $P_2$ are supplied in the initiating section from a READ statement (line 13). Line 20 calculates the suction pressure $P_1$ using $C_{vs} = 18.10^4$ with the valve half open; that is, $A_{vs} = .5$. Line 21 forms the pressure ratio PRP from the suction pressure. Line 22 is the first test for the surge condition, with a value for PRP/$N^2$ at B (Figure 11-18) of $10^{-7}$. If the total logic test on line 22 is true, the compressor is in surge, but below the maximum peak value of point B. The next check on line 29 would be to test if the compressor is below point A, indicating a return to the function, which is done on line 31 by changing W1 from 0. to 55000. and recycling the calculation to line 19 (address 4) to find the proper balance point. Line 23 also tests for the surge condition; if true (i.e., PRP/$N^2$ > point B) W1 is set $= 0.$ on line 33.

```
1*    C ***INITIATING SECTION***
2*          REAL N
3*          DIMENSION W1NX(15),W1NY(15),AVX(15),AVY(15)
4*       50 FORMAT (2F10.1)
5*       51 FORMAT (1F10.1,1E10.1)
6*       52 FORMAT (1H1,5HPBE =1F5.1,4X,5HPBX =1F5.1,4X,7HRPTMV =1F5.2,4X7HRPT
7*         1ME =1F5.2)
8*      100 FORMAT (8F10.1)
9*      300 FORMAT (1H0,4X,4HTIME,8X,2HP2,10X,2HW2,10X,2HW1,10X,1HN,11X,2HX2,
10*        110X,3HX2P,9X,3HW1T,9X,2HE1,10X,3HP2T)
11*          READ 51, (W1NX(K),W1NY(K), K=1,11)
12*          READ 50, (AVX(K),AVY(K), K=1,11)
13*          READ 100,N,P2,W1,W3,T,DT,TFIN
14*          READ 100,PBE,RPTME,P2TS,OIE,PBX,RPTMX,W1TS,OIX
15*          READ 100,P2T,E1,W1T,X2
16*      400 PRINT 52,PBE,PBX,RPTME,RPTMX
17*          PRINT 300
18*    C ***DERIVATIVE SECTION***
19*        4 W1N = W1/N
20*          P1 = 14.7 - (W1/(18.E4*.5))**2
21*          PRP = P2/P1
22*          IF ((W1 .LE. 0.) .AND. (PRP/N**2 .LT. 1.E-7)) GO TO 12
23*          IF (PRP/N**2 .GT. 1.E-7) GO TO 10
24*          PR = N**2 * FUN1(W1N,11,W1NX,W1NY)
25*          ERP = PR - PRP
26*          W1C= W1 + ERP * 2000.
27*          CALL CONV(W1,W1C,1,NC1)
28*          GO TO (5,4),NC1
29*       12 IF (PRP/N**2 .LT. 7.6E-8) GO TO 14
30*          GO TO 10
31*       14 W1 = 55000.
32*          GO TO 4
33*       10 W1 = 0.
34*        5 E2 = N**3 * 1.E-10 * (6.+4.4*W1/N)
35*          IF(T.GT.5.) W3=20000.
36*          CALL CONTR2 (W1T,X2P,0.,1.E5,W1TS,-1.,PBX,RPTMX,OIX,DIX)
37*          AV = FUN1(X2,11,AVX,AVY)
38*          W2 = 666.7 * AV * P2
39*          CALL CONTR2 (P2T,E0,0.,200.,P2TS,-1.,PBE,RPTME,OIE,DIE)
40*          DN = (E1-E2) *19150./N
41*          DP2 = (W1-W2-W3) * 8.E-4
42*          CALL PRNTF(.5,TFIN,NC,T,P2,W2,W1,N,X2,X2P,W1T,E1,P2T)
43*          GO TO (8,400),NC
44*    C ***INTEGRATION SECTION***
45*        8 CALL INTI(T,DT,2)
46*          CALL INT(N,DN)
47*          CALL TFN1(W1T,W1,,13,1.)
48*          CALL TFN1(P2T,P2,,25,1.)
49*          CALL TFN2(X2,X2P,1.,1.6,.01)
50*          CALL TFN2(E1,E0,59.,10.75,50.)
51*          CALL INT(OIX,DIX)
52*          CALL INT(OIE,DIE)
53*          CALL INT(P2,DP2)
54*          GO TO 4
55*          END
```

FIG. 11-20. Complete Program for Compressor Surge Problem

If the compressor is not in surge condition, the calculation obtains a value for PR from the FUNCTION on line 24, using W1N calculated on line 19. The function data arrays consisting of 11 coordinate pairs W1NX and W1NY, representing Figure 11-18, are entered in the data section, line 11. Lines 25 and 26 calculate an estimate for W1 defined as W1C, and line 27 calls the CONV routine to extrapolate this estimated value to the converged value W1. Line 28 redirects the calculation to continue (address 5) or recycle (address 4), depending on whether the index NC1 from the CONV routine is 1 or 2. The foregoing explanation of the compressor program is the most difficult section in this problem to understand. The remainder of the system is relatively straightforward.

### 11-7-4  Turbine

The turbine (Fig. 11-21) is driven by a steam flow having an energy of $E1(HP)$. The total work done by the turbine $E2$ can be summarized in terms of the gas flow through the compressor $W1$ and the speed $N$ by the following expression:

$$E2 = N^3 \cdot 10^{-10}\left(6. + 4.4\,\frac{W1}{N}\right)$$

The first term in the above expression includes the energy required to overcome all the friction associated with turbine and compressor. The difference between the input and output energy will control the rate of change of shaft rotational energy; that is,

$$\frac{d}{dt}(\tfrac{1}{2}JN^2) = E1 - E2$$

where $J$ = moment of inertia = $1./19150$ ft lb. This equation can be reduced to a more direct form by differentiating:

$$\frac{dN}{dt} = \frac{19150.}{N}(E1 - E2)$$

This equation is programmed on line 49, Figure 11-20.

### 11-7-5  Distribution Manifold

The distribution manifold having a relatively large diameter will be treated as a lumped vessel having an inlet flow $W1$, an outlet flow $W2$ (lb/sec) through the surge valve, and $W3$ (lb/sec) the demand load to the process (Figure 11-22). The purpose of the surge valve is to prevent surging by

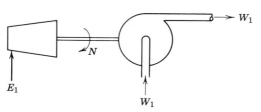

FIG. 11-21.  Turbine-Compressor

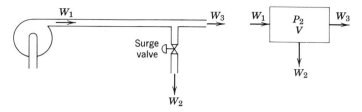

FIG. 11-22. Distribution Manifold Lumped Equivalent

opening when the compressor flow $W1$ falls to a dangerously low level. Treating the volume as an ideal gas operating isothermally, the equation for the rate of change of pressure $P2$ in the volume follows from the differential form of the ideal gas law $PV = mRT$.

Differentiating

$$\frac{d}{dt} P_2 = \frac{dm}{dt} \frac{RT}{V} = \frac{RT}{V} (W1 - W2 - W3)$$

where $V$ = volume (in.$^3$),
$R$ = gas constant,
$T$ = temperature (°K).

Substituting numerical values for the above constants, the parameter $RT/V = 8.10^{-4}$ (in.$^{-2}$). The equation for the derivative of $P_2$ is programmed in Figure 11-20, line 41.

## 11-7-6 Control System

The particular control system to be programmed is shown in Figure 11-23. The two control loops consist of a pressure controller that controls the steam to the turbine and a flow controller that operates the bypass valve. For maximum power conservation, this bypass valve is closed for steady operation and only opens during an upset if the compressor flow falls to a dangerous level. However, in order to test the computer simulation, the first run will be made with this valve partly open.

## 11-7-7 Flow Control Loop

The flow control loop consists of three units, a transmitter, a controller, and a pneumatically operated diaphragm valve. The transmitter receives the flow signal $W1$ and transmits this as $W1T$ to the controller. This transmission involves a delay that is represented by a first-order transfer function; that is,

$$\frac{W1T}{W1} = \frac{1}{\tau S + 1}$$

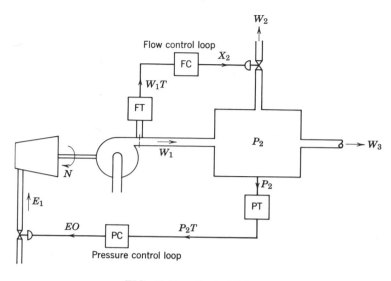

FIG. 11-23.  Control Scheme
FC = flow controller
FT = flow transmitter
PT = pressure transmitter
PC = pressure controller

The time constant $\tau$ is 0.13 sec and is typical of a pneumatic flow transmitter. This transfer function is coded in the program on line 47 and consists of a CALL TFN1 statement with a time constant of 0.13 sec and unity gain.

The controller will be a two-mode controller, that is, proportional plus reset, and will have a range of 0 to 50,000 lb/hr. It is programmed as a CALL CONTR2 statement on line 36, with a range of 0. to 1.E5 lb/hr and a set point of W1TS from the data section. The $-1$ is for reverse action; that is, the output will decrease (close valve) if the compressor flow $W1$ increases. PBE and RPTME are the proportional band and repeats/sec controller settings from the data section. OIE is the automatic reset integrator output with its derivative DIE that is integrated on line 52.

The bypass valve dynamic characteristics, that is, the response of the valve stem position $X2$ to the pressure signal from the controller $X2P$, is approximated by a second-order transfer function:

$$\frac{X2}{X2P} = \frac{1}{S^2 + 1.6S + 1}$$

This transfer function is programmed in Figure 11-20 as CALL TFN2 on

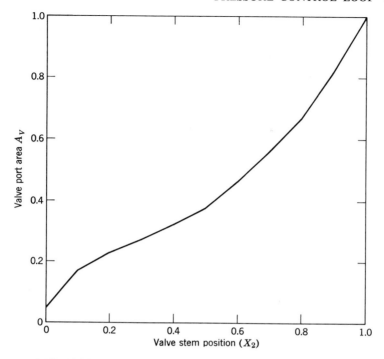

FIG. 11-24. Valve Characteristics Plotted from Measured Data

line 49, with the two parameters 1. and 1.6 and a gain of 0.01, convert the controller output range of 0 to 100 to a range of 0 to 1.0.

The relationship between the valve port area and the stem position is shown in Figure 11-24 and programmed on line 37 in Figure 11-20.

Finally, the bypass flow through this valve will be assumed to be in the critical range, since the manifold pressure $P2$ is around 90 psia and the downstream pressure is atmospheric. With reasonable control, it is unlikely that the pressure $P2$ will temporarily fall close to the critical pressure of 30 psia. Using the critical flow equation, the flow through the valve $W2 = CV2.AV.P2$, where the valve constant $CV2 = 666.7$ lb/(hr psi). This is programmed on line 38, Figure 11-20.

### 11-7-8 Pressure Control Loop

The three units to be programmed for the pressure control loop are the pressure transmitter, a two-mode controller, and the turbine steam valve. The pressure transmitter is a first-order transfer function with a 0.25 sec time

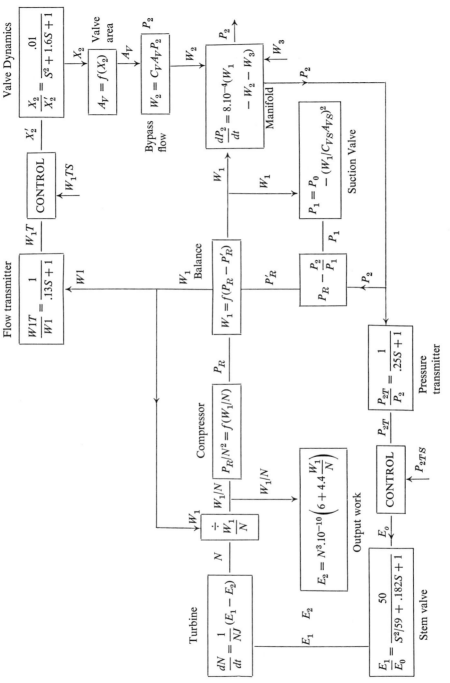

FIG. 11-25. Information Flow Diagram for Compressor Surge Control

386

constant shown on line 48. The controller is on line 39, Figure 11-20, and the steam valve, shown as a second-order transfer function on line 50, is assumed to give a linear variation of horsepower E1 with stem position of 50 HP/% stem position—hence the gain of 50 in the transfer function argument list.

### 11-7-9 System Control Test

In order to visualize the entire system, an assembled information flow diagram is shown in Figure 11-25. The procedure for determining the effectiveness of the control system is to start the digital program near an equilibrium condition found by manual calculation. The simulation continues at equilibrium for a short period, say 5 sec, then the demand load $W_3$ is suddenly changed to a new level, programmed on line 35, Figure 11-20. This constitutes the disturbance which the control system must cope with and the resulting fluctuation in pressure $P_2$. Flow $W_1$ is a measure of how rapidly the controls return the system to a new equilibrium operating level.

The numerical results for this particular run are shown in Figure 11-26 up to an elapsed time of 50 sec.

Most computer installations have a variety of plotting facilities available. One of the most common is a library subroutine that instructs the line printer to plot selected variables on designated scales. Such a plot for this run is shown in Figure 11-27 where the flow $W1$, (curve B) and pressure $P_2$ (curve A) are plotted against time. Other methods of plotting results can be achieved by displaying such curves on electronic video screens, where available. For dynamic problems, such as this example, these visual displays facilitate the evaluation of results.

The results shown numerically in Figure 11-26 and plotted in Figure 11-27 were made with the following numerical values for the READ statements on lines 11→15 of the program in Figure 11-20:

DATA, WINX/0., 1., 2., 3., 4., 5., 5.5, 6., 6.5, 7., 7.2/

DATA, $\underline{\text{WINY}}$/7.6, 8.5, 9.2, 9.8, 10., 9.8, 9.2, 8.2, 6.3, 2.8, 0./
$\qquad\quad 10^{-8}$

DATA, AVX/0., .1, .2, .3, .4, .5, .6, .7, .8, .9, 1./

DATA, AVY/0., .17, .225, .266, .3, .374, .46, .56, .67, .82, 1/

DATA N/8920./P2/90./W1/55200./W3/40200./T/0./DT/.1/TFIN/50./

DATA PBE/120./RPTME/.5/P2TS/90./0IE/46.8/PBX/100./

DATA RPTMX/.5/W1TS/55200./0IX/20./

DATA P2T/90./E1/2340./W1T/54250./X2/.2/

By supplying a variety of different controller settings in the data section, a series of runs can be made. Evaluating the plotted results permits a control engineer to arrive at suitable or even optimum controller settings. Since each run only requires a few seconds on a large computer, such a procedure is economically justified.

PBE =120.0    PBX =100.0    RPTMV = .50    RPTME = .50

| TIME | P2 | W2 | W1 | N | X2 | X2P | W1T | E1 | P2T |
|---|---|---|---|---|---|---|---|---|---|
| .00000 | .90000+02 | .13501+05 | .54248+05 | .89200+04 | .20000+00 | .20950+02 | .54250+05 | .23400+04 | .90000+02 |
| .50000+00 | .90224+02 | .13558+05 | .54309+05 | .89305+04 | .20098+00 | .21133+02 | .54298+05 | .23386+04 | .90129+02 |
| .10000+01 | .90427+02 | .13644+05 | .54294+05 | .89330+04 | .20319+00 | .21354+02 | .54301+05 | .23338+04 | .90329+02 |
| .15000+01 | .90580+02 | .13735+05 | .54243+05 | .89310+04 | .20596+00 | .21626+02 | .54259+05 | .23280+04 | .90504+02 |
| .20000+01 | .90673+02 | .13824+05 | .54182+05 | .89268+04 | .20898+00 | .21927+02 | .54200+05 | .23223+04 | .90626+02 |
| .25000+01 | .90708+02 | .13907+05 | .54125+05 | .89219+04 | .21211+00 | .22243+02 | .54142+05 | .23172+04 | .90690+02 |
| .30000+01 | .90691+02 | .13983+05 | .54079+05 | .89170+04 | .21529+00 | .22564+02 | .54092+05 | .23130+04 | .90699+02 |
| .35000+01 | .90628+02 | .14053+05 | .54046+05 | .89125+04 | .21850+00 | .22883+02 | .54055+05 | .23099+04 | .90659+02 |
| .40000+01 | .90528+02 | .14118+05 | .54028+05 | .89088+04 | .22172+00 | .23195+02 | .54033+05 | .23079+04 | .90578+02 |
| .45000+01 | .90400+02 | .14177+05 | .54026+05 | .89059+04 | .22493+00 | .23495+02 | .54025+05 | .23071+04 | .90464+02 |
| .50000+01 | .90250+02 | .14231+05 | .54038+05 | .89041+04 | .22809+00 | .23781+02 | .54033+05 | .23075+04 | .90325+02 |
| .55000+01 | .96872+02 | .15362+05 | .52281+05 | .89201+04 | .23137+00 | .25394+02 | .52809+05 | .22738+04 | .93905+02 |
| .60000+01 | .10294+03 | .16486+05 | .49848+05 | .89091+04 | .23710+00 | .28577+02 | .50501+05 | .21251+04 | .10002+03 |
| .65000+01 | .10743+03 | .17539+05 | .46404+05 | .88487+04 | .24849+00 | .33272+02 | .47342+05 | .19506+04 | .10522+03 |
| .70000+01 | .10997+03 | .18557+05 | .41781+05 | .87802+04 | .26855+00 | .40060+02 | .42968+05 | .17815+04 | .10874+03 |
| .75000+01 | .11018+03 | .19565+05 | .38089+05 | .87426+04 | .30066+00 | .47818+02 | .38849+05 | .16380+04 | .11003+03 |
| .80000+01 | .10895+03 | .21113+05 | .36650+05 | .86771+04 | .34564+00 | .54067+02 | .36961+05 | .15323+04 | .10954+03 |
| .85000+01 | .10670+03 | .22738+05 | .36139+05 | .85848+04 | .39932+00 | .59500+02 | .36206+05 | .14593+04 | .10781+03 |
| .90000+01 | .10379+03 | .24268+05 | .36442+05 | .84753+04 | .45690+00 | .64108+02 | .36349+05 | .14166+04 | .10523+03 |
| .95000+01 | .10053+03 | .25886+05 | .37561+05 | .83605+04 | .51423+00 | .68022+02 | .37067+05 | .14005+04 | .10215+03 |
| .10000+02 | .97068+02 | .27991+05 | .39233+05 | .82418+04 | .56805+00 | .70667+02 | .38751+05 | .14076+04 | .98794+02 |
| .10500+02 | .93610+02 | .29690+05 | .40862+05 | .81322+04 | .61573+00 | .72721+02 | .40565+05 | .14361+04 | .95340+02 |
| .11000+02 | .90003+02 | .30980+05 | .42040+05 | .80687+04 | .65630+00 | .75135+02 | .41681+05 | .14835+04 | .91797+02 |
| .11500+02 | .86601+02 | .31792+05 | .43846+05 | .80425+04 | .69064+00 | .76637+02 | .43360+05 | .15509+04 | .88286+02 |
| .12000+02 | .83614+02 | .32377+05 | .45372+05 | .80515+04 | .71892+00 | .77769+02 | .44975+05 | .16312+04 | .85110+02 |
| .12500+02 | .81062+02 | .32736+05 | .47037+05 | .80938+04 | .74157+00 | .78503+02 | .46598+05 | .17198+04 | .82331+02 |

| | | | | | | | | | |
|---|---|---|---|---|---|---|---|---|---|
| .13000+02 | .79078+02 | .32949+05 | .48741+05 | .81570+04 | .75905+00 | .78739+02 | .4830+05 | .18128+04 | .80067+02 |
| .13500+02 | .77626+02 | .33063+05 | .49975+05 | .82381+04 | .77169+00 | .78941+02 | .49661+05 | .19054+04 | .78359+02 |
| .14000+02 | .76616+02 | .33117+05 | .51155+05 | .83334+04 | .78030+00 | .78968+02 | .50849+05 | .19962+04 | .77121+02 |
| .14500+02 | .76048+02 | .33173+05 | .52275+05 | .84332+04 | .78571+00 | .78771+02 | .51989+05 | .20837+04 | .76333+02 |
| .15000+02 | .75882+02 | .33251+05 | .53293+05 | .85329+04 | .78842+00 | .78395+02 | .53035+05 | .21660+04 | .75967+02 |
| .15500+02 | .76062+02 | .33352+05 | .54189+05 | .86263+04 | .78881+00 | .77889+02 | .53963+05 | .22416+04 | .75975+02 |
| .16000+02 | .76532+02 | .33470+05 | .54958+05 | .87141+04 | .78724+00 | .77293+02 | .54765+05 | .23100+04 | .76300+02 |
| .16500+02 | .77236+02 | .33596+05 | .55605+05 | .87945+04 | .78404+00 | .76636+02 | .55443+05 | .23706+04 | .76886+02 |
| .17000+02 | .78123+02 | .33725+05 | .56134+05 | .88668+04 | .77954+00 | .75943+02 | .56003+05 | .24232+04 | .77682+02 |
| .17500+02 | .79150+02 | .33849+05 | .56554+05 | .89308+04 | .77405+00 | .75236+02 | .56451+05 | .24679+04 | .78638+02 |
| .18000+02 | .80276+02 | .33965+05 | .56874+05 | .89865+04 | .76784+00 | .74533+02 | .56796+05 | .25048+04 | .79714+02 |
| .18500+02 | .81465+02 | .34071+05 | .57101+05 | .90341+04 | .76118+00 | .73852+02 | .57044+05 | .25339+04 | .80872+02 |
| .19000+02 | .82690+02 | .34165+05 | .57246+05 | .90737+04 | .75430+00 | .73201+02 | .57211+05 | .25558+04 | .82079+02 |
| .19500+02 | .83921+02 | .34248+05 | .57315+05 | .91056+04 | .74738+00 | .72598+02 | .57299+05 | .25707+04 | .83306+02 |
| .20000+02 | .85135+02 | .34321+05 | .57318+05 | .91302+04 | .74061+00 | .72047+02 | .57321+05 | .25791+04 | .84529+02 |
| .20500+02 | .86312+02 | .34385+05 | .57260+05 | .91479+04 | .73412+00 | .71562+02 | .57280+05 | .25816+04 | .85724+02 |
| .21000+02 | .87431+02 | .34441+05 | .57152+05 | .91593+04 | .72805+00 | .71149+02 | .57184+05 | .25787+04 | .86872+02 |
| .21500+02 | .88476+02 | .34492+05 | .57000+05 | .91651+04 | .72250+00 | .70810+02 | .57043+05 | .25712+04 | .87954+02 |
| .22000+02 | .89434+02 | .34542+05 | .56813+05 | .91606+04 | .71755+00 | .70540+02 | .56866+05 | .25596+04 | .88955+02 |
| .22500+02 | .90291+02 | .34590+05 | .56599+05 | .91520+04 | .71329+00 | .70365+02 | .56659+05 | .25447+04 | .89863+02 |
| .23000+02 | .91039+02 | .34640+05 | .56366+05 | .91399+04 | .70974+00 | .70257+02 | .56430+05 | .25273+04 | .90666+02 |
| .23500+02 | .91671+02 | .34692+05 | .56121+05 | .91251+04 | .70694+00 | .70222+02 | .56118+05 | .25081+04 | .91356+02 |
| .24000+02 | .92183+02 | .34748+05 | .55872+05 | .91083+04 | .70490+00 | .70255+02 | .55939+05 | .24878+04 | .91927+02 |
| .24500+02 | .92572+02 | .34807+05 | .55626+05 | .90901+04 | .70361+00 | .70349+02 | .55692+05 | .24671+04 | .92377+02 |
| .25000+02 | | .34868+05 | .55339+05 | .90711+04 | .70303+00 | .70446+02 | .55545+05 | .24467+04 | .92706+02 |
| .25500+02 | .92992+02 | .34931+05 | .55169+05 | .90521+04 | .70312+00 | .70688+02 | .55227+05 | .24272+04 | .92916+02 |
| .26000+02 | .93034+02 | .34995+05 | .54969+05 | .90336+04 | .70382+00 | .70912+02 | .55022+05 | .24091+04 | .93012+02 |
| .26500+02 | .92975+02 | .35058+05 | .54794+05 | .90161+04 | .70506+00 | .71161+02 | .54840+05 | .23928+04 | .93004+02 |
| .27000+02 | .92828+02 | .35118+05 | .54648+05 | | .70676+00 | .71426+02 | .54686+05 | .23787+04 | .92901+02 |

FIG. 11-26. Numerical Results for Compressor SURGE Problem

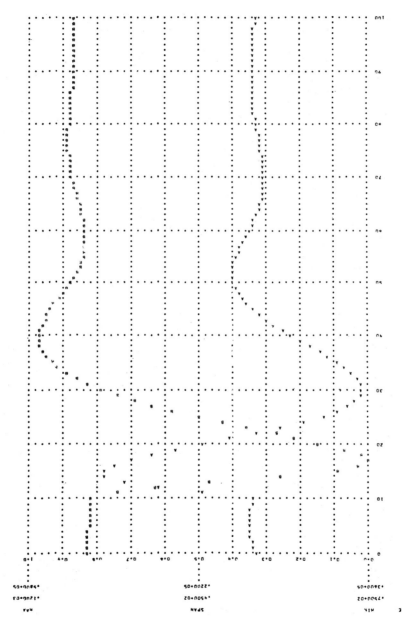

FIG. 11-27. Results Plotted on High-Speed Printer

Since the purpose of this discussion is to describe a *method* of digital simulation for control problems, and not the *solution* of the problem, this example is now complete, but additional discussion is provided in the Problem Section at the end of this chapter.

## 11-8  DISTILLATION COLUMN CONTROL

Chapter 8 develops the subroutines required for simulating the dynamics of unit operations, the most prominent of which was distillation columns. The chief interest in such dynamic simulations is the investigation of the dynamic performance of proposed control schemes. As an example, the column simulated in Chapter 8 (Figure 8-33) is controlled by sensing the

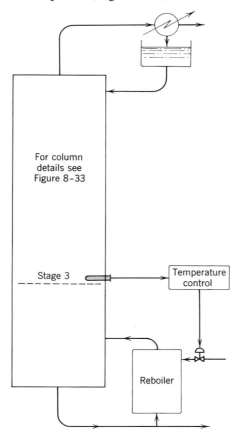

FIG. 11-28.  Distillation Column STAGE 3 Temperature Control

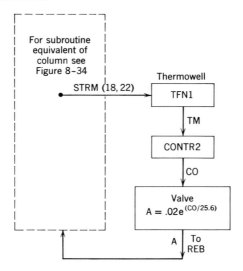

FIG. 11-29. Subroutine Equivalent of Control Loop

temperature on the third stage (STRM (18,22)) with a thermocouple in a thermowell that is connected to a two-mode controller which operates the equal-percentage valve admitting steam to the reboiler (Figure 11-28).

The thermowell can be characterized by a first-order transfer function having a time constant of 0.5 min, so it is simulated by the subroutine TFN1. The input to this subroutine is the temperature on the third-stage STRM (18,22), and the output is TM, the signal to the controller (Figure 11-29). The controller is simulated by calling the subroutine CONTR2, (Figure 11-30) which is located in the derivative section of the main program (Figure

```
TM=STRM(18,22)
OI=70.
AXN=0.



IF(T.GT.5.) AXN=-1.
CALL CONTR2(TM,CO,75.,125.,100.,AXN,100.,1.,OI,DOI)
A=.02*EXP(CO/25.6)



CALL INT(OI,DOI)
CALL TFN1(TM,STRM(18,22),.5,1.)
```

FIG. 11-30. Addition to Distillation Column Program (Figure 8-35) for Simulation of Control Action

8-35), with the companion integration for the automatic reset action located in the integration section. The controller has a scale of $75° \to 125°C$ and is reverse acting, since the reboiler steam valve must close for an increase in column temperature. The controller is to be switched to automatic after 5 min of the startup transient simulated in Chapter 8. This is most easily achieved by normally setting the manual reset $OI = 83.$ and setting the action $AXN = 0.$, making the controller inoperable. At time $T = 5.$ $AXN$ is made $-1.$, which makes the controller fully operational. The set point for the controller

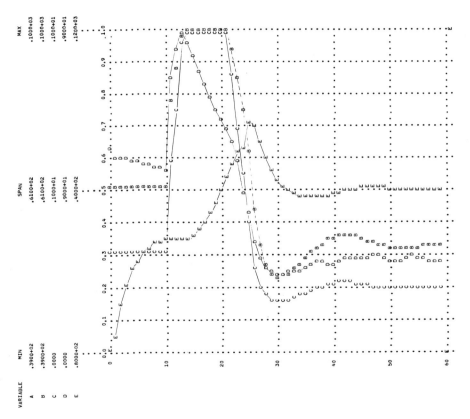

FIG. 11-31.  Automatic Control of Column Startup
Curve B  controller output CO range $39\% \to 100\%$
Curve C  valve area $a$ range $0 \to 1$
Curve D  boilup STRM $(1, 21)$ range $0 \to 9$
Curve E  temperature stage 3 STRM $(18, 22)$ range $80° \to 120°$

is 110°C, and hopefully, the temperature on stage 3 will eventually be controlled to this level. The valve area A will be related to the controller output CO by the equal-percentage valve formula explained in Section 11-5.

With nominal controller settings of 100% P.B. and 1. repeats/min, the transient results for the initial 30 min of column operation with the same initial starting values, as in Chapter 8, are shown in Figure 11-31 plotted on a line printer. Normally, a control study would now proceed to test the control system by imposing load disturbances, such as a feed change or loss of steam in the reboiler, and by adjusting the controller setting for optimum response. Other control schemes can also be tested, such as reflux control with constant boilup or feed forward controls. The net result of such a study would be the determination of the most effective control system for the column.

### Exercise Problems

1. The construction of a radiant film-drying process is represented in Figure 11-32. The film, $W$ ft wide, $t$ ft thick travels at a speed $F$ ft/min and passes under a radiant heater, receiving heat at a rate $Q$ PCU/ft² min. Moisture is expelled from the surface of the film (both sides) at a rate proportional to the temperature difference between the film and ambient $(T)$. The radiant heater is $W$ ft wide and $A$ ft long, and a distance $L$ ft separates the radiant heater from a moisture-sensing device. This device sends a signal to a two-mode controller that sends a control signal to the radiant heater control. The heat flux $Q$ is proportional to the control signal $C$ from the controller and has a first-order response $\tau_1$ (min). The response of the moisture sensor is also first-order with a time constant $\tau_2$ (min). Assume that there is negligible temperature gradient through the film and that heat loss between heater and sensor is proportional to the temperature difference between the film temperature and ambient.

Develop a mathematical model which, when programmed for computer

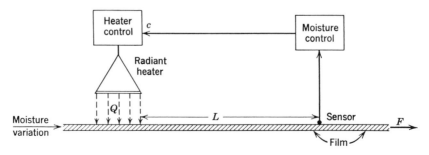

FIG. 11-32

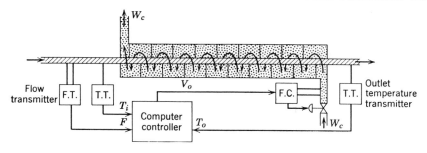

FIG. 11-33

solution, will establish the optimum controller settings for a variety of changes in moisture content in the film before it enters the heater.

2. The countercurrent, liquid-liquid heat exchanger shown in the diagram (Figure 11-33) consists of a center tube through which flows the process fluid; on the shell side, baffling causes the cooling medium to flow at an angle to the tube direction. The outlet temperature must be controlled to a very precise level, even though the inlet temperature and flow fluctuate severely.

A sophisticated control scheme is devised, consisting of a computer controller that receives input signals of flow $F$ and inlet temperature $T_i$ and outlet temperature $T_o$ and controls the coolant flow rate $W_c$ through a cascaded flow controller. Assuming a first-order response characteristic for the two thermowells and second-order characteristics for the control valve, construct a model of the system that will permit computer experiments to test out a variety of control functions for the computer controller. Also synthesize a reasonable control scheme for the computer controller that converts the input signals to the output control signal ($V_0$).

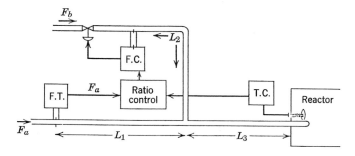

3. Two gaseous fuels are mixed together in a $T$ junction and burn in a nozzle with a flame temperature that is related to the flow rates $F_a$ and $F_b$

by the expression

$$T = \frac{a_1 F_a F_b}{a_2 F_a + a_3 F_b}$$

where $a_i$'s are constants. The flame temperature is sensed by a thermocouple with a response time $\tau_1$ and is measured by a temperature controller that sends a control signal to a ratio computer that controls the ratio of the fuel flow rates $F_a$ and $F_b$. The flow $F_a$ is free to fluctuate but is metered at a distance $L_1$ (ft) from the $T$ junction. This signal is also sent to the ratio computer whose output controls the set point of the flow control on flow $F_b$. The distance of the flow controller on $F_b$ to the $T$ junction is $L_2$ (ft), and the distance from the junction to the reactor nozzle is $L_3$. The diameter of all the pipes is $D$ (ft); the pressure in all three legs is $P$.

Assemble a model that will permit computer experiments to determine temperature and flow controller settings for a range of disturbances of the $F_a$ flow rate.

The following problems require the use of a computer.

4. A batch autoclave reactor is described in Section 6-4. The program is shown in Figures 6-11 to 6-13, and the results for a constant refrigeration temperature of 100°C is shown in Figure 6-15. Suppose that a restriction is imposed on this reaction, namely that the temperature in the reactor must be raised to 120.°C as rapidly as possible, with minimum overshoot and must be maintained at this level for the remainder of the reaction. Simulate the system with a control loop added and determine the optimum controller settings assuming that

(a) The reactor temperature is measured with a thermal system having a first-order time constant of 0.4 min.

(b) The coolant temperature through the coils is obtained by mixing two fluids with temperatures of 130°C and 70°C. Assume the mixing to be linear with valve position. The mixing valve dynamics can be simulated with a first-order time constant of 0.05 minutes.

5. The solution copolymerization problem is described in Section 6-5. The variation of instantaneous polymer composition during the batch cycle is shown in Figure 6-23. By continuously modifying the monomer addition, these compositions could be maintained to a specified level (say 0.65, 0.25, and 0.1 for A, B, C respectively). Simulate the system, and by using control methods determine these monomer feed rates as functions of time.

6. The compressor surge problem described in this chapter is simulated with the pressure controlling the steam flow to the turbine and the compressor flow controlling the bypass flow valve. Find the optimum control possible with this configuration and then determine whether a better control can be obtained by an alternate scheme, such as controlling the pressure with the

bypass valve and the turbine speed from the compressor flow. Also investigate any possible improvement using three-mode controllers.

7. The purity specification for the distillation column described in Section 8-6 is for the composition in the bottoms product, never to exceed 0.45 %C. Suppose the feed composition can change suddenly from that given in the example to A = 15%, B = 5%, C = 80%. Can the control system described in Section 11-8 maintain the bottoms parity within specification? If not, modify the control configuration to achieve a satisfactory system.

8. The fixed-bed reactor described and programmed in Chapter 10 exhibited a significant amplification factor between the input temperature and the output temperature of the reactor. Suppose a countercurrent single-pass heat exchanger is to be used to control the feed temperature, assuming that the feed is preheated to a temperature of 60 ± 2°C and that the cooling water available has a temperature of 15°C. How well will a controller be able to maintain the exit temperature to 90°C? Construct a suitable control scheme (thermowell time constant 0.4 min) and compare the performance for a single and cascade control in coping with the ±2°C fluctuations through the pre-heater.

## NOMENCLATURE

| | |
|---|---|
| $F$ | flow rate |
| $H_S$ | enthalpy of steam |
| $h$ | heat transfer coefficient |
| $H_R$ | rate of heat of reaction |
| $h_R$ | molar reaction heat |
| $k$ | reaction rate constant |
| $k_c$ | proportional gain |
| $P$ | pressure |
| $Q$ | heat flux |
| $SP$ | set point |
| $T$ | temperature |
| $t$ | time |
| $UA$ | overall heat transfer coefficient |
| $V_{max}$ | maximum valve velocity |
| $W$ | mass or water |
| $w$ | wall |
| $wc$ | heat capacity of wall |
| $X$ | mole fraction |
| $\tau$ | time constant |
| $\epsilon$ | error |
| $\alpha$ | rate amplitude ratio |

## REFERENCES

1. *Principles of Industrial Process Control*, D. P. Eckman, Wiley, New York, 1945.
2. *Automatic Process Control for Chemical Engineers*, N. H. Ceaglske, Wiley, New York, 1956.
3. *Techniques of Process Control*, P. S. Buckley, Wiley, New York, 1964.
3a. *Automatic Control of Chemical and Petroleum Processes*, T. J. Williams and V. A. Lauher, Gulf Publishing Co., 1961.


### References for Further Study

The following articles and papers describe computer simulation studies of process control problems in the chemical and petroleum industries.

4. *Simulating the Process Controller with an Analog Computer*, R. G. Franks, ISA N.J. Section, 10th Annual Symposium, 1958.
5. "Design of a Ph Control System by Analog Simulation," W. B. Field, *ISA* **6**, No. 1, 1959.
6. "Analysis and Simulation of a Chemical Feeder System," J. C. Vogt, *Simulation*, **4**, No. 4, 1964.
7. "Simulation of Process Control with an Analog Computer," F. A. Woods, *Ind. Eng. Chem.*, **50**, No. 11, 1960.
8. "Better Simulation of Process Dynamics," *Hydrocarbon Processing*, **43**, No. 12
9. "Effects of Limiting Error in Feedback Analysis," P. E. Straight and F. Michaels, *ISA J.*, **5**, No. 5, 1957.
10. "Analog Simulation for Process Design and Control," W. B. Field, *ISA*, PCT-1-58-1, 1958.
11. "Analog Simulation of a Complex Level Control System," K. I. Mumme and L. W. Zabel, *TAPPI*, **43**, No. 4.
12. "Maximizing Control Performance and Economy with Analog Simulation," *ISA J.*, **5**, No. 9, R. G. Franks, 1958.
13. "Digital Simulation for Control System Design," R. Linebarger and R. D. Brennan, *Inst. Control Systems*, October 1965.

# APPENDIX A

Tables A-1, A-2, and A-3 summarize the subroutines described in this book in both the INT and the DYFLO programs. Some of the subroutines can be used directly without any other supporting subroutines. Others require auxiliary routines, which are listed in the last column.

**Table A-1. INT Program**

| Subroutine Name | Function | Figure Number | Page Number | Auxiliary Routines Required |
|---|---|---|---|---|
| **ALGEBRAIC SUBROUTINES** | | | | |
| FUN1 | Arbitrary FUNction $Y = f(X)$ | 2-21 | 31 | None |
| FUN2 | Arbitrary FUNction $Y = f(X, Z)$ | 2-25 | 40 | None |
| CPS | Converge by Partial Substitution | 2-9 | 25 | None |
| CONV | CONVerge by Wegstein method | 2-11 | 27 | None |
| **INTEGRATION SUBROUTINES** | | | | |
| INT I | INTegration Independent variable | A. B-4 | 405 | None |
| INT | INTegration dependent variable | A. B-3 | 405 | INT I |
| PRNTF | PRiNT output and Finish | 3-14 | 64 | INT I |
| PRNTRS | Repeat PRiNT Space | 3-15 | 65 | PRNTF |
| DER | DERivative | 3-27 | 78 | INT I |
| **PROCESS CONTROL SUBROUTINES** | | | | |
| TFN1 | First-order Transfer FuNction | 11-3A | 359 | INT |
| TFN2 | Second-order Transfer FuNction | 11-3C | 360 | INT |
| CONTR1 | Proportional CONTRoller | 11-5A | 363 | None |
| CONTR2 | Proportional plus integral CONTRoller | 11-5B | 364 | INT |
| CONTR3 | Proportional plus integral plus rate CONTRoller | 11-7A | 366 | INT |
| TDL | Time DeLay | 11-7D | 368 | INT I |
| VALVE | Control VALVE | 11-10 | 372 | None |

**Table A-2. DYFLO Program for Simulation of Unit Operations**

| Subroutine Name | Function | Figure Number | Page Number | Auxiliary Routines Required |
|---|---|---|---|---|
| STATE SUBROUTINES | | | | |
| ENTHL | ENTHaLpy liquid | 5-3 | 117 | None |
| ENTHV | ENTHalpy Vapor | 5-4 | 117 | None |
| TEMP | Determines TEMPerature for enthalpy | 5-5 | 118 | CONV |
| ACTY | Liquid component ACTivitY | 5-30 | 146 | None |
| EQUIL | Vapor/liquid EQUILibrium | 5-8b | 123 | ACTY |
| DEWPT | DEW PoinT | 5-9 | 124 | ACTY |
| UNIT OPERATIONS (STEADY STATE) | | | | |
| SUM | SUM of two streams | 8-2B | 213 | TEMP |
| SPLIT | Stream SPLITter | 8-2A | 213 | None |
| HTEXCH | Enthalpy addition to stream | 5-12 | 127 | TEMP |
| CSHE | Countercurrent Steady State Heat Exchanger | 9-39-C | 301 | HT EXCH, CPS |
| FLASH | Phase splitter with enthalpy addition | 5-15 | 130 | ACTY, HT EXCH, CONV, ENTHL, ENTHV |
| CVBOIL | Constant Volume BOILer | 5-17 | 131 | DEWPT, ENTHV |
| PCON | Partial CONdenser, specified exit temperature | 5-20 | 134 | EQUIL, ACTY, ENTHV, ENTHL |

## Table A-3. DYFLO Program Control

| Subroutine name | Function | Figure Number | Page Number | Auxiliary Routines Required |
|---|---|---|---|---|
| UNIT OPERATIONS (DYNAMIC) | | | | |
| HLDP | Generalized HoLDuP, CSTR, jacketed vessel | 5-22 | 137 | INT, TEMP ENTHL, ENTHV |
| VVBOIL | Variable Volume BOILer | 5-25 | 140 | EQUIL, INT, ENTHV, ENTHL |
| STAGE | Equilibrium STAGE, distillation column | 8-20 | 235 | INT, EQUIL, ENTHV, ENTHL |
| STGF | Equilibrium Feed STaGe distillation column | 8-22 | 236 | Same as STAGE |
| STGS | Equilibrium Sidestream STaGe, distillation column | 8-23 | 237 | Same as STAGE |
| STGH | Equilibrium STaGe With Heat accumulation | 8-27b | 242 | Same as STAGE TEMP |
| BOT | Column BOTtoms | 8-25 | 239 | Same as STAGE |
| REB | Column REBoiler | 8-32 | 248 | INT, CONV |
| DHE | Countercurrent Dynamic Heat Exchanger | 10-16-a | 328 | HLDP, INT CSHE, ICHE |
| ICHE | Initial Conditions for DHE | 10-16-c | 330 | HT EXCH, ENTHL ENTHV |
| LDL | Stream time DeLay | 11-7e | 369 | INT I |
| PRL | Output PRint for streams and finish | A. B-1 | 403 | INT I |
| RPRL | Repeat PRint for streams | A. B-2 | 404 | PRL |
| NRCT | Clears RCT array | A. B-5 | 406 | — |
| START | Convenience routine (initiation) | A. B-6 | 406 | — |

# APPENDIX B

```
 1 *          SUBROUTINE PRL(PRI,FNR,NF,L1,L2,L3,L4,L5,L6,L7,L8,L9,L10,L11,L12)
 2 *          COMMON/CD/STRM(300,24),DATA(20,10),RCT(22),NCF,NCL,LSTR
 3 *          COMMON/CINT/T,DT,JS,JN,DXA(500),XA(500),IO,JS4
 4 *          LOGICAL LSTR,NF
 5 *          COMMON/CPR/NPR,NCPR,L(10,14),STP(12,24),STR(5,500),NS
 6 *          NCPR = 1
 7 *      99 FORMAT (1H0,1X,6HTIME =,1E10.4)
 8 *     100 FORMAT (1H ,1X,7HSTRM NO,5X,1I3,11I10)
 9 *     101 FORMAT (1H ,1X,4HFLOW,4X,12E10.4)
10 *     102 FORMAT (1H ,1X,4HTEMP,4X,12E10.4)
11 *     103 FORMAT (1H ,1X,6HENTHAL,2X,12E10.4)
12 *     104 FORMAT (1H ,1X,5HPRESS,3X,12E10.4)
13 *     105 FORMAT (1H ,1X,4HCOMP,1I2,2X,12E10.4)
14 *          IF (LSTR) GO TO 8
15 *          NPR = 1
16 *          IF((T.GE.TPRNT-DT/2.).AND.((JS.EQ.2).OR.(JS4.EQ.4))) GO TO 5
17 *          RETURN
18 *       5 TPRNT = TPRNT + PRI
19 *          DO 4 J = 1,NP
20 *          M = L(1,J)
21 *          DO 4 I = 1,24
22 *       4 STP(J,I) = STRM(M,I)
23 *          PRINT 99,T
24 *          PRINT 100,(L(1,J),J=1,NP)
25 *          PRINT 101,(STP(J,21),J=1,NP)
26 *          PRINT 102,(STP(J,22),J=1,NP)
27 *          PRINT 103,(STP(J,23),J=1,NP)
28 *          PRINT 104,(STP(J,24),J=1,NP)
29 *          DO 12 JC = NCF,NCL
30 *      12 PRINT 105,JC,(STP(J,JC),J=1,NP)
31 *          NPR = 1
32 *          IF (T .GE. FNR-DT/3.) GO TO 6
33 *          RETURN
34 *       8 NF = .FALSE.
35 *          LSTR=.FALSE.
36 *          L(1,1) = L1
37 *          L(1,2) = L2
38 *          L(1,3) = L3
39 *          L(1,4) = L4
40 *          L(1,5) = L5
41 *          L(1,6) = L6
42 *          L(1,7) = L7
43 *          L(1,8) = L8
44 *          L(1,9) = L9
45 *          L(1,10)=L10
46 *          L(1,11)=L11
47 *          L(1,12)=L12
48 *          DO 9 N = 1,12
49 *       9 IF (L(1,N) .EQ. 0) GO TO 10
50 *          NP = N
51 *          GO TO 5
52 *      10 NP = N - 1
53 *          GO TO 5
54 *       6 T = 0.
55 *          TPRNT = 0.
56 *          NF = .TRUE.
57 *          DO 7 J = 1,500
58 *       7 XA(J) = 0.
59 *          RETURN
60 *          END
```

FIG. B-1.  Argument list:
PRI = print interval
NF = finish index
NF = true finish
ENR = run termination Li → 12 = stream
numbers to be printed

```
  1 *           SUBROUTINE RPRL(L1,L2,L3,L4,L5,L6,L7,L8,L9,L10,L11,L12)
  2 *           COMMON/CD/STRM(300,24),DATA(20,10),RCT(22),NCF,NCL,LSTR
  3 *           COMMON/CPR/NPR,NCPR,L(10,14),STP(12,24),STR(5,500),NS
  4 *       100 FORMAT (1H0,1X,7HSTRM NO,5X,1I3,11I10)
  5 *       101 FORMAT (1H ,1X,4HFLOW,4X,12E10.4)
  6 *       102 FORMAT (1H ,1X,4HTEMP,4X,12E10.4)
  7 *       103 FORMAT (1H ,1X,6HENTHAL,2X,12E10.4)
  8 *       104 FORMAT (1H ,1X,5HPRESS,3X,12E10.4)
  9 *       105 FORMAT (1H ,1X,4HCOMP,1I2,2X,12E10.4)
 10 *           NCPR=NCPR+1
 11 *           IF (L(NCPR,1) .NE. L1) GO TO 8
 12 *           IF(NPR.NE.1) RETURN
 13 *         5 NP=L(NCPR,13)
 14 *           DO 4 J=1,NP
 15 *           M=L(NCPR,J)
 16 *           DO 4 I = 1,24
 17 *         4 STP(J,I) = STRM(M,I)
 18 *           PRINT 100,(L(NCPR,J),J=1,NP)
 19 *           PRINT 101,(STP(J,21),J=1,NP)
 20 *           PRINT 102,(STP(J,22),J=1,NP)
 21 *           PRINT 103,(STP(J,23),J=1,NP)
 22 *           PRINT 104,(STP(J,24),J=1,NP)
 23 *           DO 12 JC = NCF,NCL
 24 *        12 PRINT 105,JC,(STP(J,JC),J=1,NP)
 25 *           RETURN
 26 *         8 L(NCPR,1)=L1
 27 *           L(NCPR,2)=L2
 28 *           L(NCPR,3)=L3
 29 *           L(NCPR,4)=L4
 30 *           L(NCPR,5)=L5
 31 *           L(NCPR,6)=L6
 32 *           L(NCPR,7)=L7
 33 *           L(NCPR,8)=L8
 34 *           L(NCPR,9)=L9
 35 *           L(NCPR,10)=L10
 36 *           L(NCPR,11)=L11
 37 *           L(NCPR,12)=L12
 38 *           DO 9 N=1,12
 39 *         9 IF(L(NCPR,N).EQ.0) GO TO 10
 40 *           L(NCPR,13) =N
 41 *           GO TO 5
 42 *        10 L(NCPR,13)=N-1
 43 *           GO TO 5
 44 *           END
```

FIG. B-2. L1 12 = stream numbers to be printed

```
1  *          SUBROUTINE INT(X,DX)
2  *          COMMON /CINT/T,DT,JS,JN,DXA(500),XA(500),IO,JS4
3  *          JN=JN+1
4  *          GO TO (9,8,3,3),IO
5  *        9 X=X+DX*DT
6  *          RETURN
7  *        8 GO TO (1,2),JS
8  *        1 DXA(JN)=DX
9  *          X=X+DX*DT
10 *          RETURN
11 *        2 X=X+(DX-DXA(JN))*DT/2.
12 *          RETURN
13 *        3 GO TO(4,5,6,7),JS4
14 *        4 XA(JN)=X
15 *          DXA(JN)=DX
16 *          X=X+DX*DT
17 *          RETURN
18 *        5 DXA(JN)=DXA(JN)+2.*DX
19 *          X=XA(JN)+DX*DT
20 *          RETURN
21 *        6 DXA(JN)=DXA(JN)+2.*DX
22 *          X=XA(JN)+DX*DT
23 *          RETURN
24 *        7 DXA(JN)=(DXA(JN)+DX)/6.
25 *          X=XA(JN)+DXA(JN)*DT
26 *          RETURN
27 *          END
```

FIG. B-3.   Argument list:
X = integrated variable
DX = derivative of X

```
1  *          SUBROUTINE INTI(TD,DTD,IOD)
2  *          COMMON/CINT/T,DT,JS,JN,DXA(500),XA(500),IO,JS4
3  *          IO = IOD
4  *          JN=0
5  *          GO TO (6,5,1,1),IO
6  *        6 JS=2
7  *          GO TO 7
8  *        5 JS=JS+1
9  *          IF(JS.EQ.3)JS=1
10 *          IF(JS.EQ.2)RETURN
11 *        7 DT=DTD
12 *        3 TD=TD+DT
13 *          T=TD
14 *          RETURN
15 *        1 JS4=JS4+1
16 *          IF(JS4.EQ.5)JS4=1
17 *          IF(JS4.EQ.1) GO TO 2
18 *          IF(JS4.EQ.3) GO TO 4
19 *          RETURN
20 *        2 DT=DTD/2.
21 *          GO TO 3
22 *        4 TD=TD+DT
23 *          DT=2.*DT
24 *          T=TD
25 *          RETURN
26 *          END
```

FIG. B-4.   Listing for Subroutine INTI
Argument list:
TD = independent variable
DTD = integration step size
IOD = integration order

```
1 *          SUBROUTINE NRCT
2 *          COMMON/CD/STRM(300,24),DATA(20,10),RCT(22),NCF,NCL,LSTR
3 *          DO 5 J=NCF,NCL
4 *        5 RCT(J)=0.
5 *          RCT(21)=0.
6 *          RCT(22)=0.
7 *          RETURN
8 *          END
```

FIG. B-5.  Subroutine NRCT

```
 1 *          SUBROUTINE START(NFC,NLC,IP)
 2 *          COMMON/CD/STRM(300,24),DATA(20,10),RCT(22),NCF,NCL,LSTR
 3 *          COMMON/CINT/XT,DT,JS,JN,DXA(500),XA(500),IO,JS4
 4 *      100 FORMAT(5X,I2,5X,10E12.5)
 5 *      101 FORMAT(1H ,1X,' COMPONENT  ANTOINE C1  ANTOINE C2  ANTOINE C3  VAP
 6 *        1 ENTH.A  VAP.ENTH.B  LATENT HT.  LIQ.ENTH.A  LIQ.ENTH.B   ACTIVITY
 7 *        2    SPARE')
 8 *          LOGICAL LSTR
 9 *          LSTR=.TRUE.
10 *          IF(DT.LE.0.) DT=1.
11 *          NCF=NFC
12 *          NCL=NLC
13 *          IF(IP.EQ.1) GO TO 5
14 *          RETURN
15 *        5 PRINT 101
16 *          DO 6 J=NCF,NCL
17 *        6 PRINT 100,J,(DATA(J,I),I=1,10)
18 *,         RETURN
19 *          END
```

FIG. B-6.  Listing for Subroutine START
Argument list:
NFC = number of first component
NLC = number of last component
IP = print index (IP = 1, Print)

This subroutine is called in the initiation section of a main program. Its purpose is to take care of the following housekeeping chores:

1. It will insert NCF and NCL into COMMON/CD/.

2. It will specify DT if it is 0. to unity, thus avoiding division by zero on the first dummy pass in the subroutine BOT (if used). It is subsequently set to its proper value when INTI is called.

3. By inserting 1 in the third argument (IP), the routine will print the physical properties of all components in the program (NCF $\rightarrow$ NCL). All the examples in this book performed the first two chores in the main program. However, experience has shown that using this subroutine automatically ensures that these tasks are carried out, tasks that if omitted will cause a program abort.

# APPENDIX C

### Implementation of INT and DYFLO

The INT and DYFLO programs were developed for the Univac 1108 computer. One important feature is that the working area in core is cleared to zero before the user's program is mounted. There are many computers that do not have this convenient feature. Thus if INT or DYFLO is implemented for such a computer, it is necessary to set some of the items in the COMMON statements to zero by inserting the following DATA statement in the INTI subroutine for CINT COMMON DATA T/O./JS/O/JN/O/ DXA/O./XA/O./JS4/O/ and for DYFLO'S COMMON/CD/ DATA STRM/O./DATA/O./RCT/O./

If the size of either program is too large for a smaller computer, the capacity requirements may be reduced merely by decreasing the size of the dimension of XA and DXA in INT and the number of streams in the STRM array in DYFLO, which ensures that their dimensions are changed in all subroutines containing the particular COMMON statement. Decreasing the number of components in DYFLO (i.e., J in STRM(I,J)) would require a considerable amount of recoding within all subroutines, but could be accomplished with sufficient patience.

The program is written in FORTRAN IV but can be readily converted to FORTRAN II or some other language by a programmer familiar with the software structure of the local computer.

# INDEX